T0342103

Communication Networks and Service Management in the Era of Artificial Intelligence and Machine Learning

Communication Networks and Service Management in the Era of Artificial Intelligence and Machine Learning

Edited by Nur Zincir-Heywood, Marco Mellia, and Yixin Diao

IEEE Press
Series on
Networks and
Services Management

Dr. Veli Sahin and
Dr. Mehmet Ulema, *Series Editors*

IEEE PRESS

WILEY

Published by John Wiley & Sons, Inc., Hoboken, New Jersey.
Published simultaneously in Canada.

For general information on our other products and services or for technical support, please contact our Customer Care Department within the United States at (800) 762-2974, outside the United States at (317) 572-3993 or fax (317) 572-4002.
Wiley also publishes its books in a variety of electronic formats. Some content that appears in print may not be available in electronic formats. For more information about Wiley products, visit our web site at www.wiley.com.

Library of Congress Cataloging-in-Publication Data:

Names: Zincir-Heywood, Nur, editor. | Mellia, Marco, editor. | Diao, Yixin,
    1970- editor.
Title: Communication networks and service management in the era of
    artificial intelligence and machine learning / edited by Nur
    Zincir-Heywood, Marco Mellia, and Yixin Diao.
Description: Hoboken, New Jersey : Wiley-IEEE Press, [2021] | Series: IEEE
    Press series on networks and service management | Includes
    bibliographical references and index.
Identifiers: LCCN 2021032407 (print) | LCCN 2021032408 (ebook) | ISBN
    9781119675501 (cloth) | ISBN 9781119675440 (adobe pdf) | ISBN
    9781119675518 (epub)
Subjects: LCSH: Computer networks. | Artificial intelligence. | Machine
    learning.
Classification: LCC TK5105.5 .C5998 2021 (print) | LCC TK5105.5 (ebook) |
    DDC 004.6–dc23
LC record available at https://lccn.loc.gov/2021032407
LC ebook record available at https://lccn.loc.gov/2021032408

Cover Design: Wiley
Cover Image: © Bill Donnelley/WT Design

Set in 9.5/12.5pt STIXTwoText by Straive, Chennai, India
Printed and bound by CPI Group (UK) Ltd, Croydon, CR0 4YY

C9781119675501_160921

Contents

Editor Biographies *xv*
List of Contributors *xvii*
Preface *xxiii*
Acknowledgments *xxvii*
Acronyms *xxix*

Part I Introduction *1*

1 Overview of Network and Service Management *3*
Marco Mellia, Nur Zincir-Heywood, and Yixin Diao
1.1 Network and Service Management at Large *3*
1.2 Data Collection and Monitoring Protocols *5*
1.2.1 SNMP Protocol Family *5*
1.2.2 Syslog Protocol *5*
1.2.3 IP Flow Information eXport (IPFIX) *6*
1.2.4 IP Performance Metrics (IPPM) *7*
1.2.5 Routing Protocols and Monitoring Platforms *8*
1.3 Network Configuration Protocol *9*
1.3.1 Standard Configuration Protocols and Approaches *9*
1.3.2 Proprietary Configuration Protocols *10*
1.3.3 Integrated Platforms for Network Monitoring *10*
1.4 Novel Solutions and Scenarios *12*
1.4.1 Software-Defined Networking – SDN *12*
1.4.2 Network Functions Virtualization – NFV *14*
Bibliography *15*

2   **Overview of Artificial Intelligence and Machine Learning**   *19*
         Nur Zincir-Heywood, Marco Mellia, and Yixin Diao
2.1   Overview   *19*
2.2   Learning Algorithms   *20*
2.2.1   Supervised Learning   *21*
2.2.2   Unsupervised Learning   *22*
2.2.3   Reinforcement Learning   *23*
2.3   Learning for Network and Service Management   *24*
         Bibliography   *26*

Part II   Management Models and Frameworks   *33*

3   **Managing Virtualized Networks and Services with Machine Learning**   *35*
         Raouf Boutaba, Nashid Shahriar, Mohammad A. Salahuddin, and Noura Limam
3.1   Introduction   *35*
3.2   Technology Overview   *37*
3.2.1   Virtualization of Network Functions   *38*
3.2.1.1   Resource Partitioning   *38*
3.2.1.2   Virtualized Network Functions   *40*
3.2.2   Link Virtualization   *41*
3.2.2.1   Physical Layer Partitioning   *41*
3.2.2.2   Virtualization at Higher Layers   *42*
3.2.3   Network Virtualization   *42*
3.2.4   Network Slicing   *43*
3.2.5   Management and Orchestration   *44*
3.3   State-of-the-Art   *46*
3.3.1   Network Virtualization   *46*
3.3.2   Network Functions Virtualization   *49*
3.3.2.1   Placement   *49*
3.3.2.2   Scaling   *52*
3.3.3   Network Slicing   *55*
3.3.3.1   Admission Control   *55*
3.3.3.2   Resource Allocation   *56*
3.4   Conclusion and Future Direction   *59*
3.4.1   Intelligent Monitoring   *60*
3.4.2   Seamless Operation and Maintenance   *60*
3.4.3   Dynamic Slice Orchestration   *61*

3.4.4   Automated Failure Management   *61*
3.4.5   Adaptation and Consolidation of Resources   *61*
3.4.6   Sensitivity to Heterogeneous Hardware   *62*
3.4.7   Securing Machine Learning   *62*
          Bibliography   *63*

4   Self-Managed 5G Networks   *69*
          Jorge Martín-Pérez, Lina Magoula, Kiril Antevski, Carlos Guimarães, Jorge
          Baranda, Carla Fabiana Chiasserini, Andrea Sgambelluri, Chrysa
          Papagianni, Andrés García-Saavedra, Ricardo Martínez, Francesco
          Paolucci, Sokratis Barmpounakis, Luca Valcarenghi, Claudio EttoreCasetti,
          Xi Li, Carlos J. Bernardos, Danny De Vleeschauwer, Koen De Schepper,
          Panagiotis Kontopoulos, Nikolaos Koursioumpas, Corrado Puligheddu,
          Josep Mangues-Bafalluy, and Engin Zeydan
4.1   Introduction   *69*
4.2   Technology Overview   *73*
4.2.1   RAN Virtualization and Management   *73*
4.2.2   Network Function Virtualization   *75*
4.2.3   Data Plane Programmability   *76*
4.2.4   Programmable Optical Switches   *77*
4.2.5   Network Data Management   *78*
4.3   5G Management State-of-the-Art   *80*
4.3.1   RAN resource management   *80*
4.3.1.1   Context-Based Clustering and Profiling for User and Network
          Devices   *80*
4.3.1.2   Q-Learning Based RAN Resource Allocation   *81*
4.3.1.3   vrAIn: AI-Assisted Resource Orchestration for Virtualized Radio
          Access Networks   *81*
4.3.2   Service Orchestration   *83*
4.3.3   Data Plane Slicing and Programmable Traffic Management   *85*
4.3.4   Wavelength Allocation   *86*
4.3.5   Federation   *88*
4.4   Conclusions and Future Directions   *89*
          Bibliography   *92*

5   AI in 5G Networks: Challenges and Use Cases   *101*
          Stanislav Lange, Susanna Schwarzmann, Marija Gajic, Thomas Zinner, and
          Frank A. Kraemer
5.1   Introduction   *101*
5.2   Background   *103*
5.2.1   ML in the Networking Context   *103*

5.2.2   ML in Virtualized Networks   *104*
5.2.3   ML for QoE Assessment and Management   *104*
5.3   Case Studies   *105*
5.3.1   QoE Estimation and Management   *106*
5.3.1.1   Main Challenges   *107*
5.3.1.2   Methodology   *108*
5.3.1.3   Results and Guidelines   *109*
5.3.2   Proactive VNF Deployment   *110*
5.3.2.1   Problem Statement and Main Challenges   *111*
5.3.2.2   Methodology   *112*
5.3.2.3   Evaluation Results and Guidelines   *113*
5.3.3   Multi-service, Multi-domain Interconnect   *115*
5.4   Conclusions and Future Directions   *117*
           Bibliography   *118*

6   Machine Learning for Resource Allocation in Mobile
            Broadband Networks   *123*
            Sadeq B. Melhem, Arjun Kaushik, Hina Tabassum, and Uyen T. Nguyen
6.1   Introduction   *123*
6.2   ML in Wireless Networks   *124*
6.2.1   Supervised ML   *124*
6.2.1.1   Classification Techniques   *125*
6.2.1.2   Regression Techniques   *125*
6.2.2   Unsupervised ML   *126*
6.2.2.1   Clustering Techniques   *126*
6.2.2.2   Soft Clustering Techniques   *127*
6.2.3   Reinforcement Learning   *127*
6.2.4   Deep Learning   *128*
6.2.5   Summary   *129*
6.3   ML-Enabled Resource Allocation   *129*
6.3.1   Power Control   *131*
6.3.1.1   Overview   *131*
6.3.1.2   State-of-the-Art   *131*
6.3.1.3   Lessons Learnt   *132*
6.3.2   Scheduling   *132*
6.3.2.1   Overview   *132*
6.3.2.2   State-of-the-Art   *132*
6.3.2.3   Lessons Learnt   *134*
6.3.3   User Association   *134*
6.3.3.1   Overview   *134*
6.3.3.2   State-of-the-Art   *136*

6.3.3.3   Lessons Learnt   *136*

6.3.4   Spectrum Allocation   *136*

6.3.4.1   Overview   *136*

6.3.4.2   State-of-the-Art   *138*

6.3.4.3   Lessons Learnt   *138*

6.4   Conclusion and Future Directions   *140*

6.4.1   Transfer Learning   *140*

6.4.2   Imitation Learning   *140*

6.4.3   Federated-Edge Learning   *141*

6.4.4   Quantum Machine Learning   *142*

        Bibliography   *142*

7   **Reinforcement Learning for Service Function Chain Allocation in Fog Computing**   *147*

        José Santos, Tim Wauters, Bruno Volckaert, and Filip De Turck

7.1   Introduction   *147*

7.2   Technology Overview   *148*

7.2.1   Fog Computing (FC)   *149*

7.2.2   Resource Provisioning   *149*

7.2.3   Service Function Chaining (SFC)   *150*

7.2.4   Micro-service Architecture   *150*

7.2.5   Reinforcement Learning (RL)   *151*

7.3   State-of-the-Art   *152*

7.3.1   Resource Allocation for Fog Computing   *152*

7.3.2   ML Techniques for Resource Allocation   *153*

7.3.3   RL Methods for Resource Allocation   *154*

7.4   A RL Approach for SFC Allocation in Fog Computing   *155*

7.4.1   Problem Formulation   *155*

7.4.2   Observation Space   *156*

7.4.3   Action Space   *157*

7.4.4   Reward Function   *158*

7.4.5   Agent   *161*

7.5   Evaluation Setup   *162*

7.5.1   Fog–Cloud Infrastructure   *162*

7.5.2   Environment Implementation   *162*

7.5.3   Environment Configuration   *164*

7.6   Results   *165*

7.6.1   Static Scenario   *165*

7.6.2   Dynamic Scenario   *167*

7.7   Conclusion and Future Direction   *169*

        Bibliography   *170*

Part III   Management Functions and Applications *175*

8   **Designing Algorithms for Data-Driven Network Management and Control: State-of-the-Art and Challenges** *177*
Andreas Blenk, Patrick Kalmbach, Johannes Zerwas, and Stefan Schmid
8.1   Introduction   *177*
8.1.1   Contributions   *179*
8.1.2   Exemplary Network Use Case Study   *179*
8.2   Technology Overview   *181*
8.2.1   Data-Driven Network Optimization   *181*
8.2.2   Optimization Problems over Graphs   *182*
8.2.3   From Graphs to ML/AI Input   *184*
8.2.4   End-to-End Learning   *187*
8.3   Data-Driven Algorithm Design: State-of-the Art   *188*
8.3.1   Data-Driven Optimization in General   *188*
8.3.2   Data-Driven Network Optimization   *190*
8.3.3   Non-graph Related Problems   *192*
8.4   Future Direction   *193*
8.4.1   Data Production and Collection   *193*
8.4.2   ML and AI Advanced Algorithms for Network Management with Performance Guarantees   *194*
8.5   Summary   *194*
         Acknowledgments   *195*
         Bibliography   *195*

9   **AI-Driven Performance Management in Data-Intensive Applications** *199*
Ahmad Alnafessah, Gabriele Russo Russo, Valeria Cardellini, Giuliano Casale, and Francesco Lo Presti
9.1   Introduction   *199*
9.2   Data-Processing Frameworks   *200*
9.2.1   Apache Storm   *200*
9.2.2   Hadoop MapReduce   *201*
9.2.3   Apache Spark   *202*
9.2.4   Apache Flink   *202*
9.3   State-of-the-Art   *203*
9.3.1   Optimal Configuration   *203*
9.3.1.1 Traditional Approaches   *203*
9.3.1.2 AI Approaches   *204*
9.3.1.3 Example: AI-Based Optimal Configuration   *206*

9.3.2 Performance Anomaly Detection *207*
9.3.2.1 Traditional Approaches *208*
9.3.2.2 AI Approaches *208*
9.3.2.3 Example: ANNs-Based Anomaly Detection *210*
9.3.3 Load Prediction *211*
9.3.3.1 Traditional Approaches *212*
9.3.3.2 AI Approaches *212*
9.3.4 Scaling Techniques *213*
9.3.4.1 Traditional Approaches *213*
9.3.4.2 AI Approaches *214*
9.3.5 Example: RL-Based Auto-scaling Policies *214*
9.4 Conclusion and Future Direction *216*
Bibliography *217*

10 Datacenter Traffic Optimization with Deep Reinforcement Learning *223*
Li Chen, Justinas Lingys, Kai Chen, and Xudong Liao
10.1 Introduction *223*
10.2 Technology Overview *225*
10.2.1 Deep Reinforcement Learning (DRL) *226*
10.2.2 Applying ML to Networks *227*
10.2.3 Traffic Optimization Approaches in Datacenter *229*
10.2.4 Example: DRL for Flow Scheduling *230*
10.2.4.1 Flow Scheduling Problem *230*
10.2.4.2 DRL Formulation *230*
10.2.4.3 DRL Algorithm *231*
10.3 State-of-the-Art: AuTO Design *231*
10.3.1 Problem Identified *231*
10.3.2 Overview *232*
10.3.3 Peripheral System *233*
10.3.3.1 Enforcement Module *233*
10.3.3.2 Monitoring Module *234*
10.3.4 Central System *234*
10.3.5 DRL Formulations and Solutions *235*
10.3.5.1 Optimizing MLFQ Thresholds *235*
10.3.5.2 Optimizing Long Flows *239*
10.4 Implementation *239*
10.4.1 Peripheral System *239*
10.4.1.1 Monitoring Module (MM): *240*
10.4.1.2 Enforcement Module (EM): *240*

10.4.2   Central System   *241*
10.4.2.1   sRLA   *241*
10.4.2.2   lRLA   *242*
10.5   Experimental Results   *242*
10.5.1   Setting   *243*
10.5.2   Comparison Targets   *244*
10.5.3   Experiments   *244*
10.5.3.1   Homogeneous Traffic   *244*
10.5.3.2   Spatially Heterogeneous Traffic   *245*
10.5.3.3   Temporally and Spatially Heterogeneous Traffic   *246*
10.5.4   Deep Dive   *247*
10.5.4.1   Optimizing MLFQ Thresholds using DRL   *247*
10.5.4.2   Optimizing Long Flows using DRL   *248*
10.5.4.3   System Overhead   *249*
10.6   Conclusion and Future Directions   *251*
              Bibliography   *253*

11   The New Abnormal: Network Anomalies in the AI Era   *261*
              *Francesca Soro, Thomas Favale, Danilo Giordano, Luca Vassio, Zied Ben
              Houidi, and Idilio Drago*
11.1   Introduction   *261*
11.2   Definitions and Classic Approaches   *262*
11.2.1   Definitions   *263*
11.2.2   Anomaly Detection: A Taxonomy   *263*
11.2.3   Problem Characteristics   *264*
11.2.4   Classic Approaches   *266*
11.3   AI and Anomaly Detection   *267*
11.3.1   Methodology   *267*
11.3.2   Deep Neural Networks   *268*
11.3.3   Representation Learning   *270*
11.3.4   Autoencoders   *271*
11.3.5   Generative Adversarial Networks   *272*
11.3.6   Reinforcement Learning   *274*
11.3.7   Summary and Takeaways   *275*
11.4   Technology Overview   *277*
11.4.1   Production-Ready Tools   *277*
11.4.2   Research Alternatives   *279*
11.4.3   Summary and Takeaways   *280*
11.5   Conclusions and Future Directions   *282*
              Bibliography   *283*

12 **Automated Orchestration of Security Chains Driven by Process Learning** *289*
Nicolas Schnepf, Rémi Badonnel, Abdelkader Lahmadi, and Stephan Merz
12.1 Introduction *289*
12.2 Related Work *290*
12.2.1 Chains of Security Functions *291*
12.2.2 Formal Verification of Networking Policies *292*
12.3 Background *294*
12.3.1 Flow-Based Detection of Attacks *294*
12.3.2 Programming SDN Controllers *295*
12.4 Orchestration of Security Chains *296*
12.5 Learning Network Interactions *298*
12.6 Synthesizing Security Chains *301*
12.7 Verifying Correctness of Chains *306*
12.7.1 Packet Routing *306*
12.7.2 Shadowing Freedom and Consistency *306*
12.8 Optimizing Security Chains *308*
12.9 Performance Evaluation *311*
12.9.1 Complexity of Security Chains *312*
12.9.2 Response Times *313*
12.9.3 Accuracy of Security Chains *313*
12.9.4 Overhead Incurred by Deploying Security Chains *314*
12.10 Conclusions *315*
Bibliography *316*

13 **Architectures for Blockchain-IoT Integration** *321*
Sina Rafati Niya, Eryk Schiller, and Burkhard Stiller
13.1 Introduction *321*
13.1.1 Blockchain Basics *323*
13.1.2 Internet-of-Things (IoT) Basics *324*
13.2 Blockchain-IoT Integration (BIoT) *325*
13.2.1 BIoT Potentials *326*
13.2.2 BIoT Use Cases *328*
13.2.3 BIoT Challenges *329*
13.2.3.1 Scalability *332*
13.2.3.2 Security *333*
13.2.3.3 Energy Efficiency *334*
13.2.3.4 Manageability *335*
13.3 BIoT Architectures *335*
13.3.1 Cloud, Fog, and Edge-Based Architectures *337*

13.3.2   Software-Defined Architectures   *337*

13.3.3   A Potential Standard BIoT Architecture   *338*

13.4   Summary and Considerations   *341*

Bibliography   *342*

Index   *345*

Editor Biographies

Nur Zincir-Heywood received the PhD in Computer Science and Engineering in 1998. She is a full professor at the Faculty of Computer Science, Dalhousie University, Canada, where she directs the NIMS Research Lab on Network Information Management and Security. Her research interests include machine learning and artificial intelligence for cyber security, network, systems, and information analysis, topics on which she has published over 200 fully reviewed papers. She is a recipient of several best paper awards as well as the supervisor for the recipient of the IFIP/IEEE IM 2013 Best PhD Dissertation Award in Network Management. She is the co-editor of the book "Recent Advances in Computational Intelligence in Defense and Security" and co-author of the book "Nature-inspired Cyber Security and Resiliency: Fundamentals, Techniques, and Applications." She is an Associate Editor of the IEEE Transactions on Network and Service Management and Wiley's International Journal of Network Management. She has been a co-organizer for the IEEE/IFIP International Workshop on Analytics for Network and Service Management since 2016. She served as Technical Program Co-chair for the IEEE Symposium on Computational Intelligence for Security and Defence Applications in 2011, International Conference on Network Traffic Measurement and Analysis in 2018, International Conference on Network and Service Management in 2019, and she served as General Co-chair for the International Conference on Network and Service Management in 2020. Professor Zincir-Heywood's research record was recognized with the title of Dalhousie University Research Professor in 2021.

Marco Mellia graduated with PhD in Electronic and Telecommunication Engineering in 2001. He is a full professor at Politecnico di Torino, Italy, where he coordinates the Smartdata@PoliTO center on Big Data, Machine Learning and Data Science. In 2002, he visited the Sprint Advanced Technology Laboratories in Burlingame, CA, working at the IP Monitoring Project (IPMON). In 2011, 2012, and 2013, he collaborated with Narus Inc. in Sunnyvale, CA, working on traffic monitoring and cyber-security system design. In 2015 and 2016, he visited Cisco

Systems in San Jose, CA, working on the design of cloud monitoring platforms. Professor Mellia has co-authored over 250 papers published in international journals and presented in leading conferences, all of them in the area of communication networks. He won the IRTF ANR Prize at IETF-88 and best paper awards at IEEE P2P'12, ACM CoNEXT'13, IEEE ICDCS'15. He participated in the program committees of several conferences including ACM SIGCOMM, ACM CoNEXT, ACM IMC, IEEE Infocom, IEEE Globecom, and IEEE ICC. He is the Area Editor of ACM CCR, IEEE Transactions on Network and Service Management and Elsevier Computer Networks. He is a Fellow of IEEE. His research interests are in the area of Internet monitoring, users' characterization, cyber security, and big data analytics applied to different areas.

Yixin Diao received the PhD degree in electrical engineering from Ohio State University, Columbus, OH, USA. He is currently a Director of Data Science and Analytics at PebblePost, New York, NY, USA. Prior to that, he was a Research Staff Member at IBM T. J. Watson Research Center, Yorktown Heights, NY, USA. He has published more than 80 papers and filed over 50 patents in systems and services management. He is the co-author of the book "Feedback Control of Computing Systems" and the co-editor of the book "Maximizing Management Performance and Quality with Service Analytics." He was a recipient of several Best Paper Awards from the IEEE/IFIP Network Operations and Management Symposium, the IFAC Engineering Applications of Artificial Intelligence, and the IEEE International Conference on Services Computing. He served as Program Co-chair for the International Conference on Network and Service Management in 2010, the IFIP/IEEE International Symposium on Integrated Network Management in 2013, and the IEEE International Conference on Cloud and Autonomic Computing in 2016 and served as General Co-chair for the International Conference on Network and Service Management in 2019. He is an Associate Editor of the IEEE Transactions on Network and Service Management and the Journal of Network and Systems Management. He is a Fellow of IEEE.

List of Contributors

Ahmad Alnafessah
Department of Computing
Imperial College London
London, UK

Kiril Antevski
Telematics Engineering Department
Universidad Carlos III de Madrid
Madrid, Spain

Remi Badonnel
Université de Lorraine
CNRS
Loria, Inria, Nancy, France

Jorge Baranda
Communication Networks Division
Centre Tecnológic de
Telecomunicacions Catalunya
(CTTC/CERCA)
Barcelona, Spain

Sokratis Barmpounakis
National and Kapodistrian University
of Athens
Software Centric & Autonomic
Networking lab
Athens, Greece

Carlos J. Bernardos
Telematics Engineering Department
Universidad Carlos III de Madrid
Madrid, Spain

Andreas Blenk
Chair of Communication Networks
Department of Electrical and
Computer Engineering
Technical University of Munich
Munich, Germany

and

Faculty of Computer Science
University of Vienna
Vienna, Austria

Raouf Boutaba
David R. Cheriton School of Computer
Science
University of Waterloo
Waterloo, Ontario, Canada

Valeria Cardellini
Department of Civil Engineering and
Computer Science Engineering
University of Rome Tor Vergata
Rome, Italy

Giuliano Casale
Department of Computing
Imperial College London
London, UK

Claudio Ettore Casetti
Department of Electronics and
Telecommunications
Politecnico di Torino
Torino, Italy

Kai Chen
Department of Computer Science and
Engineering, iSING Lab
Hong Kong University of Science and
Technology
Hong Kong SAR, China

Li Chen
Department of Computer Science and
Engineering, iSING Lab
Hong Kong University of Science and
Technology
Hong Kong SAR, China

Carla Fabiana Chiasserini
Department of Electronics and
Telecommunications
Politecnico di Torino
Torino, Italy

Koen De Schepper
Nokia Bell Labs
Antwerp, Belgium

Filip De Turck
Department of Information
Technology
Ghent University – imec, IDLab
Ghent, Technologiepark-Zwijnaarde
Oost-vlaanderen, Belgium

Danny De Vleeschauwer
Nokia Bell Labs
Antwerp, Belgium

Yixin Diao
PebblePost
New York, NY, USA

Idilio Drago
University of Turin
Torino, Italy

Thomas Favale
Politecnico di Torino
Torino, Italy

Marija Gajić
Department of Information Security
and Communication Technology
Norwegian University of Science and
Technology
Trondheim, Norway

Andrés García-Saavedra
NEC Laboratories Europe
5G Networks R&D Group
Heidelberg, Germany

Danilo Giordano
Politecnico di Torino
Torino, Italy

Carlos Guimarães
Telematics Engineering Department
Universidad Carlos III de Madrid
Madrid, Spain

Zied B. Houidi
Huawei Technologies
Boulogne-Billancourt
France

Patrick Kalmbach
Chair of Communication Networks
Department of Electrical and
Computer Engineering
Technical University of Munich
Munich, Germany

Arjun Kaushik
Department of Electrical Engineering
and Computer Science
York University
Toronto, Ontario, Canada

Panagiotis Kontopoulos
National and Kapodistrian University
of Athens
Software Centric & Autonomic
Networking lab
Athens, Greece

Nikolaos Koursioumpas
National and Kapodistrian University
of Athens
Software Centric & Autonomic
Networking lab
Athens, Greece

Frank A. Kraemer
Department of Information Security
and Communication Technology
Norwegian University of Science and
Technology
Trondheim, Norway

Abdelkader Lahmadi
Université de Lorraine
CNRS
Loria, Inria, Nancy, France

Stanislav Lange
Department of Information Security
and Communication Technology
Norwegian University of Science and
Technology
Trondheim, Norway

Xi Li
NEC Laboratories Europe
5G Networks R&D Group
Heidelberg, Germany

Xudong Liao
Department of Computer Science and
Engineering, iSING Lab
Hong Kong University of Science and
Technology
Hong Kong SAR, China

Noura Limam
David R. Cheriton School of Computer
Science
University of Waterloo
Waterloo, Ontario, Canada

Justinas Lingys
Department of Computer Science and
Engineering, iSING Lab
Hong Kong University of Science and
Technology
Hong Kong SAR, China

Lina Magoula
National and Kapodistrian University
of Athens
Software Centric & Autonomic
Networking lab
Athens, Greece

Josep Mangues-Bafalluy
Communication Networks Division
Centre Tecnológic de
Telecomunicacions Catalunya
(CTTC/CERCA)
Barcelona, Spain

Jorge Martín-Pérez
Telematics Engineering Department
Universidad Carlos III de Madrid
Madrid, Spain

Ricardo Martínez
Communication Networks Division
Centre Tecnológic de
Telecomunicacions Catalunya
(CTTC/CERCA)
Barcelona, Spain

Sadeq B. Melhem
Department of Electrical Engineering
and Computer Science
York University
Toronto, Ontario, Canada

Marco Mellia
Department of Electronics and
Telecommunications
Politecnico di Torino
Torino, Italy

Stephan Merz
Université de Lorraine
CNRS
Loria, Inria, Nancy, France

Uyen T. Nguyen
Department of Electrical Engineering
and Computer Science
York University
Toronto, Ontario, Canada

Sina R. Niya
Communication Systems Group CSG
Department of Informatics IfI
University of Zürich UZH
Zürich, Switzerland

Chrysa Papagianni
Nokia Bell Labs
Antwerp, Belgium

Francesco Paolucci
Scuola Superiore Sant'Anna
Istituto TeCIP
Pisa, Italy

Francesco L. Presti
Department of Civil Engineering and
Computer Science Engineering
University of Rome Tor Vergata
Rome, Italy

Corrado Puligheddu
Department of Electronics and
Telecommunications
Politecnico di Torino
Torino, Italy

Gabriele R. Russo
Department of Civil Engineering and
Computer Science Engineering
University of Rome Tor Vergata
Rome, Italy

Mohammad A. Salahuddin
David R. Cheriton School of Computer
Science
University of Waterloo
Waterloo, Ontario, Canada

José Santos
Department of Information
Technology
Ghent University – imec, IDLab
Ghent, Technologiepark-Zwijnaarde
Oost-vlaanderen, Belgium

Eryk Schiller
Communication Systems Group CSG
Department of Informatics IfI
University of Zürich UZH
Zürich, Switzerland

Stefan Schmid
Faculty of Computer Science
University of Vienna
Vienna, Austria

Nicolas Schnepf
Department of Computer Science
Aalborg University
Aalborg, Denmark

Susanna Schwarzmann
Department of Telecommunication
Systems
TU Berlin
Berlin, Germany

Andrea Sgambelluri
Scuola Superiore Sant'Anna
Istituto TeCIP
Pisa, Italy

Nashid Shahriar
Department of Computer Science
University of Regina
Regina, Saskatchewan, Canada

Francesca Soro
Politecnico di Torino
Torino, Italy

Burkhard Stiller
Communication Systems Group CSG
Department of Informatics IfI
University of Zürich UZH
Zürich, Switzerland

Hina Tabassum
Department of Electrical Engineering
and Computer Science
York University
Toronto, Ontario, Canada

Luca Valcarenghi
Scuola Superiore Sant'Anna
Istituto TeCIP
Pisa, Italy

Bruno Volckaert
Department of Information
Technology
Ghent University – imec, IDLab
Ghent, Technologiepark-Zwijnaarde
Oost-vlaanderen, Belgium

Luca Vassio
Politecnico di Torino
Torino, Italy

Tim Wauters
Department of Information
Technology
Ghent University – imec, IDLab
Ghent, Technologiepark-Zwijnaarde
Oost-vlaanderen, Belgium

Johannes Zerwas
Chair of Communication Networks
Department of Electrical and
Computer Engineering
Technical University of Munich
Munich, Germany

Engin Zeydan
Communication Networks Division
Centre Tecnológic de
Telecomunicacions Catalunya
(CTTC/CERCA)
Barcelona, Spain

Nur Zincir-Heywood
Faculty of Computer Science
Dalhousie University
Halifax, Nova Scotia, Canada

Thomas Zinner
Department of Information Security
and Communication Technology
Norwegian University of Science and
Technology
Trondheim, Norway

Preface

Advances in artificial intelligence and machine learning algorithms provide endless possibilities in many different science and engineering disciplines including computer communication networks. Research is therefore needed to understand and improve the potential and suitability of artificial intelligence and machine learning in general for communications and networking technologies and research, but also in particular systems and networks operations and management. Approaches and techniques such as artificial intelligence, data mining, statistical analysis, and machine learning are promising mechanisms to harness the immense stream of operational data in order to improve the management and security of IT systems and networks. This will not only provide deeper understanding and better decision-making based on largely collected and available operational data but will also present opportunities for improving data analysis algorithms and methods on aspects such as accuracy, scalability, and generalization.

This book will focus on recent, emerging approaches, and technical solutions that can exploit artificial intelligence, machine learning, and big data analytics for communications networks and service management solutions. In this context, the book is intended to be a reference book for information and communications technology educators, engineers, and professionals, in terms of presenting a picture of the current landscape and discussing the opportunities and challenges of this field for the future. It is not intended as a textbook. Having said this, it can be used as a reference text for related graduate courses or high-level undergraduate courses on topic.

This book is composed of three parts and 13 chapters that provide an in-depth review of current landscape, opportunities, challenges, and improvements created by the artificial intelligence and machine learning techniques for network and service management.

The first part, Introduction, gives a general overview of the network and service management research as well as the artificial intelligence and machine learning techniques.

Chapter 1, Overview of Network and Service Management, outlines the field of network and service management that involve the setup, configuration, administration, and management of networks and associated services to ensure that network resources are effectively made available to customers and consumed as efficiently as possible by applications.

Chapter 2, Overview of Artificial Intelligence and Machine Learning, overviews the AI/ML algorithms that are most commonly used in the network and service management field, and discusses the strategic areas within network and services management that evidence growing interest of the community in developing cutting edge AI/ML solutions.

The second part of the book, Management Models and Frameworks, is dedicated to important management models and frameworks such as virtualized networks, 5G networks, and fog computing.

Chapter 3, Managing Virtualized Networks and Services with Machine Learning, exposes the state-of-the-art research that leverages Artificial Intelligence and Machine Learning to address complex problems in deploying and managing virtualized networks and services. It also delineates open, prominent research challenges and opportunities to realize automated management of virtualized networks and services.

Chapter 4, Self-Managed 5G Networks, discusses the main challenges that must be faced to successful develop 5G systems, focusing particularly on radio access networks, optical networks, data plane management, network slicing, and service orchestration, and highlights autonomous data-driven network management and federation among administrative domains that are critical for the development of 5G-and-beyond systems.

Chapter 5, AI in 5G Networks: Challenges and Use Cases, covers three representative case studies including QoE assessment, deployment of virtualized network functions, and slice management. It further points out general and use case-specific requirements and challenges and derives guidelines for network operators who plan to deploy such mechanisms.

Chapter 6, Machine Learning for Resource Allocation in Mobile Broadband Networks, provides an in-depth review of the existing machine learning techniques that have been applied to wireless networks in the context of wireless spectrum and power allocations, user scheduling, and user association.

Chapter 7, Reinforcement Learning for Service Function Chain Allocation in Fog Computing, explores the use of reinforcement learning as an efficient and scalable solution for service function chaining, especially given the dynamic behavior of the network and the need for efficient scheduling strategies, as compared to the state-of-the-art integer linear programming-based implementations.

The third part of the book, Management Functions and Applications, is focused on vital management function and applications including performance management, security management, and Blockchain applications.

Chapter 8, Designing Algorithms for Data-Driven Network Management and Control: State-of-the-Art and Challenges, provides an overview of approaches that use machine learning and artificial intelligence to learn from problem solution pairs to improve network algorithms. It discusses the applicability for different use cases and identifies research challenges within those use cases.

Chapter 9, AI-Driven Performance Management in Data-Intensive Applications, overviews recurring performance management activities for data-intensive applications and examines the role that AI and machine learning are playing in enhancing configuration optimization, performance anomaly detection, load forecasting, and auto-scaling of software systems.

Chapter 10, Datacenter Traffic Optimization with Deep Reinforcement Learning, develops a two-level deep reinforcement learning system as a scalable end-to-end traffic optimization system that can collect network information, learn from past decisions, and perform actions to achieve operator-defined goals.

Chapter 11, The New Abnormal: Network Anomalies in the AI Era, summarizes recent developments on how AI algorithms bring new possibilities for anomaly detection, and discusses new representation learning techniques such as Generative Artificial Networks and Autoencoders, and new techniques such as reinforcement learning that can be used to improve models learned with machine learning algorithms.

Chapter 12, Automated Orchestration of Security Chains Driven by Process Learning, describes an automated orchestration methodology for security chains in order to secure connected devices and their applications and illustrates how it could be used for protecting Android devices by relying on software-defined networks.

Chapter 13, Architectures for Blockchain-IoT Integration, focuses on defining and determining measures and criteria to be met for an efficient Blockchain and Internet-of-Things integration. It discusses the integration incentives and suitable use cases, as well as the dedicated metrics for scalability, security, and energy efficiency.

New York

Nur Zincir-Heywood
Marco Mellia
Yixin Diao

Acknowledgments

We sincerely thank all authors for their contributions. This book would not have been possible without their support and sharing of long-time expertise to benefit the broader audience of this book. We are especially thankful to our Book Series Editors Dr. Mehmet Ulema and Dr. Veli Sahin for inspiring us to start this book project and for providing enthusiastic support throughout. Last but certainly not least, we want to express our sincere gratitude to IEEE – Wiley editors Mary Hatcher, Teresa Netzler, and Victoria Bradshaw for their countless effort to make this book become a reality.

Acronyms

5G	Fifth generation standard for broadband cellular networks
6G	Sixth generation standard for broadband cellular networks
AD	Administrative Domain
AE	Auto Encoder
AF	Application Function
AI	Artificial Intelligence
ANN	Artificial Neural Networks
API	Application Programming Interface
AP	Access Point
ARQ	Automatic Repeat reQuest
AS	Autonomous System
ASIC	Application-Specific Integrated Circuit
AWS	Amazon Web Services
BC	Blockchain
BGP	Border Gateway Protocol
BNG	Broadband Network Gateway
C/S	Client-Server
CNN	Convolutional Neural Networks
CDN	Content Distribution Network
ConvLSTM	Convolutional Long-Short Term Memory
CQI	Channel Quality Indicator
D2D	Device-to-Device
DAG	Directed Acyclic Graphs
DASH	Dynamic Adaptive Streaming over HTTP
DC	Data Center
DDoS	Distributed Denial-of-Service
DL	Deep Learning
DLT	Distributed Ledger Technology
DNN	Deep Neural Network

E2E	End-to-End
EM	Enforcement Module
FC	Fog Computing
GBM	Gradient Boosting Machine
GCN	Graph Convolutional Network
GNN	Graph Neural Network
GP	Gaussian Process
GUI	Graphical User Interface
HDFS	Hadoop Distributed File System
HetNets	Heterogeneous Networks
IAM	Identity and Access Management
ILP	Integer Linear Programming
IoT	Internet of Things
kNN	K-Nearest Neighbors
KPI	Key Performance Indicator
LoRaWAN	Long-Range Wide-Area Network
LP-WAN	Low-Power Wide Area Network
LSTM	Long-Short Term Memory
MAC	Media Access Control
MANO	Management and Orchestration
MDP	Markov Decision Process
MEC	Multi-access Edge Computing
MIB	Management Information Base
MILP	Mixed-Integer Linear Programming
MINLP	Mixed Integer Nonlinear Programming Problems
ML	Machine Learning
MLP	Multilayer Perceptron
MM	Monitor Module
mMTC	Massive Machine Type Communications
mmWave	Millimeter Wave
MNO	Mobile Network Operator
MOS	Mean Opinion Score
MPLS	Multiprotocol Label Switching
MSE	Mean Squared Error
MTU	Maximum Transmission Unit
NFV	Network Function Virtualization
NFVI	Network Function Virtualization Infrastructure
NFVO	Network Function Virtualization Orchestrator
NIC	Network Interface Controller
NN	Neural Network
NOC	Network Operation Center

ONF	Open Networking Foundation
OTN	Optical Transport Network
OTS	Optical Transport Section
P2P	Peer-to-Peer
PK	Public Key
PoP	Point of Presence
QC	Quantum Computing
QoE	Quality of Experience
QoS	Quality of Service
RAN	Radio Access Network
RAP	Radio Access Point
RDD	Resilient Distributed Dataset
RIP	Routing Information Protocol
RL	Reinforcement Learning
RNN	Recurrent Neural Network
RRM	Radio Resource Management
RTT	Round Trip Time
SC	Smart Contract
SDN	Software Defined Networking
SFC	Service Function Chaining (updated in regards to Service Function Chain)
SINR	Signal-to-Interference-Plus-Noise Ratio
SJF	Shortest Job First
SLA	Service Level Agreement
SNMP	Simple Network Management Protocol
SNR	Signal-to-Noise Ratio
SVM	Support Vector Machine
SVR	Support Vector Regression
TO	Traffic Optimization
TPS	Transaction Per Second
TSP	Traveling Salesman Problem
V2I	Vehicle to Infrastructure
V2V	Vehicle to Vehicle
vBS	Virtual Base Station
VM	Virtual Machine
VMO	Virtual Mobile Operator
VNE	Virtual Network Embedding
VNF	Virtual Network Function
WAN	Wide Area Network
WLAN	Wireless Local Area Network
WN	Wireless Nodes

Part I

Introduction

1

Overview of Network and Service Management

Marco Mellia[1], Nur Zincir-Heywood[2], and Yixin Diao[3]

[1] *Department of Electronics and Telecommunications, Politecnico di Torino, Torino, Italy*
[2] *Faculty of Computer Science, Dalhousie University, Halifax, Nova Scotia, Canada*
[3] *PebblePost, New York, NY, USA*

1.1 Network and Service Management at Large

Nowadays the network, i.e. the Internet, has become a fundamental instrument to effectively support high value solutions that involve our daily life. Born to carry mainly data, today we use the Internet to watch high-definition videos, conduct video conferences, stay informed, participate in social networks, play games, buy goods, and do business. All these value-added services call for maintaining superior network service levels – where service disruption is not tolerated, and the quality of the service must be guaranteed. As a result of the many impacts of digitalization, the Internet has become increasingly complex and difficult to manage, with mobile broadband access networks able to connect billions of users at hundreds of megabit per seconds, backbone networks extending for thousands of kilometers with multi terabit per second channels, and huge datacenters hosting hundreds of thousands of servers, virtual machines, and applications.

The need for network and service management raised together with the first network concepts, with fundamentals that were defined within the International Organization for Standardization's Open Systems Interconnection (ISO/OSI) reference model [1, 2]. Telephone networks started moving to digital services in the 1970s, which created the need to manage these services automatically [3]. Computer communication technology radically changed the networking paradigm, with Transmission Control Protocol/Internet Protocol (TCP/IP) leading to the birth of the Internet as we know it today. Originally, computer network management was mostly a manual activity, in which the network administrator knew the configuration by hearth of each device and was able to quickly intervene

Communication Networks and Service Management in the Era of Artificial Intelligence and Machine Learning,
First Edition. Edited by Nur Zincir-Heywood, Marco Mellia, and Yixin Diao.

in case of problems. Nowadays, with networks of billions of devices, millions of nodes, thousands of applications, network and service management has evolved to be as much as possibly automated. The advances with centralized and distributed approaches have enabled the Network Operation Center (NOC) to visualize and control the network in an as much as possible automatic fashion. Today, with the abilities to collect and process large amount of data, network and service management is facing a new stimulus toward the complete automation, with machine learning and artificial intelligence approaches that start being deployed in operations.

Network and service management fundamentally implements a control loop in which data about the status of the network is collected to be then processed in a centralized or distributed fashion to detect changes, with the goal to define which actions to implement, react, and control the changes. Figure 1.1 presents a high level overview of the overall process. From the left, data about network status is collected to continuously monitor its health. Big data technologies coupled with machine learning and artificial intelligence solutions allow to collect, analyze, and derive plans to resolve issues, which are then distributed to the network devices to implement the desired changes. In the following, we present an overview of technologies to face the monitoring and execute steps. We explicitly focus on the protocols to collect and monitor the status of the network and to distribute the management decisions. We instead leave for specific chapters the description of the algorithms and approaches which are – by definition – very dependent on the use case and on the specific technologies. Our goal in this chapter is to provide a quick overview of the latest trends in the technologies for network and service management, and to give a high-level overview of solutions in dominant scenarios

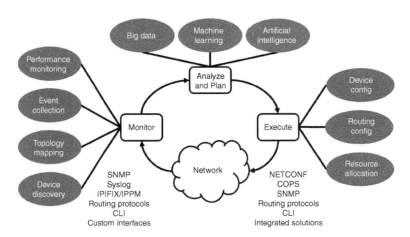

Figure 1.1 Network and service management at large.

so that the reader gets a view of the bigger picture of the problems. We leave specific solutions to the single chapters along with examples and more in-depth discussions. We focus on the Internet mainly, being it the nowadays dominant network.

1.2 Data Collection and Monitoring Protocols

Any decision process must be guided by the ability to obtain data about the status of the system. In a typical network, devices from different vendors, with different functionalities, different capabilities, different administrative domains create heterogeneous scenarios where collecting data calls for standardized instruments and tools. Often this heterogeneity produces custom solutions provided by each vendor, offering advanced and proprietary solutions to interact with the different and custom devices. Here we present an overview of the major standard protocols that allow one to collect data from network devices, leaving custom solutions out of this description.

1.2.1 SNMP Protocol Family

Original TCP/IP network management is based on the Simple Network Management Protocol (SNMP) family. SNMP standardizes the collection and organization of information about devices on an IP network. It is based on the manager/agent model with a simple request/response format. Here, the network manager issues a request and the managed agents will send responses in return. SNMP exposes management data in the form of variables organized in a Management Information Base (MIB) which describes the system status and configuration. These variables can then be remotely queried and manipulated, allowing both the collection of information and the changes in configuration – provided the manager has controlling authorization on such variables. SNMPv1 is the original version of the protocol [4]. More recent versions, SNMPv2c and SNMPv3, feature improvements in performance, flexibility, and especially security [5, 6].

Via this simple approach, an authorized agent can remotely check and change the configuration of devices under its administrative domain, propagating changes, while obtaining an updated picture of the network status. SNMP offers a means thus both to collect information from and to control the network devices, but does not provide any means to define which is the best configuration to deploy.

1.2.2 Syslog Protocol

Similarly to SNMP, the Syslog protocol family [7] offers mechanisms for collection of logging information. Initially used on Unix systems and developed since

1980, the protocol introduces a layered architecture allowing the use of any transport protocols. The Syslog protocol enables a machine to send system log messages across networks to event message collectors. It implements a push approach, where the devices send information to the collectors. The protocol is simply designed to transport and distribute these event messages, enabling the centralized collection of logs from servers, routers, and devices in general. Differently from SNMP – Syslog does not allow to distribute any configuration, which shall be achieved using other communication channels.

Messages include a facility code and a severity level. The former identifies the type of program that is logging the message (e.g. kernel, user, mail, daemon, etc.). The latter defines the urgency of the message (e.g. emergency, alert, critical, error, warning, debug, etc.). This allows for simple filtering and easy reading of the messages. When operating in a network, syslog uses a client-server paradigm, where the collector server listens for messages from clients. Born to leverage User Datagram Protocol (UDP), recent versions support TCP and Transmission Level Security (TLS) protocol for reliable and secure communications.

Syslog suffers from the lack of standard message format, so that each application supports a custom set of messages. It is common that even different software releases of the same application use different formats, thus making the parsing of the messages complicated by automatic solutions.

1.2.3   IP Flow Information eXport (IPFIX)

Both syslog and SNMP allow to collect information about the status of devices. Internet Protocol Flow Information Export (IPFIX) Protocol defines instead a means to collect in a standard way information about the traffic flowing in the network. The granularity at which it works is the flow, i.e. a group of packets having the same source and destination [8]. It defines the components involved in the measurement and reporting of information on IP flows. A Metering Process generates Flow Records; an Exporting Process transmits the information using the IPFIX protocol; and a Collecting Process receives it as IPFIX Data Records. The IPFIX protocol is a push mechanism only, and IPFIX cannot distribute configurations to the Exporters. As Syslog, it offers the means to collect information about the traffic flowing in a network, but does not provide any means to process it. Being based on traffic meters, it opens the possibility of implementing traffic profiling, traffic engineering, QoS monitoring, and intrusion detection solutions that analyze the flow-based traffic measurements and generate valuable feedback to the network managers. IPFIX is an evolution of NetFlow, a custom predecessor introduced by Cisco in 1996 to collect and monitor IP network flow information. IPFIX not only supports the Stream Control Transmission Protocol (SCTP) at the transport layer but also allows the use of the TCP or UDP to offload the meter application.

NetFlow and IPFIX protocols are examples of "metadata-based" techniques which can provide valuable operational insight for network performance, security, and other applications. For instance, in IP networks, metadata records document the flows. In each flow record, the "who" and "whom" are IP addresses and port numbers, and the "how long" is byte and packet counts. Direct data capture and analysis of the underlying data packets themselves can also be used for network performance and security troubleshooting, e.g. exporting the raw packets. This typically involves a level of technical complexity and expense that in most situations does not produce more actionable understanding vs. an effective system for the collection and analysis of metadata comprising network flow records.

The main critical point of IPFIX is its lack of scalability, for the data collection at the exporter, and the excessive the network load at the collector. This forces often to activate packet sampling options which limits visibility.

1.2.4 IP Performance Metrics (IPPM)

Internet Protocol Performance Metrics (IPPM) is an example of a successful standardization effort [9]. It defines metrics for accurately measuring and reporting the quality, performance, and reliability of the network. These include connectivity, one-way delay and loss, round-trip delay and loss, delay variation, loss patterns, packet reordering, bulk transport capacity, and link bandwidth capacity measurements. It offers a standard and common ground to define and measure performance so that even measurements performed by different vendors and implementations shall refer to the same monitored metric. In a nutshell, it opens the ability for common performance monitoring.

Among the standard protocols, the One-Way Active Measurement Protocol and Two-Way Active Measurement Protocol (OWAMP [10] and TWAMP [11], respectively) metrics specification allows delay, loss, and reordering measurements. OWAMP can be used bi-directionally to measure one-way metrics in both directions between two network elements. However, it does not natively support round-trip or two-way measurements. The TWAMP extends the OWAMP capabilities to add two-way or round-trip measurement. Two hosts are involved in the measurement. In the case of OWAMP, the sender and the receiver collaborate actively to measure the desired performance index. For instance, to compute the one-way-delay, both take a proper timestamp of the measurement packet, at the sending and receiving time, respectively. In the TWAMP, the receiver can act as a simple reflector that just sends back (or to a third party) the probe packet sent by the sender, with no additional computation effort.

Open source and proprietary implementations are readily available for both IPv4 and IPv6 protocol stacks. These are commonly integrated in monitoring platforms [12] as well, namely Perfsonar [13] or RIPE Atlas [14].

1.2.5 Routing Protocols and Monitoring Platforms

Routing protocols are among the most successful deployed solutions to manage a network. A routing protocol specifies how routers communicate each other to exchange information that allows them to get the current network topology and compute the paths to reach possible destinations. Routing protocols give the Internet the ability to dynamically adjust to changing conditions such as topology changes, links and node failures, and congestion situations. There are two main classes of routing protocols in use on IP networks. Interior gateway protocols based distance-vector routing protocols, such as Routing Information Protocol (RIP) [15], Enhanced Interior Gateway Routing Protocol (EIGRP) [16], or based on link-state routing protocols, such as Open Shortest Path First (OSPF) [17], Intermediate System to Intermediate System IS-IS [18], are used in networks that belong to the same administrator domain, i.e. within the same Autonomous System (AS). Interior gateways protocols base their decision on the minimization of the path costs, defined as the sum of link costs. As such, they aim at minimizing the cost of routing the traffic, i.e. maximizing the performance. Exterior gateway protocols aim instead at exchanging routing information between Autonomous Systems and finding the most convenient path – in terms of Autonomous Systems – to reach the destination. Here, Border Gateway Protocol (BGP) [19] is the de facto only choice. It is a path-vector routing protocol and it makes routing decisions based on network policies and rules and not based on cost functions. BGP allows network operators to define routing policies that reflects administrative costs and political decisions in terms of agreements between Autonomous Systems.

Given the importance of optimizing exterior routing policies and the partial view that each network operator can get of the global Autonomous System (AS) level topology, several mechanisms are in place to gain visibility on the current Internet routing. Among those, the University of Oregon Route Views Project [20] leverages information provided by *collectors*, vantage points that expose their partial view of the BGP data, to create interactive maps, which are historized and made browsable via an ecosystem of tools and software that simplify the management and query of the information [21]. Thanks to Routeviews and the information exposed by BGP, it is possible to observe Internet-wide outages [22, 23], routing hijacking [24], routing anomalies [25], or check the IPv4 address space utilization [26].

All the above-mentioned routing protocols implement closed loop mechanisms – from monitoring to actions. Another category of routing protocols enable traffic engineering and network management opportunities. Among those,

Multiprotocol Label Switching (MPLS) [27] is a routing technique based on the label swapping principle. Each node along the path reads the incoming packets' label and uses it to quickly route the packets to the next hop. Before the forwarding operation, the packet label is replaced with a new label that indicates the next forwarding operation to be done at the next node. Via a concatenation of labels, packets follow a pre-computed path (a so called MPLS tunnel), which is distributed to all the nodes along the path prior the actual transmission. This on the one hand avoids complex look-ups in the routing table, and on the other hand it enables the definition of explicit and well-controlled paths that traffic flows will follow. By computing explicit tunnels is then possible to implement complex traffic engineering policies [28], setup end-to-end virtual private networks (VPNs) [29], and design specific protection mechanisms that quickly recover connectivity in case of failures [30].

1.3 Network Configuration Protocol

As said, while there has been a standardized means to collect information about the status of devices and of traffic, each vendor typically offers its own mechanisms to distribute configurations. The heterogeneity of devices, vendors, and versions makes indeed it difficult to define a common and flexible structure able to support and fit different requirements. This hampered the adoption of standard protocols, which are confined to a mostly academic design, with little deployment.

1.3.1 Standard Configuration Protocols and Approaches

The NETCONF protocol is an example of a standard mechanisms that allow to install, manipulate, and delete the configuration of network devices [31]. It uses an XML-based data encoding for the configuration data as well as the protocol messages. A key aspect of NETCONF is that it allows the functionality to closely mirror the native command-line interface of the device. It provides a standard way for authentication, data integrity, and confidentiality. For this, it depends on the underlying transport protocol for this capability. For example, connections can be encrypted in TLS or SSH, depending on the device support. Along with NETCONF, a data modeling language defining the semantics of operational and configuration data, notifications, and operations has been defined via the introduction of the YANG modeling language [32]. Neither NETCONF nor YANG ever succeed in becoming an actual standard, given the difficulty to find a common and flexible ground that fits all requirements.

The Internet Engineering Task Force (IETF) defined a general policy framework for managing, sharing, and reusing policies in a vendor-independent, interoperable, and scalable manner [33]. The Policy Core Information Model (PCIM) is an object-oriented information model for representing policy information. It specifies two main architectural elements: the Policy Enforcement Point (PEP) and the Policy Decision Point (PDP). Policies allow an operator to specify how the network is to be configured and monitored by using a descriptive language. It allows the automation of management tasks, according to the requirements set out in the policy module. The IETF Policy Framework has been accepted by the industry as a standard-based policy management approach and has been adopted by the third Generation Partnership Project (3GPP) standardization as well.

The Common Open Policy Service (COPS) is a protocol that provides a client/server model to support policy control. The COPS specification is independent of the type of policy being provisioned (QoS, security, etc.) but focuses on the mechanisms and conventions used to distribute information between PDPs and PEPs. COPS has never been widely deployed because operators found its use of binary messages complicates the development of automated scripts for simple configuration management tasks.

1.3.2 Proprietary Configuration Protocols

As previously said, each vendor has implemented its own solution to collect, change, distribute configurations and system updates. Big vendors such as Cisco Systems, Juniper Networks, Huawei, etc. provide different suites that range from solutions for simple local area networks (LANs), to internet provider scale solutions. The so called Network Management Systems [34] simplify the management of the administered network offering centralized solutions that allow one to perform device discovery, monitoring and management, network performance analysis, intelligent notifications, and customizable alerts. To interact with devices, they build on standard protocols such as SNMP or syslog, but often use also custom solutions based on Command Line Interfaces (CLI) that can be reached via SSH or telnet (deprecated for security reasons). For instance, the Cisco Configuration Professional is a Graphical User Interface (GUI)-based device management tool for Cisco access routers. This tool simplifies routing, firewall, Intrusion Prevention System (IPS), VPN, unified communications, wide area network (WAN) and LAN configurations through GUI-based easy-to-use wizards.

1.3.3 Integrated Platforms for Network Monitoring

As previously said, vendors and third party companies offer a portfolio of management solutions, which range to simple network management for small deployments, to Internet Service Provider scale solutions, from LAN to Data Center Networks.

The main goal of these platforms is to offer a unified view of the network and service status. These platforms are able to collect data from devices belonging to an administration domain via SNMP, Syslog, IPFIX, and proprietary solutions. Often they implement an automatic discovery mechanism to find and add devices to their collection base so to minimize administrator intervention. Via a GUI, they present views of the status of the network, showing time series of link and CPU load, divided by applications or origin-destination of the traffic. The administrator is thus offered a unified view of the network status, with the ability to drill down into more details directly interacting with the GUI. They can also detect network node and connection health problems by using simple threshold-based algorithms. In such cases, alerts can be issued to warn the administrators. Figure 1.2 reports the Zabbix architecture as an example.

From an architecture point of view, all these platforms are similar. They have *proxy modules*, also called *agent modules*, to interact with different protocols and devices to collect data, which is then stored in a *database module*, based on open source solutions like MySQL, Postgre SQL, or commercial solutions like Oracle SQL. A typically *web-based GUI* or dashboard allows the administrator to interact and navigate through the data. The dashboard can offer also configuration

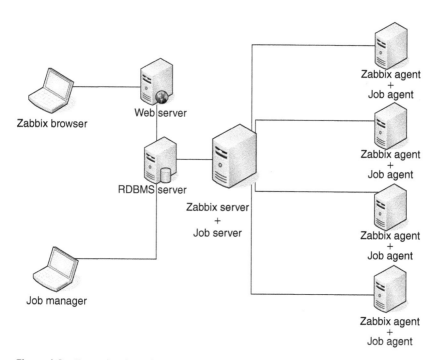

Figure 1.2 Example of monitoring architecture. Source: Courteously from Zabbix.

abilities, typically opening management connection with the devices. At last, a *media gateway* allows the system to raise and distribute alarms, via email, short message service, chat systems, ticketing systems, etc.

Some platforms are open source. They allow to integrate data collected from various deployment into a single centralized center, but rarely offer the ability to change the underlying configuration due to the difficulties in interfacing with different devices. Among those, Zabbix (https://www.zabbix.com), Nagios (https://www.nagios.org), or Cacti (https://www.cacti.net) are the oldest, with more modern solutions like LibreNMS (https://www.librenms.org) or Observium (https://www.observium.org) emerging as novel and more reactive solutions.

Proprietary solutions offer typically more options and flexibility, and include also the ability to change the network setup. Each vendor has a portfolio of solutions that fits different scenarios and deployment sizes, from small LANs to national-wide Internet Service Providers. Solutions are also available from independent vendors that have typically multi-platform support.

1.4 Novel Solutions and Scenarios

In the previous sections of this chapter, we have described the most standard approach to control and manage a network. Here we briefly present the most recent approaches which are still under investigations by the research and technical communities, with development quickly tacking grounds.

1.4.1 Software-Defined Networking – SDN

Software-defined networking (SDN) technology is an approach to network management that separate the control plane from the data plane. In the original internet design indeed, the control plane – where control protocols and management actions are performed – is tightly embedded in the data plane – where packets are routed and forwarded. SDN separates the two planes, so that switches become pure forwarding devices, while all the control and management operations are relegated to a centralized controller. The controller defines forwarding rules, which are then send to switches that use them to forward packets along the proper and desired path. This enables dynamic, programmatically efficient network configuration to improve network performance and monitoring. Martin Casado introduced the idea of relying to a centralized controller to improve network management in 2007 [35]. Since then, SDN technology has become mainstream [36], with support first for campus network, then extending its support for data center networks, and more recently in WANs via the SD-WAN [37], bringing in the WAN area the benefits of decoupling the networking hardware from its control mechanism.

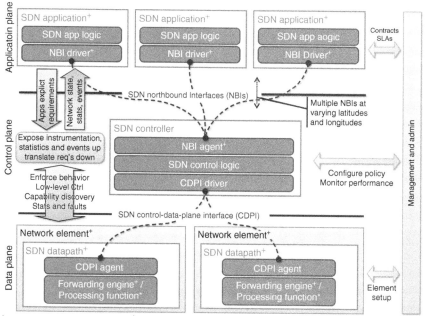

Figure 1.3 The SDN architecture. Source: Courteously from Open Networking Foundation.

The SDN architecture identifies three planes – adding an application plane on the top of the control plane. Figure 1.3 depicts the overall architecture. SDN applications are programs that directly and programmatically communicate their requirements and desired behavior via the northbound interface to the SDN network controller. Applications get an abstracted global view of the network, and suggest decisions and actions such as explicit routes, filtering rules, etc. The SDN controller sits in between. It is a logically centralized entity that translates the requirements from applications to actual action to be implemented by the control plane elements, and provides the applications an updated a common view of the network status. Logically centralized, it can be implemented in a distributed fashion to guarantee both scalability and reliability. It supports both the concept of federated controllers – each responsible of managing a portion of the network; and of hierarchical controllers – where higher hierarchy controllers summarize the information received by lower layers and make it available to applications. At the bottom, the data plane – or the Datapath – is the logical network of devices which offer forwarding and data processing capabilities. Data forwarding engines are in charge of quickly switching packets. They communicate with the SDN controller via the southbound interface, which defines standard Application Programming

Interfaces (API) to exchange information. Traffic processing functions implement decision based on packet payload. For instance, switching decision can be done considering both the sender and receiver addresses – enabling per-flow routing. Similarly, filtering decision can be based on TCP port numbers.

SDN is often associated with the OpenFlow protocol [38] that enables the remote communication with the network plane elements and the controller. However, for many companies, it is no longer an exclusive solution, and proprietary techniques are now available like the Open Network Environment and Nicira's network virtualization platform. They all offer the standard API to communicate via the southbound interface.

1.4.2 Network Functions Virtualization – NFV

Network Functions Virtualization (NFV) is a network architecture that strongly builds on the top of virtualization concepts [39]. It offers the ability to virtualize network nodes and functions into building blocks which can be connected and chained to create more complex communication services. A virtualized network function (VNF) consists of one or more virtual machines and containers that run specific software to implement networking operations in software. Firewalls, access list controllers, load balancers, intrusions detection systems, VPN terminators, etc. can thus be implemented in software – without buying and installing expensive hardware solutions.

NFV consists of three main components as sketched in Figure 1.4: On the top, the VNFs to be implemented, using a software solution; the network functions virtualization infrastructure (NFVI) sits in the middle and offers the hardware

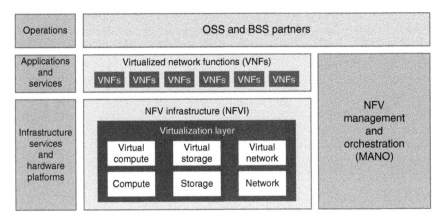

Figure 1.4 Network functions virtualization architecture. Source: Courteously from Juniper Networks.

components over which deploy the VNFs. It includes the physical servers and the network devices that build the NFV infrastructure; at last, the NFV MANagement and Orchestration (MANO) framework allows to manage the platform offering data repositories and standard interfaces to exchange information. To build a complex function, basic blocks can be chained so that a processing pipeline is built. This is called "service chaining" and allows the reuse of highly specialized and efficient blocks to build complex functionalities.

Considering the management operations, clearly NFV requires the network to instantiate, monitor, repair, and bill for the services it offers. NFV targets indeed the large carrier scenario, being it a data center manager, or an internet service providers. These functionalities are allocated to the orchestration layer, which must manages VNFs irrespective of the actual hardware and software technology sitting below.

NFV is a means to reduce cost and accelerate service development and deployment. Instead of requiring the installation of expensive hardware with dedicated functionalities, service providers rely on inexpensive network devices, storage systems, and servers to run virtual machines that implement the desired network function. When a customer asks for a net functionality, the service provider can simply spin up a new virtual machine to implement that function. This has also the benefit to reduce the dependency on dedicated hardware devices, and improve robustness via migration capabilities that move services in case of failures or maintenance operations.

Clearly, NFV calls for standard to allow interoperability of solutions. Since 2012, over 130 of the world's leading network operators have recently joined together to form a European Telecommunications Standards Institute (ETSI) Industry Specification Group (ISG) for NFV (https://www.etsi.org/technologies/nfv). NFV is also fundamental in the 5G arena, where all the advanced functionalities offered by the network like network slicing, edge computing, or decentralized radio management functions are implemented on the top of NFV.

Bibliography

1 Caruso, R.E. (1990). Network management: a tutorial overview. *IEEE Communications Magazine* 28 (3): 20–25.

2 Klerer, S.M. (1988). The OSI management architecture: an overview. *IEEE Network* 2 (2): 20–29.

3 (1984). Specification of Signalling System No. 7.

4 Case, J.D., Fedor, M., Schoffstall, M.L., and Davin, J. (1990). RFC 1157: simple network management protocol (SNMP). *Request for Comments, IETF.*

5 Case, J., McCloghrie, K., Rose, M., and Waldbusser, S. (1996). RFC 1901: introduction to community-based SNMPv2. *Request for Comments, IETF.*

6 Harrington, D., Presuhn, R., and Wijnen, B. (2002). RFC 3411: an architecture for describing simple network management protocol (SNMP) management frameworks. *Request for Comments, IETF.*

7 Gerhards, R. (2009). RFC 5424: the syslog protocol. *Request for Comments, IETF.*

8 Claise, B., Bryant, S., Sadasivan, G. et al. (2008). RFC 5101: specification of the IP flow information export (IPFIX) protocol for the exchange of IP traffic flow information. *Request for Comments, IETF.*

9 Paxson, V., Almes, G., Mahdavi, J., and Mathis, M. (1998). RFC 2330: framework for IP performance metrics. *Request for Comments, IETF.*

10 Almes, G., Kalidindi, S., and Zekauskas, M. (1999). RFC 2679: a one-way delay metric for IPPM. *Request for Comments, IETF.*

11 Hedayat, K., Krzanowski, R., Morton, Al. et al. (2008). RFC 5357: a two-way active measurement protocol (TWAMP). *Request for Comments, IETF.*

12 Bajpai, V. and Schönwälder, J. (2015). A survey on internet performance measurement platforms and related standardization efforts. *IEEE Communication Surveys and Tutorials* 17 (3): 1313–1341.

13 Hanemann, A., Boote, J.W., Boyd, E.L. et al. (2005). PerfSONAR: a service oriented architecture for multi-domain network monitoring. In: *International Conference on Service-Oriented Computing* (A. Hanemann, J.W. Boote, E. L. Boyd et al.), 241–254. Springer.

14 RIPE NCC Staff (2015). Ripe atlas: a global internet measurement network. *Internet Protocol Journal* 18 (3). http://ipj.dreamhosters.com/wp-content/uploads/2015/10/ipj18.3.pdf

15 Malkin, G. (1998). RFC 2453: RIP version 2. *Request for Comments, IETF.*

16 Savage, D., Ng, J., Moore, S. et al. (2016). RFC 7868: Cisco's enhanced interior gateway routing protocol (EIGRP). *Request for Comments, IETF.*

17 Moy, J. (1998). RFC 2328: OSPF version 2. *Request for Comments, IETF.*

18 Vasseur, J.P., Shen, N., and Aggarwal, R. (2007). RFC 4971: intermediate system to intermediate system (IS-IS) extensions for advertising router information. *Request for Comments, IETF.*

19 Shalunov, S., Teitelbaum, B., Karp, A. et al. (2006). RFC 4656: a one-way active measurement protocol (OWAMP). *Request for Comments, IETF.*

20 Meyer, D. (1997). University of Oregon Route Views Project. http://www.routeviews.org/routeviews/.

21 Orsini, C., King, A., Giordano, D. et al. (2016). BGPStream: a software framework for live and historical BGP data analysis. *Proceedings of the 2016 Internet Measurement Conference*, pp. 429–444.

22 Giotsas, V., Dietzel, C., Smaragdakis, G. et al. (2017). Detecting peering infrastructure outages in the wild. *Proceedings of the Conference of the ACM Special Interest Group on Data Communication*, pp. 446–459.

23 Luckie, M. and Beverly, R. (2017). The impact of router outages on the AS-level internet. *Proceedings of the Conference of the ACM Special Interest Group on Data Communication*, pp. 488–501.

24 Sermpezis, P., Kotronis, V., Dainotti, A., and Dimitropoulos, X. (2018). A survey among network operators on BGP prefix hijacking. *ACM SIGCOMM Computer Communication Review* 48 (1): 64–69.

25 Padmanabhan, R., Dhamdhere, A., Aben, E. et al. (2016). Reasons dynamic addresses change. *Proceedings of the 2016 Internet Measurement Conference*, pp. 183–198.

26 Livadariu, I., Elmokashfi, A., and Dhamdhere, A. (2017). On IPv4 transfer markets: analyzing reported transfers and inferring transfers in the wild. *Computer Communications* 111: 105–119.

27 Rosen, E., Viswanathan, A., and Callon, R. (2001). RFC 3031: multiprotocol label switching architecture. *Request for Comments, IETF*.

28 Xiao, X., Hannan, A., Bailey, B., and Ni, L.M. (2000). Traffic engineering with MPLS in the internet. *IEEE Network* 14 (2): 28–33.

29 Pepelnjak, I. and Guichard, J. (2002). *MPLS and VPN Architectures*, vol. 1. Cisco Press.

30 Huang, C., Sharma, V., Owens, K., and Makam, S. (2002). Building reliable MPLS networks using a path protection mechanism. *IEEE Communications Magazine* 40 (3): 156–162.

31 Enns, R., Bjorklund, M., Schoenwaelder, J., and Bierman, A. (2011). RFC 6241: network configuration protocol (NETCONF). *Request for Comments, IETF*.

32 Bjorklund, M. (2010). RFC 6020: Yang - a data modeling language for the network configuration protocol (NETCONF). *Request for Comments, IETF*.

33 Yavatkar, R., Pendarakis, D., Guerin, R. et al. (2000). RFC 2753: a framework for policy-based admission control. *Request for Comments, IETF*.

34 Martin-Flatin, J.-P., Znaty, S., and Hubaux, J.-P. (1999). A survey of distributed enterprise network and systems management paradigms. *Journal of Network and Systems Management* 7 (1): 9–26.

35 Casado, M., Freedman, M.J., Pettit, J. et al. (2007). Ethane: taking control of the enterprise. *ACM SIGCOMM Computer Communication Review* 37 (4): 1–12.

36 Kirkpatrick, K. (2013). Software-defined networking. *Communications of the ACM* 56 (9): 16–19. https://doi.org/10.1145/2500468.2500473.

37 Yang, Z., Cui, Y., Li, B. et al. (2019). Software-defined wide area network (SD-WAN): architecture, advances and opportunities. *2019 28th International Conference on Computer Communication and Networks (ICCCN)*, IEEE, pp. 1–9.

38 McKeown, N., Anderson, T., Balakrishnan, H. et al. (2008). OpenFlow: enabling innovation in campus networks. *SIGCOMM Computer Communication Review* 38 (2): 69–74. https://doi.org/10.1145/1355734.1355746.

39 Han, B., Gopalakrishnan, V., Ji, L., and Lee, S. (2015). Network function virtualization: challenges and opportunities for innovations. *IEEE Communications Magazine* 53 (2): 90–97.

2

Overview of Artificial Intelligence and Machine Learning

Nur Zincir-Heywood[1], Marco Mellia[2], and Yixin Diao[3]

[1] *Faculty of Computer Science, Dalhousie University, Halifax, Nova Scotia, Canada*
[2] *Department of Electronics and Telecommunications, Politecnico di Torino, Torino, Italy*
[3] *PebblePost, New York, NY, USA*

2.1 Overview

As the computer and network technologies improve, the ability to acquire, access, store, and process huge amounts of data from physical distant/near locations also increase. For example, people with smartphones connected all the time to different social media systems, exchanging text, voice, photos, and videos at any time and any place. This typically amounts to gigabytes to terabytes of data every day on social networking platforms. This stored data becomes useful when it is analyzed and turned into information such as for prediction, correlation, etc. To this end, artificial intelligence (AI) and machine learning (ML) have become the techniques that are increasingly employed over the years [1–3].

AI is in most part logic based [4]. AI aims to make computers do the types of things that humans' minds can do. These include but are not limited to reasoning, planning, prediction, association, perception, etc. which enable humans to achieve their goals. There are several major types of AI from classical or symbolic AI to ML, each includes many variations. Classical/symbolic AI models planning and reasoning and can also model learning. It is based on the spirit of Turing machine combined with propositional logic and the theory of neural synapses. Complex propositions are built, and deductive arguments are carried out by using logical operators to describe reasoning systems. Expert systems, knowledge bases, and case base reasoning are some examples of classical AI. Expert systems mimic the decision-making process of a human expert. The program would ask an expert in a field how to respond in a given situation, and once this was learned for a sufficient range of situations, non-experts could receive advice from that program.

Communication Networks and Service Management in the Era of Artificial Intelligence and Machine Learning,
First Edition. Edited by Nur Zincir-Heywood, Marco Mellia, and Yixin Diao.
© 2021 The Institute of Electrical and Electronics Engineers, Inc. Published 2021 by John Wiley & Sons, Inc.

These programs would be used for creating knowledge bases which then may be used for decision support systems for different application areas. Case-based reasoners solve new problems by retrieving stored "cases" describing similar prior problem-solving episodes and adapting their solutions to fit new needs. Over the years, we have seen applications of expert systems and case-based reasoning systems in network and service management [5–7].

On the other hand, ML is data driven [8]. It means programming to optimize performance criteria using examples of data or past experience. In ML, there exists a model defined by some parameters, then the learning becomes the execution of a program to optimize the parameters of the model using the training data or the past experience. Past experience case is distinct from either supervised or unsupervised learning because credit assignment is subject to delays. Thus, it is not immediately apparent which behaviors should be rewarded or penalized. This issue is specific to reinforcement learning. The model could be predictive to make predictions in the future, or it could be descriptive to gain knowledge from data, or it could be both. ML uses statistical theory to guide model building in order to infer from the training data or the past experience. In training, efficient algorithms are necessary to solve the optimization problem as well as to store and process the data/past experience. Moreover, the model that is learned at the end of training is required to have efficient representation and solution for inference purposes, possibly in real time. In some applications of ML, the efficiency of the learning and inference model, in other words the space and time complexity could be as important as its prediction accuracy. The growth of network technologies for easy access to data, cheaper access to CPU power, and fast access to data storage has enabled the use of ML algorithms in network and service management [9, 10].

2.2 Learning Algorithms

Most researchers categorize learning algorithms into three major types based on the underlying characteristics of the task: (i) supervised learning, (ii) unsupervised learning, and (iii) reinforcement learning. In supervised learning, the aim is to learn a mapping from the input data to the output data whose correct values are provided by the ground-truth (label) during training. In unsupervised learning, there is no such ground-truth, there is only input data. Thus, the aim is to find the similarities in the input data. In reinforcement learning, on the other hand, the focus is on identifying a system that is capable of maximizing the cumulative reward received when explicitly interacting with an environment. However, independent from the task, ML has three key components, namely *representation*, *cost function*, and *credit assignment*. In this context, *representation* means the learning language used to build solutions. Examples include a neural network representation vs. representation in decision tree induction or instructions from a simple instruction set. The representation may also distinguish between those capable of supporting some form of memory and those that do not. Thus, recurrent representations define the current output as a function of previous "internal" state as well as the current input (state). If the representation does not support memory mechanisms, the resulting model is limited to reactive behaviors alone. Depending on the representation assumed, solutions might be more difficult to discover or costly to estimate. Supporting memory would be beneficial under tasks that are partially observable, but might also decrease the ability to establish how decisions have been made (transparency). *Cost function* refers to how the performance of a solution is evaluated, e.g. classification or prediction accuracy, posterior probability or how simple a solution should be. *Credit assignment* guides how the representation is modified, i.e. rewarding/punishing to guide the search process. In the following, we will discuss the types of ML in more detail to gain more insight and understand their uses.

2.2.1 Supervised Learning

The goal of supervised learning is to learn a mapping from the input space to the output space where the correct values are provided by labels, called the supervisor. Figure 2.1 shows an overview of the supervised learning model. If the output data are real-valued, then such problems are also called regression. Otherwise, they are called classification, where the learning system fits a model that associates sets of (input) exemplars with labels, possibly with a corresponding measure of certainty. After training with past data, the model learns a classification rule, which may be in the form of an If-Then-Else form. Having a rule like this enables us to make predictions if the future is similar to the past. In some cases, we may want to calculate a probability, then the classification becomes learning an association between the

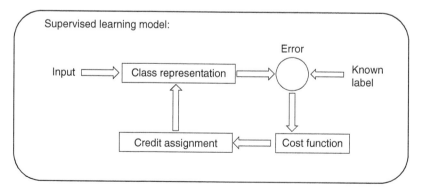

Figure 2.1 Supervised learning model.

input and output data. Learning a rule from data also allows knowledge extraction. In this case, the rule is a simple (complex) model that describes the data and therefore the learning model provides us with an insight about the process underlying the data. Moreover, a learning model also performs compression by fitting a rule to the data. This enables us to use less memory to store the data and less computation to process. Another use of supervised learning is outlier detection. In this case once the rule is learned, we focus on the parts of the data that are not covered by the rule. In other words, we identify the instances that do not follow the rule and/or are exceptions to the rule. These are outliers and imply anomalies requiring further analysis.

2.2.2 Unsupervised Learning

The goal of unsupervised learning is to identify the regularities in the input. The assumption is that there is a structure to the input space such that certain patterns occur more often than others. Figure 2.2 shows an overview of the unsupervised learning model. Thus, we aim to identify and differentiate between patterns with different underlying properties. Once this is achieved, we might also be able to distinguish between typical and atypical behaviors. One method to achieve this is clustering. Clustering aims to find groupings of input. This can be used for data exploration to understand the structure of data and/or for data preprocessing where clustering allows us to map data to a new k dimensional space, where the new dimensionality can be larger than the original dimensionality of the data. One advantage of unsupervised learning is that it does not required labeled data. Labeling data (obtaining ground-truth) is costly. Thus, we can use large amounts of unlabeled data for learning the cluster parameters. This is why unsupervised learning is widely used for "anomaly detection" in network and service management [11–13].

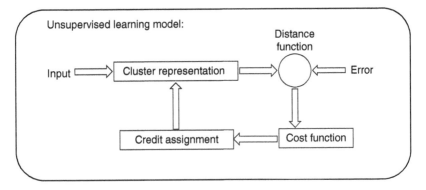

Figure 2.2 Unsupervised learning model.

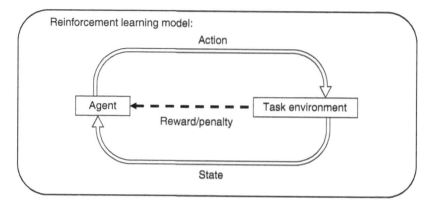

Figure 2.3 Reinforcement learning model.

2.2.3 Reinforcement Learning

The goal of reinforcement learning is to learn the best sequence of actions (policy) in a given environment to maximize the cumulative reward. Figure 2.3 shows an overview of the reinforcement learning model. In this case, reinforcement learning model acts as a decision-making agent, making actions in an environment and receives rewards/penalties while trying to solve a problem. In reinforcement learning problems, the environment is in a certain state (from a set of possible states) at any given time. The state information may be complete (Markov) or incomplete (non-Markov). The agent has a set of actions (from a set of possible actions), and when an action is taken, the state of the environment changes. Thus, unlike unsupervised or supervised learning, reinforcement learning explicitly interacts with the "task". The model is built interactively with the task, not independently from the task. At each time step, a reward signal is typically assumed, where the reward might just be "you have not failed." Indeed, there might never be any "ultimate reward" other than to maximize the duration between failures, or maximize the number of packets routed. In supervised learning, the data label explicitly tells us what to do. Conversely, reinforcement models might attempt to learn a function describing the relative "value" of being in each state. Decision-making would then simplify to identifying the action that moved the current state to the next state with most "value." Reinforcement learning is therefore also explicitly engaged in establishing the order in which it is exposed to state from the task. This is again distinct from either supervised or unsupervised learning in which the data is generally assumed to conform to the independent and identically distributed (i.i.d.) assumption. Moreover, when complete information is available, a reinforcement learning agent may make optimal decisions from the current state alone.[1] However, when

---

1 Subject to other limitations, such as the curse of dimensionality.

complete state information is not present, then the reinforcement learning agent would additionally have to develop internal models of state that extend state to previously visited values. Needless to say, this requirement has implications for the representation adopted as well as the process of credit assignment. Reinforcement learning algorithms have a wider spectrum of applications than supervised learning algorithms, however, they might take a longer time to converge given that the feedback is less explicit than with supervised and unsupervised learning. It should be noted here that the application of reinforcement learning in network and service management is developing rapidly and we see more and more impressive results in the field [14–16].

2.3   Learning for Network and Service Management

AI/ML techniques have a vital list of applications in many network and service management tasks, including (but are not limited to) traffic/service classification and prediction for performance management; intrusion, malware identification, and attribution for security management; root cause analysis and fault identification/prediction for fault management; and resource/job allocation/assignment for configuration management. As discussed in Chapter 1, the growth in connected devices as well as new communication technologies from 5G+ to SDN to NFV persuade network and service management research to explore new methodologies from the AI/ML field [17].

Given the current advances in networks/services AI/ML has found its place in performance management tasks for its ability to learn from big data to predict different conditions, to aggregate patterns, to identify triggers for operations and management actions. For example, traffic prediction has seen multiple ML-based applications from time series forecasting [18] to neural networks [19, 20] to hidden Markov models [21] to genetic algorithms [22]. Moreover, many other tasks in performance management have employed AI/ML techniques for traffic management in the cloud and mobile edge computing, network resource management and allocation, Quality of Service assurance, and congestion control. These leverage the capabilities of AI/ML techniques to learn from temporal and dynamic data [23–26]. Current examples of such developments include Deep Neural Networks [27], transfer learning [28], Deep Reinforcement Learning [15, 29], and Stream online learning [30].

Security management is another network/service management field that includes extensive and early endorsement of AI/ML techniques. Network anomaly detection is a prime example, in which ML techniques are applied for their ability to automatically learn from the data and extract patterns that can be used for identifying network anomalies in a timely manner [31]. To this end, temporal correlation [32], wavelet analysis [33], and traditional change point detection [34] approaches are applied to produce normal/malicious traffic models, where the sequence of actions in a time window are used to create profiles using clustering techniques such as Self Organizing Maps [35], K-means [36], and Gaussian Mixture Models [37]. Moreover, AI/ML techniques have been applied to network intrusion detection including, but not limited to, Decision Trees, Evolutionary Computing, Bayesian Networks, Support Vector Machines, and recently Deep and Reinforcement Learning [38–43]. Unsupervised learning and Stream online learning have been employed for security tasks as well [44, 45]. Other examples of AI/ML applications in security are moving target defence, insider threat detection, and network content filtering [46–48].

In fault management, prediction and diagnosis of faults attracted widespread use of AI/ML techniques from online learning for change point detection to Neural Networks to Hidden Markov Models to Decision Trees, and several unsupervised learning algorithms [49–53]. Additionally, other AI/ML have been introduced specifically for fault prediction, automated fault mitigation, and root cause analysis [54–57].

The application of AI/ML techniques have been slower in configuration management tasks. However, as discussed earlier, with the introduction of NFV and SDN technologies, this is changing [58–60]. Initiatives such as Intent Based Networking [61] and Zero Touch Networking [62] widespread usage of AI/ML has been seen in wireless networks. Other example tasks in configuration management employing ML are service configuration management network load balancing and routing [63–68].

In summary, AI/ML techniques have been applied to several tasks of network and service management in greater numbers over the last decade [69]. However, there are still challenges that need to be resolved for the successful usage of such techniques in production environments. One of the challenges is obtaining high quality data for training and evaluating ML techniques for network and service management functions. Even though network/service data is plenty and diverse in real world, most of the time it is difficult to obtain such data with ground truth. In return, this not only poses challenges for evaluating AI/ML techniques but also faces privacy and trust issues. Another challenge is that in today's networks/services data are generated nonstop in high volume and velocity. They include stationary as well as non-stationary behaviors superimposed. They evolve continuously as new protocols and technologies are introduced over time. All of these reflect in the data in one shape or form, as gradual drifts in user/system behaviors, or as sudden shifts maybe because of a malfunctioning device or a denial of service attack on a particular network or service. This means that AI/ML techniques require to take these dynamics and changes into account, learn under the aforementioned conditions in order to ensure successful deployment.

Yet, another challenge is the need of human experts (from network engineers to security analysts to network/service managers) to understand and trust to AI/ML based system and tools. This requires transparent AI/ML techniques for expert involvement and trust. This is of utmost importance for the widespread and successful deployment of AI/ML techniques in network and service management.

Finally, these challenges also create opportunities in the form of a need for transparent, robust, and dependable AI/ML based techniques for network and service management. To this end, we have already started to see the applications of stream learning, adversarial learning, and transfer learning to the network and service management solutions. Furthermore, research in transparent, secure, and robust AI/ML techniques have gained a big momentum in the ML community. Given the scale and dynamics of today's networks/services, we envision that the application of AI/ML techniques will become more and more ubiquitous and central for operations and management of the future services and networks. In the following Chapters 3–13, we will introduce the current state and the new trends of the AI/ML applications in network and service management.

Bibliography

1 D'Alconzo, A., Drago, I., Morichetta, A. et al. (2019). A survey on big data for network traffic monitoring and analysis. *IEEE Transactions on Network and Service Management* 16 (3): 800–813.

2 Diao, Y. and Shwartz, L. (2017). Building automated data driven systems for it service management. *Journal of Network and Systems Management* 25 (4): 848–883.

3 Alshammari, R. and Zincir-Heywood, N. (2015). How robust can a machine learning approach be for classifying encrypted VoIP? *Journal of Network and Systems Management* 23 (4): 830–869.

4 Boden, M.A. (2016). *AI Its Nature and Future*. Oxford University Press. ISBN 9780198777984.

5 Bernstein, L. and Yuhas, C.M. (1988). Expert systems in network management-the second revolution. *IEEE Journal on Selected Areas in Communications* 6 (5): 784–787.

6 Tran, H.M. and Schönwälder, J. (2015). Discaria: distributed case-based reasoning system for fault management. *IEEE Transactions on Network and Service Management* 12 (4): 540–553.

7 Fallon, L. and OSullivan, D. (2014). The Aesop approach for semantic-based end-user service optimization. *IEEE Transactions on Network and Service Management* 11 (2): 220–234.

8 Alpaydin, E. (2020). *Introduction to Machine Learning*, vol. 4. MIT Press. ISBN 9780262043793.

9 Buczak, A.L. and Guven, E. (2016). A survey of data mining and machine learning methods for cyber security intrusion detection. *IEEE Communication Surveys and Tutorials* 18 (2): 1153–1176. https://doi.org/10.1109/COMST.2015.2494502.

10 Wang, M., Cui, Y., Wang, X. et al. (2017). Machine learning for networking: workflow, advances and opportunities. *IEEE Network* 32 (2): 92–99.

11 Calyam, P., Dhanapalan, M., Sridharan, M. et al. (2014). Topology-aware correlated network anomaly event detection and diagnosis. *Journal of Network and Systems Management* 22 (2): 208–234.

12 Bhuyan, M.H., Bhattacharyya, D.K., and Kalita, J.K. (2014). Network anomaly detection: methods, systems and tools. *IEEE Communications Surveys and Tutorials* 16 (1): 303–336. https://doi.org/10.1109/SURV.2013.052213.00046. URL

13 Le, D.C. and Zincir-Heywood, N. (2018). Big data in network anomaly detection. In: *Encyclopedia of Big Data Technologies* (ed. S. Sakr and A. Zomaya), 1–9. Cham: Springer International Publishing. ISBN 978-3-319-63962-8. https://doi.org/10.1007/978-3-319-63962-8_161-1.

14 Nawrocki, P. and Sniezynski, B. (2018). Adaptive service management in mobile cloud computing by means of supervised and reinforcement learning. *Journal of Network and Systems Management* 26 (1): 1–22.

15 Bachl, M., Zseby, T., and Fabini, J. (2019). Rax: deep reinforcement learning for congestion control. *ICC 2019-2019 IEEE International Conference on Communications (ICC)*, IEEE, pp. 1–6.

16 Amiri, R., Almasi, M.A., Andrews, J.G., and Mehrpouyan, H. (2019). Reinforcement learning for self organization and power control of two-tier heterogeneous networks. *IEEE Transactions on Wireless Communications* 18 (8): 3933–3947.

17 Boutaba, R., Salahuddin, M.A., Limam, N. et al. (2018). A comprehensive survey on machine learning for networking: evolution, applications and research opportunities. *Journal of Internet Services and Applications* 9 (1): https://doi.org/10.1186/s13174-018-0087-2.

18 Syu, Y., Wang, C., and Fanjiang, Y. (2019). Modeling and forecasting of time-aware dynamic QoS attributes for cloud services. *IEEE Transactions on Network and Service Management* 16 (1): 56–71.

19 Dalmazo, B.L., Vilela, J.P., and Curado, M. (2017). Performance analysis of network traffic predictors in the cloud. *Journal of Network and Systems Management* 25 (2): 290–320. https://doi.org/10.1007/s10922-016-9392-x.

20 Hardegen, C., Pfülb, B., Rieger, S., and Gepperth, A. (2020). Predicting network flow characteristics using deep learning and real-world network traffic. *IEEE Transactions on Network and Service Management* 17 (4): 2662–2676.

21 Chen, Z., Wen, J., and Geng, Y. (2016). Predicting future traffic using Hidden Markov models. *2016 IEEE 24th International Conference on Network Protocols (ICNP)*, IEEE, pp. 1–6.

22 Zhang, Y. and Zhou, Y. (2018). Distributed coordination control of traffic network flow using adaptive genetic algorithm based on cloud computing. *Journal of Network and Computer Applications* 119: 110–120.

23 Diao, Y. and Shwartz, L. (2015). Modeling service variability in complex service delivery operations. In: *11th International Conference on Network and Service Management, CNSM 2015*, Barcelona, Spain (9–13 November 2015) (ed. M. Tortonesi, J. Schonwalder, E.R.M. Madeira et al.), 265–269. IEEE Computer Society. https://doi.org/10.1109/CNSM.2015.7367369.

24 Diao, Y. and Rosu, D. (2018). Improving response accuracy for classification-based conversational IT services. *2018 IEEE/IFIP Network Operations and Management Symposium, NOMS 2018*. Taipei, Taiwan: IEEE (23–27 April 2018), pp. 1–15. https://doi.org/10.1109/NOMS.2018.8406138.

25 Morichetta, A. and Mellia, M. (2019). Clustering and evolutionary approach for longitudinal web traffic analysis. *Performance Evaluation* 135. https://doi.org/10.1016/j.peva.2019.102033.

26 Khatouni, A.S., Seddigh, N., Nandy, B., and Zincir-Heywood, N. (2021). Machine learning based classification accuracy of encrypted service channels: analysis of various factors. *Journal of Network and Systems Management* 29 (1): 1–27.

27 Kim, H., Lee, D., Jeong, S. et al. (2019). Machine learning-based method for prediction of virtual network function resource demands. *2019 IEEE Conference on Network Softwarization (NetSoft)*, IEEE, pp. 405–413.

28 Moradi, F., Stadler, R., and Johnsson, A. (2019). Performance prediction in dynamic clouds using transfer learning. *2019 IFIP/IEEE Symposium on Integrated Network and Service Management (IM)*, IEEE, pp. 242–250.

29 Elsayed, M., Erol-Kantarci, M., Kantarci, B. et al. (2020). Low-latency communications for community resilience microgrids: a reinforcement learning approach. *IEEE Transactions on Smart Grid* 11 (2): 1091–1099. https://doi.org/10.1109/TSG.2019.2931753.

30 Khanchi, S., Vahdat, A., Heywood, M., and Zincir-Heywood, N. (2018). On botnet detection with genetic programming under streaming data label budgets and class imbalance. *Swarm and Evolutionary Computation* 39: 123–140.

31 I. Nevat, D.M. Divakaran, S. G. Nagarajan et al. (2018). Anomaly detection and attribution in networks with temporally correlated traffic. *IEEE/ACM Transactions on Networking* 26 (1): 131–144.

32 Kim, D., Woo, J., and Kim, H.K. (2016). "i know what you did before": general framework for correlation analysis of cyber threat incidents. *MIL-COM 2016 – 2016 IEEE Military Communications Conference*, pp. 782–787. https://doi.org/10.1109/MILCOM.2016.7795424.

33 Meng, M. (2008). Network security data mining based on wavelet decomposition. *2008 7th World Congress on Intelligent Control and Automation*, pp. 6646–6649. https://doi.org/10.1109/WCICA.2008.4593932.

34 Tartakovsky, A.G., Rozovskii, B.L., Blazek, R.B., and Kim, H. (2006). A novel approach to detection of intrusions in computer networks via adaptive sequential and batch-sequential change-point detection methods. *IEEE Transactions on Signal Processing* 54 (9): 3372–3382. https://doi.org/10.1109/TSP.2006.879308.

35 Bantouna, A., Poulios, G., Tsagkaris, K., and Demestichas, P. (2014). Network load predictions based on big data and the utilization of self-organizing maps. *Journal of Network and Systems Management* 22 (2): 150–173. https://doi.org/10.1007/s10922-013-9285-1.

36 Bacquet, C., Zincir-Heywood, N., and Heywood, M. (2011). Genetic optimization and hierarchical clustering applied to encrypted traffic identification. *2011 IEEE Symposium on Computational Intelligence in Cyber Security (CICS)*, April 2011, pp. 194–201. https://doi.org/10.1109/CICYBS.2011.5949391.

37 Le, D.C., Zincir-Heywood, N., and Heywood, M. (2016). Data analytics on network traffic flows for botnet behaviour detection. *IEEE Symposium Series on Computational Intelligence (SSCI '16)*, December 2016, pp. 1–7. ISBN 9781509042401. https://doi.org/10.1109/SSCI.2016.7850078.

38 Finamore, A., Mellia, M., Meo, M., and Rossi, D. (2010). KISS: stochastic packet inspection classifier for UDP traffic. *IEEE/ACM Transactions on Networking* 18 (5): 1505–1515. https://doi.org/10.1109/TNET.2010.2044046.

39 Kayacik, G., Zincir-Heywood, N., and Heywood, M. (2011). Can a good offense be a good defense? Vulnerability testing of anomaly detectors through an artificial arms race. *Applied Soft Computing* 11 (7): 4366–4383. https://doi.org/10.1016/j.asoc.2010.09.005.

40 Haddadi, F. and Zincir-Heywood, N. (2016). Benchmarking the effect of flow exporters and protocol filters on botnet traffic classification. *IEEE Systems Journal* 10 (4): 1390–1401. https://doi.org/10.1109/JSYST.2014.2364743.

41 Bronfman-Nadas, R., Zincir-Heywood, N., and Jacobs, J.T. (2018). An artificial arms race: could it improve mobile malware detectors? *Network Traffic Measurement and Analysis Conference, TMA 2018*, Vienna, Austria: IEEE (26–29 June 2018), pp. 1–8. https://doi.org/10.23919/TMA.2018.8506545.

42 Lotfollahi, M., Siavoshani, M.J., Zade, R.S.H., and Saberian, M. (2019). Deep packet: a novel approach for encrypted traffic classification using deep learning. *Soft Computing*. https://doi.org/10.1007/s00500-019-04030-2.

43 Wilkins, Z. and Zincir-Heywood, N. (2020). COUGAR: clustering of unknown malware using genetic algorithm routines. In: *GECCO '20: Genetic and Evolutionary Computation Conference*, Cancún Mexico (July 8-12, 2020) (ed. C.A.C. Coello), 1195–1203. ACM. https://doi.org/10.1145/3377930.3390151.

44 Ahmed, S., Lee, Y., Hyun, S., and Koo, I. (2019). Unsupervised machine learning-based detection of covert data integrity assault in smart grid networks utilizing isolation forest. *IEEE Transactions on Information Forensics and Security* 14 (10): 2765–2777. https://doi.org/10.1109/TIFS.2019.2902822.

45 Le, D.C. and Zincir-Heywood, N. (2020). Exploring anomalous behaviour detection and classification for insider threat identification. *International Journal of Network Management*. https://doi.org/e2109.

46 Dietz, C., Dreo, G., Sperotto, A., and Pras, A. (2020). Towards adversarial resilience in proactive detection of botnet domain names by using MTD. *NOMS 2020 - 2020 IEEE/IFIP Network Operations and Management Symposium*, pp. 1–5. https://doi.org/10.1109/NOMS47738.2020.9110332.

47 Le, D.C., Zincir-Heywood, N., and Heywood, M. (2020). Analyzing data granularity levels for insider threat detection using machine learning. *IEEE Transactions on Network and Service Management* 17 (1): 30–44.

48 Bag, T., Garg, S., Rojas, D.F.P., and Mitschele-Thiel, A. (2020). Machine learning-based recommender systems to achieve self-coordination between son functions. *IEEE Transactions on Network and Service Management* 17 (4): 2131–2144. https://doi.org/10.1109/TNSM.2020.3024895.

49 Makanju, A., Zincir-Heywood, N., and Milios, E.E. (2013). Investigating event log analysis with minimum apriori information. *2013 IFIP/IEEE International Symposium on Integrated Network Management (IM 2013)*, pp. 962–968.

50 Jiang, H., Zhang, J.J., Gao, W., and Wu, Z. (2014). Fault detection, identification, and location in smart grid based on data-driven computational methods. *IEEE Transactions on Smart Grid* 5 (6): 2947–2956. https://doi.org/10.1109/TSG.2014.2330624.

51 Uriarte, R.B., Tiezzi, F., and Tsaftaris, S.A. (2016). Supporting autonomic management of clouds: service clustering with random forest. *IEEE Transactions on Network and Service Management* 13 (3): 595–607. https://doi.org/10.1109/TNSM.2016.2569000.

52 Fadlullah, Z.M., Tang, F., Mao, B. et al. (2017). State-of-the-art deep learning: evolving machine intelligence toward tomorrow's intelligent network traffic control systems. *IEEE Communication Surveys and Tutorials* 19 (4): 2432–2455. https://doi.org/10.1109/COMST.2017.2707140.

53 Messager, A., Parisis, G., Kiss, I.Z. et al. (2019). Inferring functional connectivity from time-series of events in large scale network deployments. *IEEE Transactions on Network and Service Management* 16 (3): 857–870. https://doi.org/10.1109/TNSM.2019.2932896.

54 Tiwana, M.I., Sayrac, B., and Altman, Z. (2010). Statistical learning in auto-mated troubleshooting: application to lte interference mitigation. *IEEE Transactions on Vehicular Technology* 59 (7): 3651–3656. https://doi.org/10.1109/TVT.2010.2050081.

55 Ahmed, J., Josefsson, T., Johnsson, A. et al. (2018). Automated diagnostic of virtualized service performance degradation. *NOMS 2018 – 2018 IEEE/IFIP Network Operations and Management Symposium*, pp. 1–9. https://doi.org/10.1109/NOMS.2018.8406234.

56 Renga, D., Apiletti, D., Giordano, D. et al. (2020). Data-driven exploratory models of an electric distribution network for fault prediction and diagnosis. *Computing* 102 (5): 1199–1211. https://doi.org/10.1007/s00607-019-00781-w.

57 Steenwinckel, B., Paepe, D.D., Hautte, S.V. et al. (2021). FLAGS: a methodology for adaptive anomaly detection and root cause analysis on sensor data streams by fusing expert knowledge with machine learning. *Future Generation Computer Systems* 116: 30–48. https://doi.org/10.1016/j.future.2020.10.015.

58 Xie, J., Yu, F.R., Huang, T. et al. (2018). A survey of machine learning techniques applied to software defined networking (SDN): research issues and challenges. *IEEE Communication Surveys and Tutorials* 21 (1): 393–430.

59 Zhang, C., Patras, P., and Haddadi, H. (2019). Deep learning in mobile and wireless networking: a survey. *IEEE Communication Surveys and Tutorials*. 21 (3)

60 Park, S., Kim, H., Hong, J. et al. (2020). Machine learning-based optimal VNF deployment. *21st Asia-Pacific Network Operations and Management Symposium, APNOMS 2020*, Daegu, South Korea (22–25 September 2020), IEEE, pp. 67–72. https://doi.org/10.23919/APNOMS50412.2020.9236970.

61 Lerner, A. (2017). Intent-based networking. *Gartner Blog*: https://blogs.gartner.com/andrew-lerner/2017/02/07/intent-based-networking/ (accessed 15 April 2021).

62 ETSI (2020). Zero-touch network and Service Management. https://www.etsi.org/technologies/zero-touch-network-service-management (accessed 13 April 2021).

63 Tsvetkov, T., Ali-Tolppa, J., Sanneck, H., and Carle, G. (2016). Verification of configuration management changes in self-organizing networks. *IEEE Transactions on Network and Service Management* 13 (4): 885–898. https://doi.org/10.1109/TNSM.2016.2589459.

64 Zhang, Y., Yao, J., and Guan, H. (2017). Intelligent cloud resource management with deep reinforcement learning. *IEEE Cloud Computing* 4 (6): 60–69. https://doi.org/10.1109/MCC.2018.1081063.

65 Mismar, F.B., Choi, J., and Evans, B.L. (2019). A framework for automated cellular network tuning with reinforcement learning. *IEEE Transactions on Communications* 67 (10): 7152–7167. https://doi.org/10.1109/TCOMM.2019.2926715.

66 Yao, H., Mai, T., Jiang, C. et al. (2019). Ai routers network mind: a hybrid machine learning paradigm for packet routing. *IEEE Computational Intelligence Magazine* 14 (4): 21–30. https://doi.org/10.1109/MCI.2019.2937609.

67 Zhang, Q., Wang, X., Lv, J., and Huang, M. (2020). Intelligent content-aware traffic engineering for SDN: an Ai-driven approach. *IEEE Network* 34 (3): 186–193. https://doi.org/10.1109/MNET.001.1900340.

68 Zhang, J., Ye, M., Guo, Z. et al. (2020). CFR-RL: traffic engineering with reinforcement learning in SDN. *IEEE Journal on Selected Areas in Communications* 38 (10): 2249–2259. https://doi.org/10.1109/JSAC.2020.3000371.

69 Le, D.C. and Zincir-Heywood, N. (2020). A frontier: dependable, reliable and secure machine learning for network/system management. *Journal of Network and Systems Management* 28 (4): 827–849.

Part II

Management Models and Frameworks

3

Managing Virtualized Networks and Services with Machine Learning

Raouf Boutaba[1], Nashid Shahriar[2], Mohammad A. Salahuddin[1], and Noura Limam[1]

[1] *David R. Cheriton School of Computer Science, University of Waterloo, Ontario, Canada*
[2] *Department of Computer Science, University of Regina, Saskatchewan, Canada*

3.1 Introduction

Virtualization is instigating a revolutionary change in the networking industry, similar to that of the computer industry in the 1980s. Indeed, before IBM compatibles and Windows, the mainframe computer industry in the late 1970s and early 1980s was closed with vertically integrated specialized hardware, operating system and applications – all from the same vendor. A revolution happened when open interfaces started to appear, the industry became horizontal and innovation exploded. A similar revolution is happening in the networking industry, which previously had the "mainframe" mindset relying on vendor specific, proprietary and vertically integrated solutions. Network Virtualization (NV) and the provision of open interfaces for network programming are expected to foster innovation and rapid deployment of new network services.

The idea of NV gained momentum to address the Internet ossification problem by enabling radically different architectures [1]. The current Internet suffers from ossification, as the Internet size and rigidity make it difficult to adopt new networking technologies [2]. For example, the transition from Internet Protocol version 4 (IPv4) to IPv6 has started more than a decade ago, while IPv6 adoption rate is still significantly low as reported by major service providers (i.e. less than 30% of Google users have adopted IPv6 [3]). It is becoming increasingly cumbersome to keep up with emerging applications quality of service (QoS) requirements of bandwidth, reliability, throughput, and latency in an ossified Internet. NV solves the ossification problem by allowing the coexistence of multiple virtual networks (VNs), each customized for a specific purpose on the shared Internet. Although the

Communication Networks and Service Management in the Era of Artificial Intelligence and Machine Learning,
First Edition. Edited by Nur Zincir-Heywood, Marco Mellia, and Yixin Diao.
© 2021 The Institute of Electrical and Electronics Engineers, Inc. Published 2021 by John Wiley & Sons, Inc.

idea of NV originated to address the Internet ossification, NV has been adopted as a diversifying attribute of different networking technologies, including wireless [4], radio access [5], optical [6], data center (DC) [7], cloud computing [8], service-oriented [9], software-defined networking (SDN) [10, 11], and Internet of Things (IoT) [12].

Another prolific application of virtualization in networking is the adoption of virtualized network services through network functions virtualization (NFV). NFV decouples network or service functions from underlying hardware, and implements them as software appliances, called virtual network functions (VNFs), on virtualized commodity hardware. Numerous state-of-the-art VNFs have already shown the potential to achieve near-hardware performance [13, 14]. Moreover, NFV provides ample opportunities for network optimization and cost reduction. First, hardware-based network or service functions come with high capital expenditures, which can be reduced by deploying VNFs on commodity servers. Second, hardware appliances are usually placed at fixed locations, whereas in NFV, a VNF can be deployed on any server in the network. VNF locations can be determined intelligently to meet dynamic traffic demand and better utilize network resources. NFV opens-up the opportunity to simultaneously optimize VNF locations and traffic routing paths, which can significantly reduce the network operational expenditure. Finally, hardware-based functions are difficult to scale, whereas NFV offers to cost-efficiently scale VNFs on-demand. A service-function chain (SFC) is an ordered sequence of VNFs composing a specific service [15]. For example in a typical DC network, traffic from a server passes through an intrusion detection system (IDS), a firewall, and a network address translator (NAT) before reaching the Internet.

Virtualizing networks and services facilitate a new business model, namely Network-as-a-Service (NaaS), which provides a separation between the applications and services, and the networks supporting them [16]. Network operators can adopt the NaaS model to partition their physical network resources into multiple VNs (also called *network slices*) and lease them to service providers [17]. In turn, service providers use VNs to offer services with diverse QoS requirements, without any investment in establishing and managing a physical infrastructure. A perfect incarnation of the NaaS model is network slicing for fifth generation (5G) mobile networks. Using network slicing, a single 5G physical network can be sliced into multiple isolated logical networks of varying sizes and structures, dedicated to different types of services. These "self-contained" VNs should be flexible enough to simultaneously accommodate diverse business-driven use cases from multiple service providers on a common network infrastructure, and created on-demand according to the service providers' requirements.

The benefits of virtualized networks and services come at the cost of additional management challenges for network operators. First, a network operator has to orchestrate VNs/network slices in such a way that they can coexist in a single infrastructure, without affecting each other. Hence, smart orchestration decisions need to be carried out to provision VNs satisfying requirements from diverse users and applications, while ensuring desired resource utilization. This also involves configuring a large number of virtual instances and their operating parameters. The initial orchestration and configuration need to be adapted to cope with time-varying traffic demands and change in network states. Second, the added virtualization layer introduces new attack and failure surfaces across different administrative and technological domains. For instance, any failure in the underlying physical resource can propagate to the hosted virtual resources, though the reverse is not always true. Similarly, remediation and mitigation mechanism for one VN should not jeopardize the operation of coexisting VNs. These diverse challenges call for automated management that cannot be satisfied with the traditional, reactive human-in-the loop management approach. The management of VNs should be intelligent to leverage the sheer volume of operational data generated within a live network, and take automated decisions for different operational and management actions. Therefore, Artificial Intelligence (AI) and Machine Learning (ML) can play pivotal roles for realizing the automation of control and management for VNs and their services [18, 19].

AI and ML techniques have been widely used in addressing networking problems in the last few decades [18, 19]. However, when it comes to virtualized network management, the lack of real-world deployment of virtualized services impedes the application of AI and ML techniques. Despite that, there has been a recent surge in research efforts that aim to leverage ML in addressing complex problems in NV environment. This chapter summarizes state-of-the-art research and outlines potential avenues in the application of AI and ML techniques in virtualized network and service management. The rest of the chapter is organized as follows. We provide a detailed technology overview of virtualized networks and services in Section 3.2. We present state-of-the-art research that apply AI and ML in three core sub-areas of virtualized networks and services, namely NV, NFV, and network slicing in Section 3.3. We conclude the chapter in Section 3.4 with a brief summary, and outline possible research avenues to advance the state-of-the-art in applying AI and ML for managing virtualized networks and services.

3.2 Technology Overview

Virtualization in networking is not a new concept. Virtual channels in X.25-based telecommunication networks (e.g. ATM networks) allow multiple users to share a large physical channel. Virtual Local Area Networks (VLANs) partition a physical Local Area Network (LAN) among multiple logical LANs with elevated levels of trust, security, and isolation. Similarly, virtual private networks (VPNs) offer dedicated communications that connect multiple geographically distributed sites through private and secure tunnels over public communication networks (e.g. the Internet). Overlay networks (e.g. PlanetLab) create virtual topologies on top of the physical topology of another network. Overlays are typically implemented in the application layer, though various implementations at lower layers of the network stack do exist. These technologies deploy narrow fixes to specific problems without a holistic view of the interactions between coexisting VNs. Therefore, in this section, we provide a comprehensive overview of different technologies that enable virtualization of networks and services.

3.2.1 Virtualization of Network Functions

An Network Function (NF) is a functional block within a network infrastructure that has well-defined external interfaces and functional behavior [13]. NFs in traditional wired networks can be classified in two categories: forwarding functions and value-added functions. Forwarding functions, such as routers, switches, and transponders, provide the functionality to forward data along a network path. On the other hand, value-added functions, such as Dynamic Host Configuration Protocol (DHCP), Network Address Translation (NAT), Universal Plug and Play (UPnP), Firewall, Optimizer, and Deep Packet Inspectors (DPI), offer additional capabilities to the data forwarding path. Similarly, NFs in mobile networks are categorized in two classes: Radio Access Network (RAN) functions and core functions. We will discuss more about RAN and core functions later when we discuss network slicing. In this subsection, we discuss two popular methods of virtualizing NFs as follows (a summary is depicted in Figure 3.1).

3.2.1.1 Resource Partitioning

Partitioning is a convenient method to create multiple virtual entities on a single networking device (e.g. routers and switches) that provide forwarding functions. Resource partitioning can be achieved either by hard partitioning (i.e. dedicated switch ports, CPU cores, cards) or by soft partitioning (i.e. CPU execution capping, routing, and forwarding table partitioning). Hard partitioning provides excellent isolation, but it requires abundant hardware to implement. In contrast, soft partitioned instances may not provide the highest level of isolation and security due to their shared nature.

A hard partitioned router, called a Logical Router (LR), can run across processors on different cards of a router device. All the underlying hardware and software resources, including network processors, interfaces, and routing and forwarding tables, are dedicated to an LR. Examples of LR are "protected system domains" by Juniper Networks, or "logical routers" by Cisco Systems. Hardware partitioned

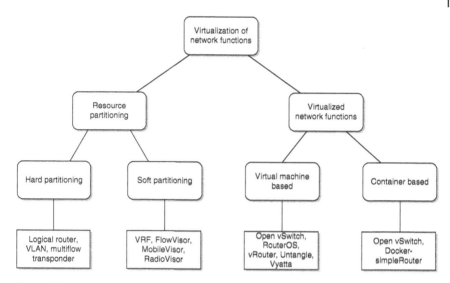

Figure 3.1 Technologies for virtualizing network functions with examples.

routers are mainly deployed in Points of Presence (PoP) of network carriers to save space and power, and reduce management overhead. Similarly, VLANs divide a physical switch into multiple logical switches by grouping ports on a switch. How a switch does grouping is implementation dependent, but a common solution is for the switch to tag each frame with a VLAN ID as it arrives on a port. When the frame is sent to another port, only the ports configured with the VLAN ID carried in the frame will output the packet. A VLAN can also span multiple inter-connected switches using the IEEE standard 802.1Q. The limitation of VLAN is its low scalability, primarily due to a maximum of 4094 VLANs in a layer-2 network. To support a larger number of VLANs in a broadcast domain, VXLAN has been developed for large multi-tenant DC environments. In the optical domain, mul-tiflow transponders can be used to create a number of subtransponders from the hardware resource pool [6]. These subtransponders can be used to carry different flows arriving from a single router interface by using flow identifiers.

Examples of soft partitioning include Virtual Routing and Forwarding (VRF) that allow multiple instances of routing and forwarding tables to co-exist within the same router. The various routing and forwarding tables may be maintained by a single process or by multiple processes (e.g. one process for each routing and forwarding table). Routing protocols should understand that certain routes may be placed only in certain VRFs. The routing protocols manage this by peering within a constrained topology, where a routing protocol instance in a VRF peers with other instances in the same VN. Another example of soft partitioning is FlowVisor that slices the flowspace of OpenFlow switches based on OpenFlow match fields, such as switch port, MAC addresses, and IP addresses. FlowVisor basically acts as a proxy between OpenFlow switches and controllers, and intercepts messages between them. By abstracting the OpenFlow control channel, FlowVisor provides mechanisms for bandwidth, switch CPU, and flowspace isolation.

3.2.1.2 Virtualized Network Functions

The main idea of VNFs is to decouple the physical network equipment from the functions that run on them. A VNF is an implementation of an NF that is deployed on virtual resources, such as a virtual machine (VM) or container [13]. A single VNF may be composed of multiple internal components, and hence it could be deployed over multiple VMs/containers, in which case each VM/container hosts a single component of the VNF.

For instance, a virtual router (vRouter) is a software function that implements the functionality of a layer 3 IP routing in software. The underlying physical resources are shared with other co-hosted VMs. In a well-implemented vRouter, users can see and change only the configuration and statistics for "their" router. Examples of vRouter include Alpine Linux, Mikrotik RouterOS, Brocade vRouter, Untangle, and Vyatta. Similarly, a virtual switch (vSwitch) is a software emulation of a physical switch that performs functions, such as traffic switching, multiplexing, and scheduling. It detects which VMs are logically connected to each of its virtual ports and uses this information to forward traffic to the correct VMs. Examples of vSwitch include Open vSwitch, Cisco Nexus 1000v, and VMware virtual switch. Due to the diversity of value-added NFs, different kinds of VNFs may exist based on different network layers. Even for each kind of NF, there may be multiple implementations with different features by various vendors. For example, the virtual NAT implemented by VMware provides a way for VMs to communicate with the host, while the one implemented by NFWare is extended to the carrier-grade level. For a comprehensive list of VNF products, the reader is referred to [14].

There are pros and cons of deploying a VNF on top of a VM or container. In case of VMs, the entire operational function of a VM is completely isolated from that of the host and other guest VMs. Hence, VM-based virtualization enforces a stronger isolation among VMs and the physical machine, and is regarded as a more secure and reliable solution. However, VM-based virtualization suffers from scalability and performance issues, due to the overhead of emulating a full computer machine within a VM. In contrast, containers do not need hardware indirection and run more efficiently on top of host OS, whereas each VM runs as an independent OS. Hence, containers can be used to deploy VNFs in a more flexible and agile manner but with a reduced level of isolation and security. Recently, unikernels have emerged as lightweight alternatives that take the best of both VM- and container-based virtualization. Unikernels usually package the VNFs with only the required libraries, unlike VMs that provide an entire guest OS.

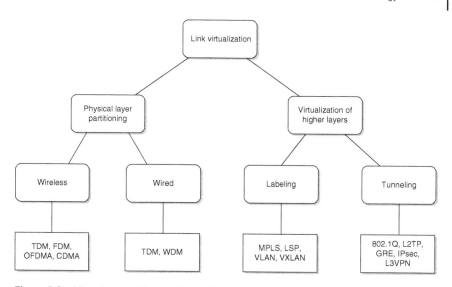

Figure 3.2 Link virtualization technologies with examples.

3.2.2 Link Virtualization

Link virtualization technologies enable creation of virtual links that can connect physical or virtual NFs. A virtual link can consist of a single physical link or can encompass a sequence of physical links forming a path. In this subsection, we discuss two popular technologies of virtualizing network links as follows (a summary is depicted in Figure 3.2).

3.2.2.1 Physical Layer Partitioning

Using various multiplexing technologies, a wired (e.g. fiber, copper cable) or wireless (e.g. wireless spectrum) physical medium can be split into distinct channels or time slots. A set of channels or time slots are then assigned to a virtual link with a specific bit rate such that the sender and receiver of the virtual link get the illusion that they own the physical medium. The type of multiplexing technique depends on the physical medium properties, the associated constraints and impairments. For example, a wireless link can be partitioned using time division multiplexing (TDM), frequency division multiplexing (FDM), or code division multiple access (CDMA). A combination of different multiplexing techniques can also be applied to achieve higher bandwidth, such as for broadband wireless networks. For example, orthogonal frequency-division multiple access (OFDMA) can be described as a combination of FDM and TDM multiple access, where the resources are partitioned in both time and frequency domains, and slots are assigned along the OFDM symbol index, as well as OFDM subcarrier index.

In fiber-optic communications, wavelength-division multiplexing (WDM) is a technology that multiplexes a number of optical carrier signals onto a lightpath (i.e. a set of concatenated optical fiber links) by using different wavelengths (i.e. colors) of laser light. This is similar to FDM, since wavelength and frequency communicate the same information. Physical layer multiplexing provides hard partitioning and better isolation among virtual links, since physical medium resources are assigned in a dedicated manner to virtual links.

3.2.2.2 Virtualization at Higher Layers

At higher layers (e.g. link, network, or application layers), link resource partitioning is achieved by allocating a specific bandwidth (i.e. transmission bit rate, link capacity) to a virtual link. Such partitioning can be enforced by rate-limiting or allocating an appropriate amount of link queues and link buffers. Since virtualization at higher layers is achieved through soft-partitioning of link resources, isolation between virtual links is especially critical. To ensure isolation among virtual links, two popular methods include: (i) labeling and (ii) tunneling.

Labeling involves specifying certain fields (e.g. tags, IDs, etc.) in the packet header that serve for identification and isolation of virtual links. For instance, VLANs apply tags to network packets and handle these tags in switches – creating the appearance and functionality of network traffic that is physically on a single network but acts as if it is split between separate VNs. VLANs can be used to distinguish data from different VLANs and to help form data paths for the broadcasting domain. Similarly, Multiprotocol Label Switching (MPLS) and label switched path (LSP) technologies can be used to specify the path that data packets take. In MPLS, labels identify virtual links (paths) between nonadjacent NFs. This requires MPLS capable routers (e.g. label-switched routers) to forward packets to outgoing interface based only on label value, unlike using IP addresses in traditional routers.

Tunneling is a popular method for link virtualization that has been adopted by many different technologies, such as VPN and VLAN. It ensures isolation of traffic from multiple VNs transported over a shared network. It also provides direct connection between network devices that are not physically adjacent. Tunneling is performed by using encapsulation and occasionally encryption techniques. A number of different tunneling technologies exist, including IEEE 802.1Q, Layer 2 Tunneling Protocol (L2TP), Generic Routing Encapsulation (GRE), Internet Protocol security (IPsec), and layer 3 virtual private network (L3VPN).

3.2.3 Network Virtualization

As discussed in the previous two subsections 3.2.2.1 and 3.2.2.2, both NFs and links can be independently virtualized while being oblivious to each other. It is also possible to virtualize only NFs and use non-virtualized links to connect VNFs and vice versa. In contrast, NV seeks to create slices of a network, i.e. VNs at the particular networking layer. For instance, a VN in the IP layer comprises of vRouters/vSwitch and overlay IP links connecting them, whereas a VN in the optical layer connects multiflow transponders through optical lightpaths. It should be noted that a given VN should have its own resources, including its own view of the network topology, its own portions of link bandwidths, dedicated CPU resources in NFs, and its own slices of CPU, forwarding/routing tables in switches and routers. Such a holistic NV can be achieved through network hypervisors that abstract the physical network (e.g. communication links, network elements, and control functions) into logically isolated VNs [11]. A number of network hypervisors, such as OpenVirteX, FlowVisor, OpenSlice, MobileVisor, RadioVisor, and Hyper-Flex, have been developed for different network technologies. The reader is referred to [11] for a more comprehensive survey of NV hypervisors.

3.2.4 Network Slicing

Network slicing extends the concept of NV in the context of 5G mobile networks from two perspectives. First, a 5G network slice is an end-to-end (E2E) VN that spans multiple technological and administrative network segments (e.g. wireless radio, access/core transport networks, Multi-access Edge Computing [MEC] and central DCs), whereas a traditional VN concerns only one particular network technology, such as wired transport or wireless network. Examples of network slices are shown in Figure 3.3 where the dark gray network slice goes all the way to the central DC, and dotted light gray network slice terminates at the central office of a mobile network. E2E perspective of network slices offer more opportunities to optimize the deployment of network slices, and meet fine-grained QoS requirements. Second, network slicing allows to virtualize RAN and core NFs, and include

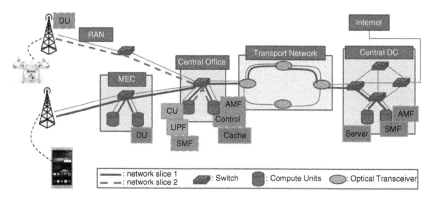

Figure 3.3 Examples of network slices.

them within a network slice that are typically not considered by conventional Vns. Virtualizing RAN and core NFs enable a more flexible way of creating, operating, managing, and deleting network slices on-demand. It also allows to deploy these VNFs with the appropriate capacity at the right place, to meet stringent requirements (e.g. E2E latency) imposed by 5G services.

Let us now discuss more about RAN and core NFs. The most common RAN functions responsible for baseband processing are: Service Data Adaptation Protocol (SDAP), Radio Resource Control (RRC), Packet Data Convergence Protocol (PDCP), Radio Link Control (RLC), Medium Access Control (MAC), and Physical (PHY) layer functions. In traditional mobile networks, Baseband Units (BBUs), co-located with antennas, are responsible for performing RAN NFs. However, in 5G RAN architecture, these NFs are envisioned to be virtualized and placed on commodity servers deployed either at antenna sites or MECs. Due to the strict timing requirements of some NFs, the RAN NFs are grouped in two entities: Central Unit (CU) and Distributed Unit (DU) [20]. DU hosts time-critical functions, such as MAC, RLC, and PHY, and serves a number of mobile users within the DU's coverage. On the other hand, CU may host time-tolerant functions, such as SDAP, PDCP, and RRC, and can serve multiple DUs. Both DU and CU can also be considered as aggregated VNFs and deployed on VMs/containers on servers located at antenna sites or MECs.

Similarly, a new core network architecture for 5G mobile networks, namely the Next Generation (NG) core, that separates the current Evolved Packet Core (EPC) functions into more fine-grained NFs has been proposed [20]. The most prominent NFs in NG core are as follows: Access and Mobility Management Function (AMF), Session Management Function (SMF), Policy Control Function (PCF), User Plane Function (UPF), and Unified Data Management (UDM). These core NFs can also be considered as VNFs and easily deployed in a virtualized environment. The benefit of this service oriented RAN and core architecture is that it allows for sharing of fine-grained VNFs among network slices without compromising the performance and QoS requirements. For instance in Figure 3.3, dark gray and dotted light gray network slices share CU and UPF NFs while using completely dedicated RAN and core NFs and their application functions (e.g. cache, control, or server). Similarly, the control plane functions, such as RLC, MAC, AMF, and PCF, can be shared between slices while using dedicated UPFs, including PDCP and UPF. Finally, the network slices that require the highest level of security (e.g. public safety or first responder's slice) may use dedicated VNFs not shared with others.

3.2.5 Management and Orchestration

SDN has the potential to simplify network configuration and reduce management complexity. In contrast to today's networks, where control and forwarding functions are tightly coupled and embedded within each network device (i.e. switches and routers), SDN accumulates the control functionality in a logically centralized and programmable control plane, which is decoupled from the forwarding plane. The control plane is implemented in software (i.e. SDN controller) on one or more dedicated computer servers, has a global network view, and provides a unified interface to configure and control the network. On the other hand, packet forwarding remains the responsibility of the switches/routers and is implemented on commodity hardware.

Management and Orchestration (MANO) is quintessential to unlock the full potential of NV, which includes seamless operation and efficient delivery of services. OpenStack is an open source cloud computing platform that controls large pools of virtual resources to build and manage private/public clouds. However, with the advent of NFV, OpenStack has become a crucial component in NFV MANO, as a Virtualized Infrastructure Manager (VIM). It is responsible for dynamic management of network function virtualization infrastructure (NFVI) hardware resources (i.e. compute, storage, and networking) and software resources (i.e. hypervisors), offering high availability and scalability. OpenStack also facilitates additional features in NFVI, including service function chaining and network slicing. Open Platform for Network Function Virtualization (OPNFV), a carrier-grade, open source platform also leverages OpenStack as its VIM solution [21].

Open Network Automation Platform (ONAP) and Open Source MANO (OSM) are two prominent NFV MANO initiatives. ONAP, a open source project hosted by the Linux Foundation, offers real-time, policy-driven orchestration of both physical and virtualized NFs, to facilitate efficient and automated delivery of on-demand services and support their lifecycle management. All ONAP components are offered as Docker containers, allowing for custom integration in different operator environments. It also allows for integration with multiple VIMs, VNF managers, and SDN controllers. ONAP primarily consists of two components: (i) design-time and (ii) run-time, each having subcomponents.

ONAP's design-time component offers a service design and creation (SDC) environment, that supports OASIS Topology and Orchestration Specification for Cloud Applications (TOSCA), for describing resources and services (i.e. assets), along with their associated policies and processes. Its run-time component executes the policies prepared in the design-time, which pertain to monitoring, data collection, analytics, service orchestration, etc. ONAP leverages the Closed Loop Automation Management Platform (CLAMP), to enable lifecycle management of VNFs and automate E2E deployment processes. In contrast, OSM is an European Telecommunications Standards Institute (ETSI) initiative to offer cost-effective and automated delivery of services. Both ONAP and OSM conform to the ETSI NFV Reference architecture. A comparative evaluation of ONAP and OSM, with respect to features and performance gaps, is provided in [22]. Authors in [23, 24] propose an architecture for network slice management on top of ONAP, while [25] enhances OSM (along with OpenStack and OpenDaylight SDN controller) to enable service deployment across a multi-domain infrastructure.

3.3 State-of-the-Art

3.3.1 Network Virtualization

The embedding of VNs into substrate networks is a critical aspect of NV. The virtual network embedding (VNE) is a resource allocation problem that involves embedding virtual nodes and links to substrate nodes and links, respectively. For successful network embedding, it is paramount that resources are allocated efficiently. VNE is a well-studied problem that has been proved to be NP-hard [26, 27]. As a result, several linear programming algorithms, mixed integer programming algorithms, as well as heuristic algorithms have been proposed in the research literature. Most of the proposed heuristic algorithms solve the problem in two stages: (i) node embedding first and (ii) link embedding next. In the first stage, substrate nodes are ranked based on a specific metric (e.g. availability) and a greedy node mapping strategy is applied where mapping is decided by rank results. In the second stage, the virtual links are usually mapped to the shortest path that has enough bandwidth resources between nodes. On the other hand, linear programming and mixed integer programming algorithms are used to solve the VNE problem in a single stage, by simultaneously mapping nodes and links.

The majority of VNE solutions perform static mappings and resource allocations i.e. they do not consider the remapping of embedded VNs by migrating virtual nodes and/or links or adjusting the resource allocated to the VN, as new requests are received, or the network load, traffic pattern changes. Indeed, this is counterintuitive, considering the dynamic nature of Internet traffic. The proven inefficiency of static resource allocation motivated the emergence of dynamic solutions. ML, in particular reinforcement learning (RL), have been proven particularly efficient for solving the dynamic resource allocation problem, considering the higher complexity of the problem compared to static VNE. Table 3.1 provides a summary of the state-of-the-art that addresses VNE and resource allocation.

Mijumbi et al. [28] address the dynamic resource allocation problem using an RL-based approach. They model the substrate network as a decentralized system of Q-learning agents, associated to substrate nodes and links. The agents use Q-learning to learn an optimal policy to dynamically allocate network resources to virtual nodes and links. The reward function encourages high virtual resource utilization, while penalizing packet drops and high delays. The agents ensure that

Table 3.1 Summary of the state-of-the-art for virtual network embedding.

References	Problem/objective	Features	ML technique
[28]	Dynamic resource allocation to achieve high resource utilization and QoS	Virtual resource substrate resource	RL with Q-learning
[29]	Dynamic resource allocation to achieve high resource utilization and QoS	Virtual resource substrate resource	RL with ANN
[30, 31]	Node mapping to achieve high revenue-to-cost ratio	CPU bandwidth topological features	RL with ANN
[32]	Node mapping to achieve high revenue-to-cost ratio	CPU bandwidth degree	RL with RNN
[33]	VNE admission control	CPU bandwidth topological features	RNN
[34]	Substrate subgraph extraction to speed up VNE process	CPU bandwidth topological features	Hopfield network
[35]	Node mapping to achieve high acceptance ratio, high revenue-to-cost ratio, and load balancing	CPU bandwidth embedding status	Deep RL with GCN

while the VNs have the resources they need, at any given time only the required resources are reserved for this purpose. Simulations show that the RL-based dynamic resource allocation significantly improves the VN acceptance ratio, and the maximum number of accepted VN requests at any time, in comparison to the static approach. The approach also ensures that VN's QoS requirements, such as packet drop rate and virtual link delay, are not affected.

In a subsequent work [29], Mijumbi et al. leverage artificial neural networks (ANNs) and propose an adaptive resource allocation mechanism, which unlike the Q-learning-based solution in [28], does not restrict the state-action space. Similar to [28], resource allocation decisions are made in a decentralized fashion by RL agents associated to each substrate node and link. Each agent relies on an ANN whose input is the status of the substrate node (respectively link) and embedded virtual nodes (respectively links), and that outputs an allocation action. An error function that evaluates the desirability of the ANN output is used for training purposes. The objective of the error function is to encourage high virtual resource utilization, while penalizing packet drops and high delays. Simulations show that the ANN-based RL solution outperforms the Q-learning-based solution, which is attributed to a state-action space expressed at a finer granularity.

In [30, 31], Yao et al. build on the intuition that network requests follow an invariable distribution, such that if an embedding algorithm works well for historical VN requests, it is likely to have the same performance for incoming VN requests. They propose in [30] a two-phased VNE algorithm i.e. a policy gradient RL-based node-mapping phase, followed by a breadth-first search for the shortest paths between the selected host nodes in the link-mapping phase. The node-mapping agent is implemented as an ANN. It is trained with historical network data and tuned using policy gradient based on the average revenue-to-cost ratio metric. The agent's goal is to observe the current status of the substrate network and output node mapping decisions. The status of the substrate network is represented by a matrix that combines topological features and resource usage extracted from every substrate node. In [31], this matrix is further reduced using a spectrum analysis method. The reduced matrix is combined with a reduced form of the substrate network adjacency matrix. Perturbation is applied to the resulting matrix every time an embedding occurs, in lieu of systematic updates for reduced complexity. Simulations show that the model devised in [31] outperforms the original model from [30].

More recently, Yao et al. [32] explore replacing the ANN node-mapping agent with a Recurrent Neural Network (RNN), after formulating VNE as a time-series problem. The intuition is that node embedding is a continuous decision process. The RNN agent, implemented as a seq2seq model, is trained with historical network data and fine tuned using the policy gradient algorithm based on the long-term average revenue-to-cost ratio metric. Simulation results show an improvement compared to the original model from [30] in terms of request acceptance ratio, long-term revenue and long-term revenue-to-cost ratio.

In [33], Blenk et al. study the online VNE satisfiability problem. They propose an RNN-based classifier that, for a given VN request, outputs whether the embedding is possible or not. The model is meant to run prior to the VNE algorithm per se, as an admission control procedure. The goal is to save time and resources that might be wasted trying to satisfy an embedding request that cannot be satisfied, at least not in an acceptable time, in the current state of the substrate network. The authors additionally devise a novel, relatively low-complexity representation of the substrate network, as well as VN requests that combine topological features and resource usage. Simulations show that their classifier is highly accurate and significantly reduces the overall computational time for the online VNE problem, without severely impacting the performance of embedding.

In their continued effort to speed up and improve rigorous online VNE algorithms, Blenk et al. [34], leverage Hopfield networks to devise a VNE preprocessing mechanism that performs search space reduction and candidate subgraph extraction. More precisely, the designed Hopfield network computes a probability for each substrate node to be part of the candidate subgraph for a given embedding request. A rigorous VNE algorithm is then used to find the final embedding solution within the extracted subgraph. Simulations show that the proposed preprocessing step improves the runtime and/or performance of most of the tested online VNE algorithms, depending on the parameters of the Hopfield network, which have to be determined beforehand.

Yan et al. [35] build on recent advancements in deep learning and propose a deep RL solution to the node mapping problem, to reduce the overall runtime of the VNE algorithm. The authors focus on the static allocation of substrate resources. They use Graph Convolutional Networks (GCN), for the learning agent to extract spatial features in the substrate network and find the optimal node mapping. The learning agent is trained using a parallel policy gradient approach, which is shown to converge faster and perform better than sequential training. In addition to rewarding higher acceptance ratio and revenue-to-cost ratio, the used reward signal also encourages policy exploration and is shown to lead to higher performance than more traditional reward functions. The proposed deep RL solution is shown to outperform state-of-the-art non-ML embedding algorithms.

3.3.2 Network Functions Virtualization

3.3.2.1 Placement

Placement of SFCs can have varying objectives, such as minimizing the cost of placement, cost of operation (e.g. licensing fee, energy consumption), service-level agreement (SLA) and QoS requirements. This problem is known to be NP-hard, making it difficult or even prohibitive to solve it optimally for large problem instances. Furthermore, heuristics tend to be inefficient in the face of high number of constraints and changes in network dynamics [36, 37]. Recently, RL has been explored to facilitate SFC placement in virtualized environments. Traditional RL maintains a Q-table to store policies (i.e. Q-values), and the RL agent uses feedback from the environment to learn the best sequence of actions or policy to optimize a cumulative reward. However, it does not scale for large state-action space [38]. In contrast, deep RL leverages Neural Networks (NNs) to learn the Q-function that map states, actions to Q-values. Deep RL can be classified into value-based, such as deep Q-learning network (DQN), and policy-based approaches. Table 3.2 provides a summary of the state-of-the-art that addresses NFV placement.

In [39], Pei et al. translate QoS requirements as a penalty when failing to serve a service-function chain request (SFCR) in VNF placement. They employ double deep Q-learning network (DDQN) that includes two NNs, one for selecting state, action and the other for evaluating the Q-value. Once the DDQN has been trained,

Table 3.2 Summary of the state-of-the-art for ML-based placement in NFV.

References	Problem/objective	Features	ML technique
[39]	Minimize operational cost and penalty for rejecting SFCR	CPU, memory, bandwidth	Deep RL with DDQN
[40]	Minimize cost of provisioning VNFs on multi-core servers for SFCRs	CPU	RL with Q-learning and ϵ-greedy policy
[36]	Maximize the number of SFCs based on QoS requirements	CPU, memory, storage, bandwidth	Deep RL with DDPG and MCN
[37]	Minimize infrastructure power consumption	CPU, storage, bandwidth, propagation delay	NCO with stacked LSTM and policy gradient
[38]	Minimize operational cost and maximize QoS w.r.t. total throughput of accepted SFCR	CPU, memory, bandwidth, latency	Deep RL with policy gradient
[41]	Minimize discrepancy in predicted and actual total response time	Transmission, propagation, processing times, CPU, storage	RL with Q-learning and ϵ-greedy policy

it can be used for VNF placement. Each action has an associated reward that reflects the influence of the action on the network. After deployment, the DDQN evaluates the performance of the actions and selects the highest reward action according to a threshold-based policy, to trigger horizontal scaling. After VNF placement, the authors use SFC-MAP [42] to construct the routing paths for the ordering required in the SFCRs.

In contrast, to avoid expensive bandwidth consumption, Zheng et al. [40] jointly optimize the cost of provisioning VNFs on multi-core servers (i.e. VNF assignment to CPU core). However, there is still unpredictability in VNF deployment, such as the random arrival of SFCRs, resources consumed and cost of provisioning. The authors leverage Q-learning to alleviate the need to know the state transitions a priori. They employ value iteration to select a uniform, random action, implement it, and evaluate the reward. In this way, their approach updates the Q-table to identify the state transitions and be resilient in the face of changing rate of SFCRs. The authors leverage an ϵ-greedy algorithm that strikes a balance between exploration and exploitation, and control the influence of historical experience. On the other hand, Quang et al. [36] employ deep Q-learning (DQL) to maximize the number of SFCs on a substrate network, while abiding by infrastructure constraints. They leverage deep deterministic policy gradient (DDPG), where deep NNs (DNNs) i.e. actor and critic, separately learn the policy and Q-values, respectively. The authors improve DDPG by using multiple critic network (MCN) for an action, where the actor NN is updated with the gradient of the best critic in the MCN, thus improving convergence time.

The Neural Combinatorial Optimization (NCO) paradigm is extended by Solozabal et al. [37], to optimize VNF placement. Their NCO leverages an NN to model the relationship between problem instances (i.e. states) and corresponding solutions (i.e. actions), where the model weights are learnt iteratively via RL, specifically policy gradient method. Once the RL agent converges, given a problem instance, it returns a solution. This allows to infer a placement policy for a given SFCR that minimizes the overall power consumption of the infrastructure (i.e. the cost function or reward), given constraints, such as availability of virtual resource and service latency thresholds. The constraints are incorporated into the cost function using Lagrange relaxation, which indicates the degree of constraint dissatisfaction. For NN, the authors employ stacked Long Short-Term Memory (LSTM), which allows to accommodate for SFCs of varying sizes. The authors show that the proposed agent when used in conjunction, improves the performance of the greedy First-Fit heuristic.

In [38], Xiao et al. jointly address the following SFC deployment challenges: (i) capturing the dynamic nature of service request and network state, (ii) handling the different network service request traffic characteristics (e.g. flow rate) and QoS requirements (e.g. bandwidth and latency), and (iii) satisfying both provider and customer objectives i.e. minimize operation cost and maximize QoS, respectively. For the first challenge, the authors leverage Markov Decision Process (MDP) to model the dynamic network state transitions, where a state is represented as the current network resource utilization (i.e. CPU, memory, and bandwidth) and impact of current SFCs, while the action corresponds to the SFC deployment corresponding to an arriving service request. For the second challenge, the authors employ policy gradient based deep RL to automatically deploy SFCs. After RL convergence, it provides SFC deployment solution to each arriving request, abiding by resource constraints. They address the third challenge by jointly maximizing the weighted total throughput of accepted service requests (i.e. income) and minimizing the weighted total cost of occupied servers (i.e. expenditure), as the MDP reward function (i.e. income minus expenditure). Via trace-driven simulation, the authors show their approach to outperform greedy and Bayesian learning-based approaches, providing higher throughput and lower operational cost on average.

Bunyakitanon et al. [41] define E2E service level metrics (e.g. VNF processing time, network latency, etc.) in support of VNF placement. They account for heterogeneous nodes with varying capabilities and availability. The authors purport that their Q-learning based model generalizes well across heterogeneous nodes and dynamically changing network conditions. They predict the service level metrics and take actions that maximize the reward for correct predictions. The Q-values are updated using a weighted average of new and historical Q-values. The reward incorporates an acceptable margin of error, with the highest reward for predicting a value that equals the actual value. They employ an ϵ-greedy policy to strike a balance between exploration and exploitation, starting with an equal probability to explore or exploit. Then, they generate a random number, and compare it to the ϵ-greedy value to steer toward exploration or exploitation. The authors show that their model has the best performance with approximately 94% in exploration and 6% in exploitation.

3.3.2.2 Scaling

VNF resource scaling assumes an initial deployment of SFCs, with the primary objective of accommodating for the changes is service demand. Static threshold-based scaling is relatively simple to implement, where predefined thresholds are used per performance metric, such as CPU utilization, bandwidth utilization, etc. For example, Ceilometer, in OpenStack Heat, can be used to create alarms based on CPU utilization thresholds to spin up or terminate virtual network function instances (VNFIs) [43]. However, it is not only nontrivial to choose these thresholds, they may also require frequent updates to accommodate for the varying service requirements. Table 3.3 provides a summary of the state-of-the-art that addresses NFV scaling.

Static threshold-based scaling is reactive and unable to cope with sudden changes in service demand, leading to resource wastage and SLA violations. Moreover, over provisioning can lead to low resource utilization and high operational costs, while under provisioning can result in service disruption and even outage. Tang et al. [44] propose an alternative to static threshold-based scaling mechanisms, which is SLA-aware and resource efficient. They model VNF scaling as an MDP and leverage Q-learning to decide on the scaling policy. In the evaluation on daily busy-and-idle and bursty traffic scenarios, their approach outperforms static threshold-based and voting policy-based (e.g., majority of the performance metrics have to agree to a scaling action, based on their respective thresholds) approaches, while striking a trade-off between SLA guarantee for network services and VNF resource consumption.

Proactive scaling leverages service demand and/or threshold predictions to dynamically allocate resources to SFCs. ML is an ideal technique to perform predictions based on historical data, while ML features play a pivotal role in its performance. Cao et al. [45] use novel ML features for scaling, which include VNF and infrastructure level metrics. They train an NN on labeled data, to capture the complex relationships between resource allocation, VNF performance, and

Table 3.3 Summary of the state-of-the-art for ML-based scaling in NFV.

References	Problem/objective	Features	ML technique
[44]	Trade-off between SLA and VNF resource consumption	CPU, memory, storage, bandwidth, network users and requests	RL with Q-learning
[45]	Learning resource allocation and VNF performance relationship	VNF internal statistics (e.g. request queue size) and resource utilization	NN, decision table, random forest, logistic regression, naïve bayes
[46]	Meet service demands	Performance measurements (e.g. max sustainable traffic load) and resource requirements (e.g., CPU, memory)	Support vector regression, decision tree, multi-layer perceptron, linear regression, ensemble
[47]	Predict VNFC resource reliability	QoS requirements	Bayesian learning
[43]	Predict VNFC resource reliability	CPU, memory, link delay	GNN with FNNs
[48]	Predict VNFIs, and minimize QoS violations and operational cost	Time of day, measured traffic load at different time units, and changes in traffic	Multi-layer percep., bayesian network, reduced error pruning tree, random and C4.5 decision trees, random forest, decision table
[49]	Minimize average oper. cost, SLA violation and VNF latency w.r.t. resizing, deployment, off-loading	CPU, memory, QoS	Deep RL with twin delayed DDPG and DNN

service demand. However, labeling is not only cumbersome, tedious, and error prone but it also requires NFV domain expert knowledge. The authors prioritize resource allocation for VNFs based on urgency and attempt to distribute load across all instances of the VNF, using traffic forwarding rules. However, if existing instances of a VNF cannot meet the service demand, new instances must be spawned using VNF placement algorithms. While Cao et al. [45] show the benefit of composite features (i.e. VNF and infrastructure level), Schneider et al. [46] promote the use of ML for creating performance profiles that precisely capture the complex relationships between VNF performance and resource requirement. On the other hand, Shi et al. [47] leverage MDP to scale virtual network function components (VNFCs). To improve MDP performance, the authors employ

Bayesian learning and use historical resource usage of VNFCs to predict future resource reliability. These predictions are leveraged in an MDP to dynamically allocate resources to VNFCs, and facilitate system operation without disruption. Their approach outperforms greedy methods in overall cost.

Mijumbi et al. [43] draw logical relationships among VNFCs in a SFC, to forecast future resource requirements. The novelty lies in identifying relationships among VNFCs that may or may not be ordered within a VNF. The authors leverage graph NN (GNN) to model each VNFC in the SFC as two parametric functions, each modeled as a feedforward NN (FNN). These pairs of FNN are responsible for learning the resource requirements of the VNFC, using historical resource utilization information from the VNFC and its neighboring VNFCs (i.e. using the first FNN), followed by prediction of future resource requirements of the VNFC (i.e. using the second FNN). The authors employ backpropagation-through-time to update the NN weights and improve prediction performance. Similar to [45], they also leverage VNF (e.g. CPU utilization, memory, processing delay) and infrastructure level (e.g. link, capacity, latency) features. Their model yields the lowest mean absolute percentage error, when the prediction window size is within the training window size. Otherwise, the prediction accuracy suffers, requiring model retaining.

In [48], Rahman et al. use traffic measurements and scaling decisions across a time period to extract features and define classes for ML classifiers (e.g. random forest, decision table, multi-layer perceptron, etc.). The features represent measured service demand and its change from recent history, while classes represent the number of VNFIs. These features and classes are used to train ML classifiers and predict future scaling decisions. The authors leverage two classifiers, the first predicts scaling to avoid QoS violations, while the other predicts scaling to reduce operational cost. In the face of inaccurate scaling predictions and/or delays in VM startup time, ML classifiers trained to reduce QoS violations stay in a state of degraded QoS for shorter periods of time. Containerization has been shown to reduce startup times for VNFIs and significantly improve QoS.

Roig et al. [49] use unlabeled data to decide on vertical, horizontal scaling or offloading to a cloud, based on service requests, operational cost, QoS requirements and end-user perceived latency. The authors employ a parameterized action MDP, where a set of continuous parameters are associated with each action. The actions correspond to the user-server assignment, while the parameters identify the scaling of VNF server resources (i.e. compute and storage). This allows for selecting different servers for users requesting the same VNF service to increase sensitivity to end-user perceived latency and enable asynchronous manipulation of server resources. The authors leverage deep RL that parameterizes the policy, and employ actor and critic NNs to learn the policy using a twin delayed DDPG. A DNN is used to approximate the policy that optimizes the weighted average of latency, operational cost, and QoS. Since the weights can be adjusted and used to update the policy, it not only performs well under constant service demand but it also quickly adapts to variation in service requests and is resilient to changes in network dynamics.

3.3.3 Network Slicing

3.3.3.1 Admission Control

Admission control dictates whether a new incoming slice request should be granted or rejected based on available network resources, QoS requirements of the new request and its consequence on the existing services, and ensuring available resources for future requests. Evidently, accepting a new request generates revenue for the network provider. However, it may degrade the QoS of existing slices, due to scarcity of resources, consequentially violating SLA and incurring penalties, loss in revenue. Therefore, there is an inherent trade-off between accepting new requests and maintaining or meeting QoS. Admission control addresses this challenge and aims to maximize the number of accepted requests without violating SLA. Several research efforts, as outlined below, have addressed the slice admission control problem from different perspectives using ML. Table 3.4 provides a summary of the state-of-the-art for ML-based admission control approaches in network slicing.

Bega et al. [50] present a network slice admission control algorithm that maximizes the monetization of the infrastructure provider, while ensuring slice SLAs. The algorithm achieves the objective by autonomously learning the optimal admission control policy, even when slice behavior is unknown and data is unlabeled. The authors consider two types of slices: (i) inelastic, whose throughput should always be above the guaranteed rate, and (ii) elastic, whose throughput is allowed to fall below the guaranteed rate during some periods, as long as the average stays above the specified rate. Since the type of the slice, its arrival and

Table 3.4 Summary of the state-of-the-art for ML-based admission control approaches in network slicing.

References	Problem/objective	ML technique
[50]	Maximize monetization of infrastructure provider, while ensuring slice SLAs	Deep RL framework with two different NNs
[51]	Minimize loss of revenue and loss due to penalties in service degradation	Resource prediction and RL
[52]	Maximize resource utilization while respecting slice priorities	RL with Q-learning

departure are unknown in advance, it is impossible to establish the ground truth for the admission control problem. Therefore, the authors propose a deep RL approach that interacts with the environment and takes decision at a given state, while receiving feedback from past experiences. Their deep RL framework uses two different NNs, one to estimate the revenue for each state when accepting the slice request, and another to reject the request. The framework then selects the action with the highest expected revenue, and the reward for the action is fed back to RL. Through evaluation, the authors show that their proposed algorithm performs close to the optimal under a wide range of configurations, and outperforms naïve approaches and smart heuristics.

Raza et al. [51] address the network slice admission control problem by taking into account revenues of accepted slices, and penalties proportional to performance degradation, if an admitted slice cannot be scaled up later due to resource contention. The authors propose a supervised learning (SL)- and a RL-based algorithm for slice admission control. The SL-based solution leverages prediction for the incoming slice requirement, and future changes in requirement for the incoming slice and all other slices currently provisioned. This facilitates identification of possible degradation in performance upon admission, for the incoming slice or currently provisioned slices, which results in slice rejection. On the other hand, the RL-based algorithm learns the relationship between slice requirement and current resource allocation, along with the overall profit. This relationship guides slice admission policy, allowing to only accept slices that are likely to experience/create minimal to no degradation in performance. The objective of the admission policy is loss minimization, where the loss has two components: (i) loss of revenue due to rejecting slice requests, and (ii) the loss incurred due to penalties in service degradation, as described in [53].

An RL-based solution for cross-slice congestion control problem in 5G networks, which impacts the slice admission control process, is proposed by Han et al. [52]. Their solution identifies active slices with loose requirements i.e. their amount of allocated resources can be reduced based on resource availability, slice requirements, and the queue state. The identified slices' resources are then scaled down, to make room for a larger number of higher priority slices. To achieve this, the authors use Q-learning that can learn optimal resource reallocation strategy, by jointly maximizing resource utilization and respecting slice priorities. The evaluation results show that the proposed solution is able to increase the percentage of accepted slice requests, without negatively affecting the performance of high priority slices.

3.3.3.2 Resource Allocation

An E2E network may simultaneously require radio, network, computing, and storage resources from multiple network segments. An emerging challenge for

Table 3.5 Summary of the state-of-the-art for ML-based resource allocation approaches in network slicing.

References	Resource type	Problem/objective	ML technique
[54]	Virtual protocol stack functions, RRU association, sub-channel and power allocation	Maximize service utility in terms of the difference between revenue and expense	RL with ϵ-greedy Q-learning
[55]	VMs and bandwidth	Minimize processing delays for received requests and resource usage costs	RL with policy gradient methods
[56]	Service capacity requirement	Maximize revenues in short- and long-term resource reallocation	ANN-based deep learning prediction
[57]	Slice bandwidth allocation and scheduling of SFC flows	Maximize the weighted sum of spectrum efficiency and QoE	RL with Deep Q-Learning
[58]	Bandwidth or time-slots	Maximize SSR and spectrum efficiency	Dueling GAN-DDQN
[59]	Computing, storage, and radio resources	Maximize the long-term average reward	RL (Q-learning, DQL, deep double Q-learning, and deep dueling)

the network provider is how to concurrently manage multiple interconnected resources. Due to the dynamic demand of services, the frequency of slice requests, their occupation time, and requirements are not known a priori, while the resources are limited. Hence, dynamically allocating resources in real-time to maximize a specific objective is another challenge for the network provider. Table 3.5 provides a summary of the state-of-the-art for ML-based resource allocation approaches in network slicing.

Wang and Zhang [54] propose a two-stage network slice resource allocation framework based on RL (i.e. Q-learning). The first stage performs the mapping of virtual protocol stack functions of a network slice to physical server node. The second stage manages remote radio unit (RRU) association, sub-channel, and power allocation. Instead of applying one Q-learning model to solve the joint problem, the authors use two ϵ-greedy Q-learning models sequentially, to keep the model scalable. The optimization goal of the proposed model is to maximize the service utility (i.e. difference between revenue and expenditure) of the whole network, where the revenue comes from the service rate, and the expenditure comes from the virtual function deployment cost and E2E delay loss. Simulation results show that compared to the baseline schemes (e.g. minimum cost function deployment and radio resource allocation maximizing signal to noise ratio), the proposed algorithm can increase the utility of the whole system. However, there is an upper limit, due to the limited node resources, while simulation is performed only on a few tens of users in the system.

A deep RL approach is proposed by Koo et al. [55], which addresses the network slice resource allocation problem by considering unknown slice arrival characteristics, and heterogeneous SLA and resource requirements (e.g. VMs, bandwidth, and memory). The slice resource allocation pertains to allocating VMs and bandwidth for each slice, with the objective of minimizing processing delays for received requests and resource usage costs. The authors formulate the resource allocation problem as an MDP, where the constrained multi-resource optimization problem is formulated for each service upon arrival and a batch of services. For both types of request, RL models are trained offline to learn efficient resource allocation policies, which are used in real-time resource allocation. The policies are stochastic, and determine real valued resource allocations for each slice that has large and continuous action space. The authors use policy gradient methods as opposed to Q-learning, which cannot represent stochastic and continuous action spaces. Simulations using both simulated and real traces show that the model outperforms a baseline of equal-slicing strategy, which fairly divides the resources among each slice.

Bega et al. [56] present DeepCog, a data analytics tool for cognitive management of resources in 5G network slices. DeepCog forecasts the capacity needed to accommodate future traffic demands of individual network slices, while accounting for the operator's desired balance between resource over provisioning (i.e. allocating resources exceeding the demand) and SLA violations (i.e. allocating less resources than required). DeepCog uses an ANN-based deep learning prediction mechanism that consists of an encoder with three layers of three-dimensional convolutional NNs and a decoder implemented by multi-layer perceptrons. The encoder–decoder structure is shown to predict service capacity requirement with high accuracy, based on measurement data collected in an operational mobile network. The authors claim that the structure is general enough to be trained to solve the capacity forecast problem for different network slices with diverse demand patterns. The capacity forecast returned by DeepCog, can then be used by operators to take short- and long-term resource reallocation decisions and maximize revenues.

In [57], Li et al. address resource management for network slicing independently for radio resource slicing and priority-based core network slicing. In the radio part, resource management pertains to slice bandwidth allocation to maximize the weighted sum of spectrum efficiency and quality of experience (QoE). For the core network slicing, the goal is to schedule flows to SFCs that incur acceptable waiting times. For both of these problems, the authors leverage DQL to find the optimal resource allocation policies, which enhance effectiveness and agility of network slicing in a resource-constrained scenario. However, their approach does not consider the effects of random noise on the calculation of spectrum efficiency and QoE for radio resource slicing. To overcome this limitation, Hua et al. [58] combine distributional RL and Generative Adversarial Network (GAN), to propose GAN-powered deep distributional Q network (GAN-DDQN). Furthermore, the authors adopt a reward-clipping scheme and introduce a dueling structure to GAN-DDQN (i.e. Dueling GAN-DDQN), to separate the state-value distribution and the action advantage function from the action-value distribution. This circumvents the inherent training problem of GAN-DDQN. Simulation results show the effectiveness of GAN-DDQN and Dueling GAN-DDQN over the classical DQL algorithms.

Van Huynh et al. [59] propose a resource management model that allows the network provider to jointly allocate computing, storage, and radio resources to different slice requests in a real-time manner. To deal with the dynamics, uncertainty, and heterogeneity of slice requests, the authors adopt semi-MDP. Then, several RL algorithms, i.e. Q-learning, DQL, deep double Q-learning, and deep dueling, are employed to maximize the long-term average reward for the network provider. The key idea of the deep dueling algorithm is to use two streams of fully connected hidden layers to concurrently train the value and advantage functions, thus improving the training process. Simulation results show that the proposed model using deep dueling can yield up to 40% higher long-term average reward, and is a few thousand times faster compared to other network slicing approaches. The advantage of the proposed model is that it can accommodate adding more resources or removing some resources (i.e. scaling out or scaling in, respectively) by considering some new events in the system state space. However, the work of [59] overlooks the network resources that is needed for an E2E slice provisioning.

3.4 Conclusion and Future Direction

Virtualized networks and services bring inherent challenges for network operators, which calls for automated management that cannot be satisfied with the traditional, reactive human-in-the loop management approach. Furthermore, the requirement for higher QoS and ultra-low latency services necessitates intelligent management that should harness the sheer volume of data within a network, and take automated management decisions. Therefore, AI and ML can play a pivotal role to realize the automation of management for virtualized networks and services. In Section 3.3, we discuss the state-of-the-art in employing AI and ML techniques to address various challenges in managing virtualized networks and services, specifically in NV, NFV, and network slicing. In this section, we delineate open, prominent research challenges and opportunities for holistic and automated management of virtualized networks and services.

3.4.1 Intelligent Monitoring

Monitoring requires the identification of Key Performance Indicators (KPIs), such as perceived latency, alarms, and utilization of virtualized network components [60]. These play a crucial role in analytics to facilitate automated decision-making for managing virtualized networks and services. It is quintessential that the employed measurement techniques collect telemetry data with high accuracy, while minimizing overhead. However, measurement can add significant overhead (e.g. consumed network bandwidth, switch memory due to probing, and storage) when a large number of virtualized network components are monitored at regularly occurring intervals. This instigates the need for adaptive measurement schemes that can dynamically tune monitoring rate and decide what to monitor. ML techniques, such as regression, can facilitate adaptive monitoring by predicting telemetry data that would have otherwise been measured. Another challenge is to devise mechanisms for timely and high precision instrumentation to monitor KPIs for virtualized networks with demanding QoS requirements, especially for ultra-low latency services.

3.4.2 Seamless Operation and Maintenance

ML-based predictive maintenance can enable seamless operation of virtualized networks [19]. It involves inferring future events based on measured KPIs, identifying causes of performance degradation, and proactively taking preventive measures. An example of inference is to determine if a performance degradation (e.g. increased packet loss, prolonged downtime) would lead to future QoS violations. It is also crucial to infer causes (e.g. misconfiguration, failure) of performance degradation in correlation with potential alarms. However, realizing this from the enormous volume of telemetry data and stochastic nature of network events is challenging. Data-driven approaches, including ML, can be explored to address these problems. Once the cause for performance degradation is identified, mitigation workflows are needed to minimize the impact on KPIs. Deducing these workflows and optimally scheduling their execution with minimal interruption to the existing traffic is nontrivial. However, RL seems well suited to the problem and should be investigated to find optimal mitigation workflows.

3.4.3 Dynamic Slice Orchestration

In 5G mobile networks, an E2E VN slice spans multiple network segments, each of which can have different technological and physical constraints. For instance, the access network may have limited bandwidth and scalability to minimize cost and energy, while the core network may not have these issues of capacity or scalability. However, the core network may have higher latency and energy footprint due to long geographical distances and more complex network devices. Similar trade-offs exist between edge and central DCs, with respect to processing capacity, latency, and energy consumption. Therefore, it will be impractical to provision a network slice for its peak traffic demand. Hence, dynamic slice provisioning algorithms must be investigated, where resource orchestration decisions are facilitated by ML models for slice traffic volume prediction with temporal, spatial considerations and QoS requirements. Such dynamic slice provisioning will be enabled by NFV that allows for spawning on-demand virtualized NFs, and SDN controllers that can route traffic to newly spawned NFs.

3.4.4 Automated Failure Management

Even with predictive maintenance, some failures, such as fiber cuts and device burns are inevitable. Ability of a network provider to quickly repair a failure is crucial to keep the network operational. Failure management involves three steps: failure detection, localization, and identification. The goal of failure detection is to trigger an alert after the failure has occurred. Once detected, the failed element (e.g. the node or link responsible for the failure) must be localized in the network to narrow down the root cause of failure. Even after localization, it might still be complex to understand the exact cause of the failure. For example, inside a network node, the degradation can be due to misconfiguration or malfunction. To speed up the failure repair process, all three steps of failure repair should be automated. An interesting avenue of research is to develop ML models and algorithms for automated failure detection, localization, and identification based on the data generated in production networks. These models will decrease the mean time to repair after failure events, thus improving the availability of a network slice or a virtualized network/service.

3.4.5 Adaptation and Consolidation of Resources

The traffic demand and/or QoS requirement of a virtualized network or a network slice may evolve over time, due to change in number of users and communication patterns [61]. Hence, the initial resource allocation need to be adapted to accommodate for such changes, while causing minimal to no disruption to existing traffic. This calls for ML models to predict change in requirements in a timely and accurate manner, and facilitate dynamic adaptation of resource allocation. Furthermore, over time, arrival and departure of virtualized networks or network slices can lead to fragmentation and skewed utilization of links and processing servers. These, in turn, can impact the acceptance of future requests and result in unnecessary energy consumption. One way to mitigate this is by re-optimizing bandwidth allocation and periodically consolidating VMs or containers. The solution should also output the sequence of operations (e.g. VM migration, virtual link migration, and bandwidth reallocation) that lead to a load-balanced state. RL is an ideal technique to generate the sequence of operations needed to reach the optimized state.

3.4.6 Sensitivity to Heterogeneous Hardware

In NFV deployment or in network slices, VNFIs that reside on VMs or containers are scaled to meet the service demands. However, the performance of VNFs is sensitive to the underlying hardware [45, 46]. For example, traffic processing capabilities of virtual CPUs on Intel Xeon processor differ from AMD Opteron processor [45]. Similarly, boot up time for VMs differ across VIMs, such as OpenStack, Eucalyptus, and OpenNebula [43]. Nevertheless, most research assumes homogeneous hardware, being oblivious to its impact on VNF performance. This is an oversimplification, which can lead to inferior ML models and inaccurate scaling decisions in practice. Therefore, it is quintessential to develop performance profiles [46], which incorporate the sensitivity of VNF performance on different hardware. In case of horizontal scaling, these profiles can be leveraged to accurately gauge the impact on performance for new VNFIs on different physical servers. Indeed, incorporating these profiles will increase the dimensionality of the scaling problem. A naïve option is to incorporate hardware-sensitivity as a cost. However, building VNF performance profiles for different hardware is cumbersome. It remains to be evaluated how these hardware-specific performance profiles will impact the accuracy of ML models and VNF scaling decisions.

3.4.7 Securing Machine Learning

Evidently, there has been a surge in the application of ML for managing virtualized networks, ranging from placement and scaling of VNFs to admission control in network slices. However, numerous research assumes ML itself to be invincible. This is an unrealistic assumption, as adversaries can poison the training data, or compromise the RL agent by manipulating system states and policies, leading to inferior actions [62]. For example, impeding actual resource consumption of substrate network can result in suboptimal SFC placement, leading to resource wastage and/or SLA violations. Inherently, ML models lack robustness against adversarial attempts. Adversarial learning addresses this concern by leveraging carefully crafted adversarial (i.e. fake) samples, with minor perturbations to regular inputs [63, 64]. These can be used to inculcate robustness into ML models against data poisoning attacks. GANs have been widely used to generate such adversarial samples. GANs are a class of deep learning techniques that use two neural networks, discriminator and generator, to compete with each other for model training. However, GANs can suffer from training instability, due to fake training data that degrades model performance [65]. Therefore, ensuring convergence of GANs is an open research problem. Furthermore, the use of GANs to harden RL agents against complex threat vectors is rather unexplored. Adversarial deep RL with multi-agents [66, 67], trained across distributed virtualized environments, can also help alleviate the impact of adversarial attempts.

Bibliography

1 Anderson, T., Peterson, L., Shenker, S., and Turner, J. (2005). Overcoming the internet impasse through virtualization. *Computer* 38 (4): 34–41.

2 Turner, J.S. and Taylor, D.E. (2005). Diversifying the internet. *IEEE Global Telecommunications Conference (GLOBECOM)*, Volume 2, IEEE, p. 6.

3 Google IPv6 adoption statistics. https://www.google.com/intl/en/ipv6/statistics .html#tab=ipv6-adoption.

4 Liang, C. and Yu, F.R. (2014). Wireless network virtualization: a survey, some research issues and challenges. *IEEE Communication Surveys and Tutorials* 17 (1): 358–380.

5 Costa-Pérez, X., Swetina, J., Guo, T. et al. (2013). Radio access network virtualization for future mobile carrier networks. *IEEE Communications Magazine* 51 (7): 27–35.

6 Jinno, M., Takara, H., Yonenaga, K., and Hirano, A. (2013). Virtualization in optical networks from network level to hardware level. *Journal of Optical Communications and Networking* 5 (10): A46–A56.

7 Bari, Md.F., Boutaba, R., Esteves, R. et al. (2012). Data center network virtualization: a survey. *IEEE Communication Surveys and Tutorials* 15 (2): 909–928.

8 Jain, R. and Paul, S. (2013). Network virtualization and software defined networking for cloud computing: a survey. *IEEE Communications Magazine* 51 (11): 24–31.

9 Duan, Q., Yan, Y., and Vasilakos, A.V. (2012). A survey on service-oriented network virtualization toward convergence of networking and cloud computing. *IEEE Transactions on Network and Service Management* 9 (4): 373–392.

10 Drutskoy, D., Keller, E., and Rexford, J. (2012). Scalable network virtualization in software-defined networks. *IEEE Internet Computing* 17 (2): 20–27.

11 Blenk, A., Basta, A., Reisslein, M., and Kellerer, W. (2015). Survey on network virtualization hypervisors for software defined networking. *IEEE Communication Surveys and Tutorials* 18 (1): 655–685.

12 Alam, I., Sharif, K., Li, F. et al. (2020). A survey of network virtualization techniques for internet of things using SDN and NFV. *ACM Computing Surveys (CSUR)* 53 (2): 1–40.

13 Mijumbi, R., Serrat, J., Gorricho, J.-L. et al. (2015). Network function virtualization: state-of-the-art and research challenges. *IEEE Communication Surveys and Tutorials* 18 (1): 236–262.

14 Yi, B., Wang, X., Li, K. et al. (2018). A comprehensive survey of network function virtualization. *Computer Networks* 133: 212–262.

15 Ghaznavi, M., Shahriar, N., Kamali, S. et al. (2017). Distributed service function chaining. *IEEE Journal on Selected Areas in Communications* 35 (11): 2479–2489.

16 Carapinha, J. and Jiménez, J. (2009). Network virtualization: a view from the bottom. *ACM Workshop on Virtualized Infrastructure Systems and Architectures*, ACM, pp. 73–80.

17 Nikaein, N., Schiller, E., Favraud, R. et al. (2015). Network store: exploring slicing in future 5G networks. *International Workshop on Mobility in the Evolving Internet Architecture*, ACM, pp. 8–13.

18 Ayoubi, S., Limam, N., Salahuddin, M.A. et al. (2018). Machine learning for cognitive network management. *IEEE Communications Magazine* 56 (1): 158–165.

19 Boutaba, R., Salahuddin, M.A., Limam, N. et al. (2018). A comprehensive survey on machine learning for networking: evolution, applications and research opportunities. *Journal of Internet Services and Applications* 9 (1): 16.

20 Afolabi, I., Taleb, T., Samdanis, K. et al. (2018). Network slicing and softwarization: a survey on principles, enabling technologies, and solutions. *IEEE Communication Surveys and Tutorials* 20 (3): 2429–2453.

21 China Mobile Technology. (2016) Accelerating Business with OpenStack and OPNFV. https://object-storage-ca-ymq-1.vexxhost.net/swift/v1/ 6e4619c416ff4bd19e1c087f27a43eea/www-assets-prod/marketing/presentations/ OpenStack-OPNFVDatasheet-A4.pdf (accessed 04 August 2020).

22 Yilma, G.M., Yousaf, Z.F., Sciancalepore, V., and Costa-Perez, X. (2020). Benchmarking open source NFV MANO systems: OSM and ONAP. *Computer Communications*. 161 86–98.

23 Rodriguez, V.Q., Guillemin, F., and Boubendir, A. (2020). 5G E2E network slicing management with ONAP. *Conference on Innovation in Clouds, Internet and Networks and Workshops (ICIN)*, IEEE, pp. 87–94.

24 Rodriguez, V.Q., Guillemin, F., and Boubendir, A. (2020). Network slice management on top of ONAP. *IFIP/IEEE Network Operations and Management Symposium (NOMS)*, IEEE, pp. 1–2.

25 Karamichailidis, P., Choumas, K., and Korakis, T. (2019). Enabling multi-domain orchestration using Open Source MANO, OpenStack and Open-Daylight. *IEEE International Symposium on Local and Metropolitan Area Networks (LANMAN)*, IEEE, pp. 1–6.

26 Cao, H., Hu, H., Qu, Z., and Yang, L. (2018). Heuristic solutions of virtual network embedding: a survey. *China Communications* 15 (3): 186–219.

27 Fischer, A., Botero, J.F., Beck, M.T. et al. (2013). Virtual network embedding: a survey. *IEEE Communication Surveys and Tutorials* 15 (4): 1888–1906.

28 Mijumbi, R., Gorricho, J., Serrat, J. et al. (2014). Design and evaluation of learning algorithms for dynamic resource management in virtual networks. *IEEE Network Operations and Management Symposium (NOMS)*, pp. 1–9.

29 Mijumbi, R., Gorricho, J.-L., Serrat, J. et al. (2014). Neural network-based autonomous allocation of resources in virtual networks. *European Conference on Networks and Communications (EuCNC)*, IEEE, pp. 1–6.

30 Yao, H., Chen, X., Li, M. et al. (2018). A novel reinforcement learning algorithm for virtual network embedding. *Neurocomputing* 284: 1–9.

31 Yao, H., Zhang, B., Zhang, P. et al. (2018). RDAM: a reinforcement learning based dynamic attribute matrix representation for virtual network embedding. *IEEE Transactions on Emerging Topics in Computing* 1. https://ieeexplore.ieee .org/document/8469054

32 Yao, H., Ma, S., Wang, J. et al. (2020). A continuous-decision virtual network embedding scheme relying on reinforcement learning. *IEEE Transactions on Network and Service Management*. 17 (2): 864–875

33 Blenk, A., Kalmbach, P., Van Der Smagt, P. et al. (2016). Boost online virtual network embedding: using neural networks for admission control. *International Conference on Network and Service Management*, Montreal, Canada, October 2016, pp. 10–18.

34 Blenk, A., Kalmbach, P., Zerwas, J. et al. (2018). NeuroViNE: a neural pre-processor for your virtual network embedding algorithm. *IEEE International Conference on Computer Communications (INFOCOM)*, IEEE, pp. 405–413.

35 Yan, Z., Ge, J., Wu, Y. et al. (2020). Automatic virtual network embedding: a deep reinforcement learning approach with graph convolutional networks. *IEEE Journal on Selected Areas in Communications* 38 (6): 1040–1057.

36 Quang, P.T.A., Hadjadj-Aoul, Y., and Outtagarts, A. (2019). A deep reinforcement learning approach for VNF forwarding graph embedding. *IEEE Transactions on Network and Service Management* 16 (4): 1318–1331.

37 Solozabal, R., Ceberio, J., Sanchoyerto, A. et al. (2019). Virtual network function placement optimization with deep reinforcement learning. *IEEE Journal on Selected Areas in Communications* 38 (2): 292–303.

38 Xiao, Y., Zhang, Q., Liu, F. et al. (2019). NFVdeep: adaptive online service function chain deployment with deep reinforcement learning. *International Symposium on Quality of Service*, pp. 1–10.

39 Pei, J., Hong, P., Pan, M. et al. (2019). Optimal VNF placement via deep reinforcement learning in SDN/NFV-enabled networks. *IEEE Journal on Selected Areas in Communications* 38 (2): 263–278.

40 Zheng, J., Tian, C., Dai, H. et al. (2019). Optimizing NFV chain deployment in software-defined cellular core. *IEEE Journal on Selected Areas in Communications* 38 (2): 248–262.

41 Bunyakitanon, M., Vasilakos, X., Nejabati, R., and Simeonidou, D. (2020). End-to-end performance-based autonomous VNF placement with adopted reinforcement learning. *IEEE Transactions on Cognitive Communications and Networking*. 6 (2): 534–547

42 Pei, J., Hong, P., Xue, K., and Li, D. (2018). Efficiently embedding service function chains with dynamic virtual network function placement in geo-distributed cloud system. *IEEE Transactions on Parallel and Distributed Systems* 30 (10): 2179–2192.

43 Mijumbi, R., Hasija, S., Davy, S. et al. (2017). Topology-aware prediction of virtual network function resource requirements. *IEEE Transactions on Network and Service Management* 14 (1): 106–120.

44 Tang, P., Li, F., Zhou, W. et al. (2015). Efficient auto-scaling approach in the telco cloud using self-learning algorithm. *IEEE Global Communications Conference (GLOBECOM)*, IEEE, pp. 1–6.

45 Cao, L., Sharma, P., Fahmy, S., and Saxena, V. (2017). ENVI: elastic resource flexing for network function virtualization. *USENIX Workshop on Hot Topics in Cloud Computing (HotCloud)*.

46 Schneider, S.B., Satheeschandran, N.P., Peuster, M., and Karl, H. (2020). Machine learning for dynamic resource allocation in network function virtualization. *IEEE Conference on Network Softwarization (NetSoft)*.

47 Shi, R., Zhang, J., Chu, W. et al. (2015). MDP and machine learning-based cost-optimization of dynamic resource allocation for network function virtualization. *IEEE International Conference on Services Computing*, IEEE, pp. 65–73.

48 Rahman, S., Ahmed, T., Huynh, M. et al. (2018). Auto-scaling VNFs using machine learning to improve QoS and reduce cost. *IEEE International Conference on Communications (ICC)*, IEEE, pp. 1–6.

49 Roig, J.S.P., Gutierrez-Estevez, D.M., and Gündüz, D. (2019). Management and orchestration of virtual network functions via deep reinforcement learning. *IEEE Journal on Selected Areas in Communications* 38 (2): 304–317.

50 Bega, D., Gramaglia, M., Banchs, A. et al. (2019). A machine learning approach to 5G infrastructure market optimization. *IEEE Transactions on Mobile Computing*. 19 (3): 498–512

51 Raza, M.R., Natalino, C., Wosinska, L., and Monti, P. (2019). Machine learning methods for slice admission in 5G networks. *OptoElectronics and*

Communications Conference (OECC) and 2019 International Conference on Photonics in Switching and Computing (PSC), IEEE, pp. 1–3.

52 Han, B., DeDomenico, A., Dandachi, G. et al. (2018). Admission and congestion control for 5G network slicing. *IEEE Conference on Standards for Communications and Networking (CSCN)*, IEEE, pp. 1–6.

53 Raza, M.R., Natalino, C., Öhlen, P. et al. (2019). Reinforcement learning for slicing in a 5G flexible RAN. *Journal of Lightwave Technology* 37 (20): 5161–5169.

54 Wang, X. and Zhang, T. (2019). Reinforcement learning based resource allocation for network slicing in 5G C-RAN. *Computing, Communications and IoT Applications (ComComAp)*, IEEE, pp. 106–111.

55 Koo, J., Mendiratta, V.B., Rahman, M.R., and Walid, A. (2019). Deep reinforcement learning for network slicing with heterogeneous resource requirements and time varying traffic dynamics. *arXiv preprint arXiv:1908.03242*.

56 Bega, D., Gramaglia, M., Fiore, M. et al. (2019). DeepCog: cognitive network management in sliced 5G networks with deep learning. *IEEE Conference on Computer Communications (IEEE INFOCOM)*, Paris, France, May 2019, pp. 280–288.

57 Li, R., Zhao, Z., Sun, Q. et al. (2018). Deep reinforcement learning for resource management in network slicing. *IEEE Access* 6: 74429–74441.

58 Hua, Y., Li, R., Zhao, Z. et al. (2019). GAN-powered deep distributional reinforcement learning for resource management in network slicing. *IEEE Journal on Selected Areas in Communications*. 38 (2): 334–349

59 Van Huynh, N., Hoang, D.T., Nguyen, D.N., and Dutkiewicz, E. (2019). Optimal and fast real-time resource slicing with deep dueling neural networks. *IEEE Journal on Selected Areas in Communications* 37 (6): 1455–1470.

60 Chowdhury, S.R., Bari, Md.F., Ahmed, R., and Boutaba, R. (2014). Payless: a low cost network monitoring framework for software defined networks. *IEEE Network Operations and Management Symposium (NOMS)*, IEEE, pp. 1–9.

61 Hadi, M., Pakravan, M.R., and Agrell, E. (2019). Dynamic resource allocation in metro elastic optical networks using Lyapunov drift optimization. *Journal of Optical Communications and Networking* 11 (6): 250–259.

62 Behzadan, V. and Munir, A. (2017). Vulnerability of deep reinforcement learning to policy induction attacks. *International Conference on Machine Learning and Data Mining in Pattern Recognition*, Springer, pp. 262–275.

63 Grosse, K., Papernot, N., Manoharan, P. et al. (2017). Adversarial examples for malware detection. In: *Computer Security – ESORICS 2017* (ed. S.N. Foley, D. Gollmann, and E. Snekkenes), 62–79. Springer International Publishing.

64 Pawlicki, M., Choraś, M., and Kozik, R. (2020). Defending network intrusion detection systems against adversarial evasion attacks. *Future Generation Computer Systems*. 110 148–154.

65 Kodali, N., Abernethy, J., Hays, J., and Kira, Z. (2017). On convergence and stability of gans. *arXiv preprint arXiv:1705.07215.*

66 Nguyen, T.T. and Reddi, V.J. (2019). Deep reinforcement learning for cyber security. *arXiv preprint arXiv:1906.05799.*

67 Zhang, K., Yang, Z., and Başar, T. (2019). Multi-agent reinforcement learning: a selective overview of theories and algorithms. *arXiv preprint arXiv:1911.10635.*

4

Self-Managed 5G Networks[1]

Jorge Martín-Pérez[1], Lina Magoula[2], Kiril Antevski[1], Carlos Guimarães[1], Jorge Baranda[3], Carla Fabiana Chiasserini[4], Andrea Sgambelluri[5], Chrysa Papagianni[6], Andrés García-Saavedra[7], Ricardo Martínez[3], Francesco Paolucci[5], Sokratis Barmpounakis[2], Luca Valcarenghi[5], Claudio EttoreCasetti[4], Xi Li[7], Carlos J. Bernardos[1], Danny De Vleeschauwer[6], Koen De Schepper[6], Panagiotis Kontopoulos[2], Nikolaos Koursioumpas[2], Corrado Puligheddu[4], Josep Mangues-Bafalluy[3], and Engin Zeydan[3]

[1] *Telematics Engineering department, Universidad Carlos III de Madrid, Madrid, Spain*
[2] *National and Kapodistrian University of Athens, Software Centric & Autonomic Networking lab, Athens, Greece*
[3] *Communication Networks Division, Centre Tecnològic de Telecomunicacions Catalunya (CTTC/CERCA), Barcelona, Spain*
[4] *Department of Electronics and Telecommunications, Politecnico di Torino, Torino, Italy*
[5] *Scuola Superiore Sant'Anna, Istituto TeCIP, Pisa, Italy*
[6] *Nokia Bell Labs, Antwerp, Belgium*
[7] *NEC Laboratories Europe, 5G Networks R&D Group, Heidelberg, Germany*

4.1 Introduction

Besides the performance enhancements (i.e. lower latency, higher bandwidth, increased reliability, among others) and new advances in radio technologies (i.e. new spectrum by the introduction of sub-6 GHz and mmWave), the 5th generation (5G) of mobile communications aims at extending the general purpose connectivity design of earlier generations to support a wide variety of use cases with a disparate set of requirements and capabilities. The traditional one-size-fits-all approach to network infrastructure is no longer suitable for supporting such

1 This work has been partially supported by EC H2020 5GPPP 5Growth project (Grant #856709). It has also been (partially) funded by the H2020 EU/TW joint action 5G-DIVE (Grant #859881).

vision; a shift is required toward a custom-fit approach that supports different virtual and isolated networks within the same shared network infrastructure. As such, the design of 5G networks moves toward a highly modular, highly and programmable architecture, integrating innovative and disruptive mechanisms from other technology domains, including service-based architectures, Information Technology (IT)-centric cloud services and artificial intelligence/machine learning (AI/ML).

This enables the creation of novel services and use cases that were not possible with the previous generations of mobile communication. A wide variety of use cases are identified in [1], that require the support of 5G networks: video entertainment, mass media, automotive, industry 4.0, and eHealth related services. Each one of these use cases have disparate requirements not only in terms of latency, bandwidth, reliability, and mobility but also in terms of service priority. As an example, the traffic of a remote surgery service must be prioritized with its communication requirements being met at all times, not impacted by any lower priority services, such as a 4K TV live broadcast service.

This diversity of services supported by 5G networks is changing the way the mobile network is managed and operated. It is required to ensure that different services can coexist and be provisioned over the same network infrastructure, while satisfying the requirements of each one of these services. To prioritize service traffic in order to meet service key performance indicators (KPIs), efficiently allocate wireless resources, fiber optics wavelengths, and computational resources; these are management tasks that must be automated in 5G systems. For example, a natural disaster might require an eHealth service for remote surgery in the affected area. Under such circumstances, the 5G network should automatically prioritize and allocate resources for the eHealth services, while supporting existing services. Zero touch network and service management (ZSM) is needed as adding the human in the loop increases the service response time with direct impact on human lives. Moreover, the use of static or predefined set of rules and policies may not be sufficient for unexpected events, leading to undesired and/or nonoptimal decisions. Thus, novel and automated network and service management mechanisms are required.

5G enables different virtual and isolated networks (hereinafter referred as *network slices*) to be supported within the same shared network infrastructure. An important property is that each network slice is isolated from one another, meaning that each slice's performance must not have any impact on the performance of coexisting slices. Following the previous example, a network slice could be dedicated to eHealth services, having the top priority among all other slices, and even wider slots in the radio transmission intervals; whilst video entertainment services would run on top of a slice with less priority meeting bandwidth capacity to provide 4K video streaming.

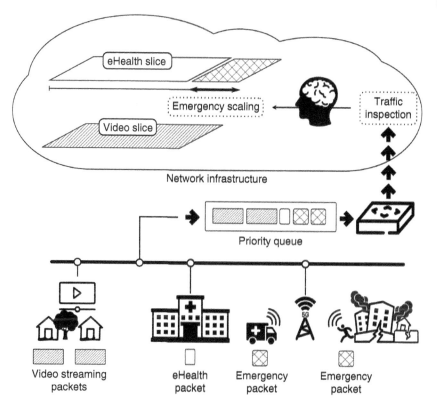

eHealth slice

Emergency scaling

Traffic Inspection

Video slice

Network infrastructure

Priority queue

Video streaming packets

eHealth packet

Emergency packet

Emergency packet

Figure 4.1 Slicing and prioritization of 5G network traffic. An eHealth network slice scales up after the occurrence of an emergency.

In [2], the 3rd generation partnership project (3GPP) organization provides a detailed specification on how to incorporate *network slicing* to the standard 5G architecture specification. The specification states that network slices might differ "as they deliver a different committed service," as shown in Figure 4.1. In Figure 4.1, it is illustrated how the infrastructure resources dedicated to an eHealth slice scale up after an emergency happens. Moreover, Figure 4.1 depicts[2] an intelligent agent that takes the scaling decision based on the inspection of the received traffic. The intelligent agent is a software-based management tool which can be based on AI/ML as discussed latter in this chapter, and it enhances the network with self-management that, in the depicted scenario, increases the resources of the eHealth slice to accommodate the incoming traffic of the emergency event.

2 Icons made by https://freeicons.io/profile/[3335,823,726,2257,3063,722,3031] from www.freeicons.io

Network Function Virtualization (NFV) appeared as a key enabler for the realization of network slicing, by allowing network functions and services to be virtualized on standard hardware instead of running on dedicated hardware. As such, NFV adds increased flexibility to run the virtualized network functions (VNFs) across servers in different locations. Independently of the virtualization technology being used [3], NFV offers the possibility of instantiating new network services (NSs), decommissioning NSs when no longer required, or scaling and migrating NSs to cope with changes on demands. Furthermore, a network slice can be deployed through NFV defined NS, and consist of one or more such NSs. As discussed in this chapter, an agent can automate the network provisioning of the aforementioned slices.

Satisfying service level agreements (SLAs) is a key component in contractual agreements between verticals and service providers. Given that mobile traffic demand will grow up to 43.9 Mbps by 2023,[3] the risk of bottlenecks and SLAs violation is high unless there are changes in the network management paradigm. An immediate consequence of increasing traffic demand is the scarce of computational, and network resources owned by service providers. In case a service provider's infrastructure runs out of resources, it could rely on the *federation* paradigm, an approach proposing a share of services or resources among infrastructure owners, or service providers. For example, two infrastructure owners would expose interfaces to rent its resources among each other, so both of them can immediately rent disk space, radio coverage in a specific area, or a located-caching service. The renting happens under the umbrella of agreed SLAs, using peering infrastructure/service provider's interfaces. *Federation* in 5G networks will not only enhance service providers and infrastructure owners with a wider pool of resources and services but will also allow service providers to satisfy service requirements by delegating the deployment to peering domains.

Network *slicing*, *virtualization*, and *federation* are three key concepts that will enable self-management in 5G networks. To make the most out of the three of them, it is necessary to understand what is the status of current network technology. Thus, Section 4.2 provides a technology overview on radio access network (RAN), NFV, data plane, and optical switches programmability, and network data management. This chapter covers all these aspects and collects the current state-of-the-art (SoA) on how to use AI/ML techniques to asses RAN resource management, orchestration of NSs, slicing in the data plane, allocation of fiber optics' wavelengths, and federation. Section 4.3 reviews how the SoA incorporates AI/ML techniques to implement agents that deal with network management tasks. This chapter finishes with Section 4.4 stating future directions on how to enhance self-management, and improve 5G networks performance.

---

3 Numbers obtained at [4]

4.2 Technology Overview

This section gives a brief overview of the enabling technologies that are fundamental to build and manage the resources end-to-end. Through such technologies, 5G networks can automate the resource handling, and isolation between logical and physical resources. In this sense, Section 4.2.1 discusses about the latest approaches in the RAN domain with open radio access network (O-RAN) as a prominent example. Section 4.2.2 focuses on relevant controllers based on the European Telecommunications Standards Institute Management and Orchestration (ETSI MANO) stack, and in particular, how computing resources are handled (e.g. through Openstack). The following two Sections 4.2.3–4.2.4 discuss about the fundamental ideas and latest trends in data plane programmability at the packet level (Section 4.2.3) and at the optical switching level (Section 4.2.4), hence focusing on transport network resource management. Finally, Section 4.2.5 covers the data pipeline creation to provide the management stack with network monitoring and analytics.

4.2.1 RAN Virtualization and Management

Radio access virtualization (*a.k.a.* virtual radio access network (vRAN)), with the promise of considerable operational/capital expenditure (OPEX/CAPEX) savings, high flexibility, and openness to foster innovation and competition, is the last milestone in the NFV revolution and will be a key technology for beyond-5G (B5G) systems. Harnessing the strengths of NFV into the RAN arena, however, entails many challenges that are the objects of study by multiple initiatives such as Rakuten's greenfield deployment in Japan,[4] Cisco's Open vRAN Ecosystem,[5] telecom infra project (TIP) TIP's vRAN Fronthaul Project Group,[6] and O-RAN Alliance.[7] Arguably, O-RAN is the most promising among these efforts.

Figure 4.2 depicts a high-level view of the O-RAN architecture. Doubtlessly, the most important functional components introduced by O-RAN are the non-real-time (Non-RT) radio intelligent controller (RIC) and the near-RT RIC. While the former is hosted by the Service Management and Orchestration (SMO) framework of the system (e.g. integrated within Open Network Automation Platform [ONAP]), the latter may be co-located with 3GPP gNB functions, namely, O-RAN-compliant cloud unit (O-CU) and/or O-RAN-compliant distributed unit (O-DU), or fully decoupled from them as long as latency constraints

---

4 https://global.rakuten.com/corp/news/press/2019/0605_01.html
5 https://blogs.cisco.com/sp/cisco-multi-vendor-open-vran-ecosystem-for-mobile-networks
6 https://telecominfraproject.com/vran/
7 https://www.o-ran.org/

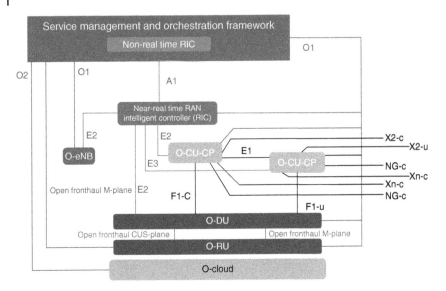

Figure 4.2 O-RAN high-level architecture. Source: O-RAN Alliance [5].

are respected. Furthermore, the O-CU might be decoupled in two VNFs for the control (O-CU-CP) and data-plane (O-CU-DP). Figure 4.2 also depicts the O-Cloud, an O-RAN compliant cloud platform that uses hardware accelerator add-ons when needed (e.g. to speed up fast fourier transform (FFT) or decoding workflows) and a software stack that is decoupled from the hardware to deploy evolved NodeB/next generation NodeB (eNBs/gNBs) as VNFs in vRAN scenarios (Open RAN NodeB (O-gNBs)).

Service Management and Orchestration (SMO): The SMO consolidates several orchestration and management services, which may go beyond pure RAN management such as 3GPP next generation (NG-) core management or end-to-end network slice management. In the context of O-RAN, the main responsibilities of SMO are (i) fault, configuration, accounting, performance, and security (FCAPS) interface to O-RAN network functions; (ii) large timescale RAN optimization; and (iii) O-Cloud management and orchestration via O2 interface, including resource discovery, scaling, FCAPS, software management, create, read, update, and delete (CRUD) O-Cloud resources.

Non-RT RAN Intelligent Controller (Non-RT RIC): As mentioned earlier, this logical function resides within the SMO and provides the A1 interface to the near-RT RIC. Its main goal is to support large timescale RAN optimization (seconds or minutes), including policy computation, ML model management (e.g. training), and other radio resource management (RRM) functions within this timescale. Data management tasks requested by the non-RT RIC should be converted into the O1/O2 interface; and contextual/enrichment information can be provided to the near-RT RIC via A1 interface.

Near-RT RAN Intelligent Controller (Near-RT RIC): Near-RT RIC is a logical function that enables near-real-time optimization and control and data monitoring of O-CU and O-DU nodes in near-real-time timescales (between 10 ms and 1 second). To this end, near-RT RIC control is steered by the policies and assisted by models computed/trained by the Non-RT RIC. One of the main operations assigned to the near-RT RIC is RRM but near-RT RIC also supports third party applications (so-called xApps).

vRAN orchestration is a challenging research problem as the performance of vBSs depends on numerous factors like channel conditions and users' demand. Moreover, the bit-rate is influenced by the SNR, RAP configuration, and the pooling of computational resources [6]. Additionally, CPU dimensioning and radio orchestration decisions (such as modulation and coding scheme selection) impact not only performance but also cost on the computational loads. In Section 4.3.1.3, we present a vRAN orchestration solution with assistance of ML to jointly decide the radio and CPU allocation policy in different context and network conditions.

4.2.2 Network Function Virtualization

The NFV concept introduced by ETSI enables that specific networking functions are entirely deployed in software exploiting virtualization techniques [7] in cloud computing environment. By doing so, telecom networks attain higher networking flexibility as well as other appealing benefits, such as reduced operational and capital costs when compared to traditional dedicated hardware solutions supporting such network functions. In this context, an end-to-end service (referred to as NS) is formed by a sequence of VNFs hosted in cloud (e.g. Data Centers, data center (DCs)) or edge facilities being interconnected over different segments/domains providing the required NS functionality.

In [8], ETSI proposed the architectural framework for the network function softwarization specifying the key control and management elements, functions, and interfaces supporting the deployment of any NS made up of different VNFs. In a nutshell, this architecture covers the following functionalities:

- Supporting VNFs operation across different hypervisors and computing resources to provide access to shared storage, computation, and physical/virtual networking.
- Construction of VNF forwarding graphs.
- NSs with different requirements exploit virtualization techniques.
- Use of DC technology.

In this NFV architectural framework, the core element is the MANO which handles the orchestration and lifecycle management of the NSs/VNFs. In other words, it takes over of all the virtualization-specific management tasks. The MANO is divided into a set of functional modules which interact among them via well-defined interfaces:

- *Virtual infrastructure manager* (VIM): This element provides control and management of the virtualized resources available in the Network Function Virtualization infrastructure (NFVI) such as storage, network, and compute. This means the allocation, modification, release of the NFVI resources (i.e. association of virtual and physical resources), etc.
- *Virtualized network function manager* (VNFM): This function handles the lifecycle management of the deployed VNFs for a NS.
- *NFV orchestrator (NFVO)*: It coordinates (i.e. orchestrates) the selection and allocation of the resources (both compute and networking) across multiple VIMs. Moreover, it is the responsible for managing the lifecycle of the deployed NSs.

Among the different VIM solutions available in the open source world, Open-Stack [9] is one of the most popular software for NFVI management and VNF deployment in the form of virtual machines (VMs). It offers a mature cloud computing platform for resource orchestration taking advantage of the support of a vast community of developers and industry members. Other container-oriented VIMs have recently received increasing attention (e.g. Kubernetes). On the other hand, for the available MANO solutions, a set of community driven approaches exist gaining notable relevance in the last years. In this context, there is the Open Source MANO (OSM) project [10], the ONAP [11], the Open Baton framework [12] and the Open Platform for Network Function Virtualization (OPNFV) [13]. All of them present their own development stage but in general they tackle mostly the functions of the NFVO and VNFMs, as well as supporting as said before the most popular VIM implementation, i.e. OpenStack.

4.2.3 Data Plane Programmability

Originally conceived for intra-data center scenarios, the SDN paradigm is experiencing rapid development thanks to its extreme flexibility of use and the availability of standard control protocols which have made them implementable on an ever increasing number of switching devices and in different scenarios ranging from 5G fronthaul, edge, up to metro aggregation networks and optical transport. However, SDN rules are stateless and applicable to a well-defined traffic flow. This implies that more intelligent network behaviors, based on context or stateful information, are delegated to the central controller, subject to significant scalability limitations.

In addition, so far, SDN switch manufacturers have implemented closed and proprietary solutions of the switch pipeline (i.e. the functional structure implemented in ASIC), selecting, due to architectural tradeoff reasons, some functions directly in hardware and others in software, therefore affecting the switch performance in a rigid way.

A new SDN phase is opening the way to define and program a switch pipeline for implementing stateful workflows without the need to query the controller. The programming protocol-independent packet processors (P4) language has been developed by the P4.org consortium [14] as a high-level language, platform-independent, used to program control, actions, coding and decoding of protocol headers and tables of a programmable SDN switch, operating on general purpose bare metal switches, field-programmable gate array (FPGAs), and Smart NICs. In particular, P4 allows definition and management of metadata, i.e. additional data that can be associated to a packet, along with stateful objects (e.g. registers, counters, and meters). Such features allow to implement state machines inside the pipeline itself, able to identify and manage complex events, and make dynamic per-packet decisions. In addition, P4 allows to define and implement proprietary headers.

In a P4 switch, incoming packets are first passed to P4-defined parsers (either standard or proprietary). Then, two programmable pipelines sections are applied, before output port selection (i.e. ingress pipeline) and after output port selection (i.e. egress pipeline). Each pipeline is programmable with different flow tables and actions. Moreover, control sections define the order and conditional execution of selected flow tables, enabling the implementation of simple finite state machines.

Application examples are ranging from advanced multi-layer Traffic Engineering policies (e.g. dynamic switching and traffic aggregation based on statistical traffic features) [15], to the implementation of dedicated network protocols transparent to the end user, active in-band telemetry applications [16], sliceable optical networking based on NICs [17], multi-tenancy [18] up to context sensitive cyber security functions[19], feature extraction for AI engines [20], and potentially direct processing of the application layer. In the 5G context, some functions delegated to specific system VNFs have been implemented directly in the infrastructure bare metal switches, for example multi-access edge computing (MEC) GRPS-tunneling terminations and Broadband Network Gateways (BNG) [21], with clear benefits in terms of performance and latency.

4.2.4 Programmable Optical Switches

A Reconfigurable Add Drop Multiplexer (ROADM, i.e. an optical switch) is a device that allows the switching of the optical signals without any electrical conversion (i.e. all-optical). It presents the add/drop ports, connected to the tunable Dense Wavelength Division Multiplexing (DWDM) optical line interfaces of the transponder, and a number (according to the nodal degree) of line interfaces connected to other ROADMs. In general, the control of ROADMs is performed via proprietary GUI/LCT. The SDN paradigm has been for optical networks, the main enabler for the development of new functionalities. Many vendors are opening the blackbox of their transmission solutions, exposing new application programming interface (APIs) to the SDN controller. Different working groups (i.e. OpenConfig [22], OpenROADM [23], and TIP [24]), led by the main network operators, are putting their effort on the definition of sets of vendor-agnostic yet another next generation (YANG) data models, allowing the representation and the control of the optical devices. The network configuration protocol (NETCONF) protocol is adopted for the communication among the reconfigurable optical add-drop multiplexer (ROADMs) and the SDN optical controller instance(s), relying on YANG models and allowing both the control of the device and the monitoring of the main transmission parameters.

Considering the OpenROADM YANG models, a ROADM device is represented in a hierarchical structure, composed by six main sections: (i) the general information, (ii) the list of shelves, (iii) the list of circuit-packs, (iv) the list of interfaces and the lists of both (v) internal, and (vi) external links. Each "shelf," includes a list of slots, where the different components are installed. Each "circuit-pack" consists on a list of physical ports. The virtual-interfaces, at the different optical transport network (OTN) layers (i.e. optical transport section (OTS), optical multiplex section (OMS)), are reported in the "list of interfaces." Each virtual interface is defined on top of a physical interface or on top of another virtual interface (following the OTN stack). The "internal link list" includes all the connections among the installed components, while, all the external connections with other optical switches are included in the "external link list," reporting the network topology information.

Nowadays the YANG models have been enhanced with telemetry functionalities, enabling the real-time monitoring of the optical switches. This functionality consists in the streaming of the key transmission parameter values to a specific data collector (i.e. monitoring platform), allowing the detection of possible failures (i.e. both soft and hard failures) and anomalies.

4.2.5 Network Data Management

In a high-level network data management architecture, five main components/ modules exist and are interconnected with each other in various forms: (i) data connect, (ii) data ingest, (iii) data analysis, (iv) data storage, and (v) data visualization. Data connect is used as a trigger to connect to a data source, which can be either on web, mobile/IoT device, or data store. Data sources can be in various forms, such as static data source from files, databases (MySQL, Mongobb, etc.), or streaming data sources from third party APIs or frameworks. This stage can be configured to transfer data into the data ingestion module.

The data ingestion module acts as the intermediary stage for the incoming data from the connection module. Data ingestion is generally used to transfer data between external systems and the big data cluster (e.g. Hadoop based). The streaming data can be ingested in real or near-real time into the cluster. As it is loaded, data can be used for later processing (e.g. using Apache Spark) or storage purposes (e.g. using hadoop distributed file system (HDFS) [25]). During data ingestion, data enrichment, aggregation, transformation, etc. can also be performed. Some examples in this category are Apache Kafka [26], Apache Spark Streaming [27], and Apache Flink (using DataStream API) [28].

After data is ingested, the data processing/analytics module ensures to accomplish tasks, such as data transformation, cleaning, staging, integrity verification and combining over both streaming and batch data. There are many frameworks that can work both in batch and streaming analytics modes (Spark, Flink, Drill [29], etc.). For example, Apache Spark can be used to perform standard extract, transform, and load (ETL) processes on big data. In the data storage stage, various SQL/NoSQL databases, such as Postgre SQL, MongoDB, Cassandra, CouchDB, Hbase, and CosmosDB can be used for short-term or long-term high volume storage purposes. The data visualization stage reports on performance to explain and present the data analysis results or the data itself in a comprehensive form. It can also serve as an interface for users to execute or compose analytics on data processing and analytics frameworks and to visualize the results. Some example tools include ElasticSearch's Kibana [30], Tableau [31], Grafana [32], and Apache Superset [33]. A simple example of how such a pipepline can be integrated in a 5G network management and orchestration framework may be found in [34].

For orchestration, there are various distributed open-source frameworks available to assess resource management and scheduling of clusters, data centers or cloud environments various distributed open-source frameworks options are available. These frameworks are mainly used with applications (e.g. Kafka, Hadoop, Elasticseach, Spark, etc.) and provide APIs for resource management, orchestration, and scheduling. Some of the most prominent ones to define and schedule tasks programmatically are Apache Mesos [35] and Apache YARN (for operating the cluster and monitoring the executed jobs), Apache Zookeeper, Apache AirFlow [36] and Apache Oozie (for distributed coordination, scheduling of the workflow in the cluster), Apache TEZ, Apache Ambari (providing management interface), Kubernetes [37], and Docker Swarm.

4.3 5G Management State-of-the-Art

In this section, we will present the main approaches available in the literature for the management of various architectural components of a 5G network. In the course of the discussion, we note how the diversity of the problems at each architectural level of course calls for different approaches.

In Section 4.3.1, we will examine RAN resource management through the detection of user behavioral patterns and RAN resource allocation leveraging reinforcement learning (RL) techniques, such as Q-learning approaches. To conclude Section 4.3.1, we present traditional optimization techniques, and AI-based approaches to solve network orchestration problems involving any compute resources. Later on, Section 4.3.3 presents SoA techniques to achieve data plane slicing via isolation of coexisting traffic. Taking a wider look at the different network segments of a 5G network, Section 4.3.4 covers the different programmable techniques and algorithms for the management of optical resources. Finally, Section 4.3.5 addresses the all-important issue of resource federation and the AI techniques used to assist the realization of the federation procedures.

4.3.1 RAN resource management

4.3.1.1 Context-Based Clustering and Profiling for User and Network Devices

There is a plethora of proposed studies that attempt to analyze, exploit, and manage user and network behavioral patterns as well as introduce innovative mechanisms based on both supervised and unsupervised approaches [38–42] toward efficient and proactive RAN resource management. In [38], a Knowledge Discovery in Databases (KDD) framework is introduced catering for the extraction and exploitation of user behavioral patterns from network and service information using K-means and spectral clustering, Naïve Bayes, and decision tree learning algorithms. The authors in [39], introduce a framework for traffic clustering and forecasting using K-Means, Neural Networks (NNs), and stochastic processes so as to manage traffic behaviors for a huge number of base stations. Valtorta et al. [40], propose a methodology for radio and network behavior-aware IoT devices profiling using K-Means clustering. Another work [41] applies K-means and hierarchical clustering techniques on mobile network data for anomaly detection on mobile wireless networks. In [42] the authors propose location and traffic-aware hierarchical clustering as well as an improved version of affinity propagation for cell towers and introduce three location aware algorithms enhanced with mobility and handovers for packing baseband units (BBUs). Apart from the well-established and improved machine learning techniques, there are numerous studies that introduce new and innovative clustering algorithms in order to exploit the aggregated traffic data from the network in [43–45]. In [43], the authors propose both a static and a dynamic network selection clustering algorithm for distributing traffic flows into different network interfaces based on the characteristics of the flow. In [44], the authors propose a user specific and adaptive cell mobility-aware clustering with non-coherent Coordinated Multi-Point – Joint Transmission (CoMP-JT), targeting to customize the cell cluster size separately for each user.

4.3.1.2 *Q*-Learning Based RAN Resource Allocation

With the rise of Machine Learning algorithms and techniques, the focus of resource allocation problems has shifted in many occasions to the use of NNs and their adaptability to cover a wide target area and different set of communication scenarios or architectures (device to device (D2D) communications, Cloud-RAN, etc.); some indicative examples are provided in [46–48], in which the authors attempt to optimize specific RAN resource allocation KPIs, taking into consideration different parameters or constraints, such as channel availability, interference, adaptive power control approaches, service coverage, number of tenants, etc.

In conjunction with NNs, the adoption of the *Q*-Learning approach, i.e. a model-free reinforcement learning algorithm, is a promising solution, which can be tailored in order to overcome the complexity and computational requirements of such optimization problems [49–54]. In [49], the interference is used as a key indicator to allocate the available radio resources and–in conjunction with caching and computing resources–targets to minimize the end-to-end delay. Following the interference path, the [50] tries to minimize it by controlling the transmitting power of the base stations. Also, the authors in [52], taking into account the inter-cell interference and cell range expansion, target to increase the user throughput and reduce the handover failure, by managing the power transmission of the heterogeneous network base stations. Another approach is taken by Do and Koo [53], where authors minimize power consumption with the use of renewable energy powered base stations. Abiko et al. [54] presents a more dynamic approach by introducing resource blocks to quantify the changing number of slices needed and use time and frequency division to allocate the minimum number of resource blocks. In [51], the authors use Deep-*Q*-Learning (DQL) to aid the resource allocation of LoRaWAN where memory capacity for the computation of the channel allocation is limited.

4.3.1.3 vrAIn: AI-Assisted Resource Orchestration for Virtualized Radio Access Networks

This section gives an example use case of applying AI/ML for RAN resource management, in particular optimizing dynamic and joint selection of radio and compute orchestration policies, as introduced in Section 4.2.1. This optimization problem can be formulated as a contextual bandit (CB) problem, which is a particular case of reinforcement learning (RL). There, one observes a *context* vector, chooses an action, and receives a reward signal as feedback, sequentially at different time stages. The goal is to find a policy that maps input contexts into compute/radio control policies or actions that maximize the expected reward. An example of this approach is introduced in [6], which defines the following aspects of the CB model.

- *State or context space*: Given the information of user equipment (UEs), vrAIn creates a large state vector with buffer size, and SNR information, that is later compressed into a context vector using a sparse auto-encoder.
- *Action space*: Both computing control and radio control actions are taken. For the first one, [6] allocates a fraction of CPU time c_i for each vBS i. For the latter, the agent sets an upper-bound on the eligible MCS index m_i.
- *Rewards*: The design of a reward function depends on the system's goal. Here, a two-fold objective is considered: (i) minimizing operational costs due to CPU reservations and (ii) maximizing performance by reducing decoding error rates and latency. The reward function is designed accordingly, so as (i) long buffer occupancy (which is a primary source of delay) penalizes the reward, (ii) nonzero decoding errors due to CPU deficit (which induce throughput losses) penalizes the reward, and (iii) smaller CPU assignments increase the reward.

Figure 4.3 illustrates the decision-making closed-loop process implementing the RL formulation above. Each orchestration period or *stage* is divided into T *monitoring slots*. At the end of each slot t, vrAIn aggregates *radio samples*, such

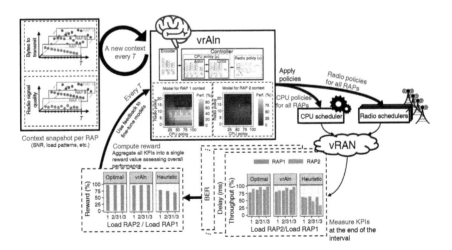

Figure 4.3 vrAIn's control loop.

as mean SNR and new data pending to be transmitted, collected during the last slot across all users in each vBS. As a result, at the beginning of each stage, vrAIn encodes a context vector containing T samples of radio information such as SNR and data load. Then, vrAIn maps an encoded representation of such context vector into a pair of control actions, as defined above to be applied as scheduling policies during the next stage. At the end of this stage, performance indicators are collected and encoded into a reward signal which is fed back into vrAIn to optimize vrAIn's internal NNs.

4.3.2 Service Orchestration

Several research works have been proposed in the direction of applying the concept of NFV to 5G networks. Vertical services can indeed be defined as a set of connected VNFs, often represented as a graph, which can be deployed on the computing, storage, and network resources within the MNO infrastructure. To efficiently support the various vertical services, the network slicing paradigm has also emerged, so that the mobile operator's can deploy different services, ensuring for each of them their isolation requirements as well as their required KPIs, in spite of the limited resources available. Network slicing also supports composed services, i.e. such that the VNF set includes subsets, each corresponding to a sub-slice service.

To create a slice, MNOs must decide where to place each VNF and allocate the necessary resources (e.g. virtual machines or containers, and virtual links connecting them). The dynamic placement of VNFs close to the Edge of the network, as well as across the Cloud, Edge, and Fog, have been examined thoroughly. The optimal selection of a VNF placement solution is known to be non deterministic polynomial time (NP)-hard. Therefore, it has become increasingly important to exploit approximate linear models so as to efficiently support the decision-making in the service orchestration process.

The typical scenario considered for NFV-based 5G networks foresees that the VNF placement decisions are made by NFV Orchestrator (NFVO), defined in the ETSI MANO framework [7]. However, ETSI [55, section 8.3.6] specifies four granularity levels for placement decisions: individual host, zone (i.e. a set of hosts with common features), zone group, and point-of-presence (PoP) (e.g. a datacenter). Real-world mobile network implementations, including [56], assume that the NFVO, or similar entities, make PoP-level decisions. Placement and sharing decisions within individual PoPs, instead, can be made by other entities under different names and with slight variations between the solutions of IETF, the NGMN alliance, and 5G PPP. The latter, in particular, includes a Software-Defined Mobile Network Coordinator (SDM-X), as depicted in Figure 4.4. The SDM-X operates at a lower level of abstraction than the NFVO and makes intra-PoP VNF placement

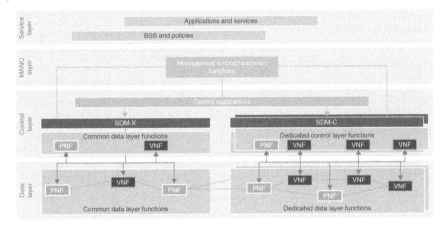

Figure 4.4 5G-PPP network architecture. Source: Rost et al. [57].

decisions. Specifically, for each service newly requested by a vertical, the SDM-X makes decisions on: (i) whether any subset of the service VNFs can reuse an existing sub-slice; (ii) if not possible, which PoP should host the virtual machine (VM) implementing the VNF; (iii) how to allocate (including scaling up/down) the computational capability of the VMs within the PoP.

As for existing solutions to the VNF placement problem, there is a plethora of scientific papers formulating this problem as an integer linear programming (ILP), considering a variety of KPIs and imposing resource-related constraints in order to meet the targeted performance [58–61]. More specifically, in [58], Sun et al. propose a time-efficient heuristic offline algorithm which is extended so as to predict future VNF demands and reduce the setup delay of the service function chains (SFCs). Davit Harutyunyan et al. [59] propose an heuristic approach demonstrating a trade-off between the optimality and scalability of the VNF placement solution in large-scale environments. In [60] Richard Cziva et al. employed optimal stopping theory principles to define when to re-evaluate their placement solution, considering the changes in latency values of a real-world scenario and migrate VNFs if necessary. In [61], Luizelli et al. incorporate a variable neighborhood search (VNS), targeting to minimize the number of VNF instances mapped on the physical nodes (PNs) and the operational costs, respectively.

Since the ILP becomes computationally intractable as the size of mobile network increases, several papers propose different solutions such as novel complexity-aware heuristic, genetic algorithms, and decision tree learning [62–65]. Importantly, both [65, 66] also address the problem of sub-slices sharing by different service instances, in order to minimize the deployment cost while meeting possible isolation requirements.

Additionally, many recent research works address the VNF placement problem using deep reinforcement learning (DRL) models due to their efficiency and applicability in circumstances where the response of an environment to an action is not a priori known, but needs to be learned [67–70].

Another major research NFV challenge is the fluctuation of traffic load that needs to be efficiently handled by the NS to meet the required SLAs. This translates into the need of dynamic scaling actions to adapt the allocated resources for certain VNF instances belonging to the NS to meet the required performance. In the literature, auto-scaling techniques are mainly classified in two main groups, namely, reactive and proactive approach. The reactive approach is based on the application of threshold rules, and it is currently considered as built-in mechanism in current open-source MANO management and orchestration (MANO) platforms like OSM. Nonetheless, the selection of thresholds is complicated and it may be not be effective in the case of sudden traffic surges. To solve this problem, several research works propose novel heuristic and adaptive solutions [71–74], as well as neural network models [75, 76], which assist in proactive auto-scaling decisions by periodically estimating the required number of VNF instances operating on each host PN.

4.3.3 Data Plane Slicing and Programmable Traffic Management

Centralization of the network's intelligence in SDN is an advantage for applications that do not have strict real-time requirements and depend on global network state. However, when the service uses local state information, e.g. to support QoS, the same level of flexibility must be supported at the data plane. In modern programmable networking devices, the traffic management logic is not programmable. Previous research on extending programmability to the data plane [77] stressed the importance of customizing traffic management algorithms, such as queuing strategies and scheduling techniques, to application requirements. Specifically, Sivaraman et al. [77] argue that there is no "one-size-fits-all" algorithm by analyzing different active queue management (AQM) approaches. Furthermore, they enable programmability at the data plane by adding an FPGA to the fast path of a hardware switch with a simple interface to packet queues, implementing controlled delay active queue management (CODEL) and random early detection (RED) as a proof of concept. Toward the same direction, Kundel et al. [78] demonstrated that is possible to implement such algorithms for P4 programmable data planes, illustrating P4 capabilities and constraints. In [79], a PI2 [80]implementation in P4 is provided, which also showcase ways to overcome P4 limitations toward the development of AQM algorithms. Finally, in [81] both proportional integral controller enhanced (PIE) and RED AQM schemes are implemented and tested within the P4 context. These approaches enhance queue utilization within common network infrastructures (links) but they cannot fully provide per tenant (slice) bandwidth and delay guarantees.

Previous efforts on supporting multiple network contexts within the same data plane, leverage approaches similar to hypervisor-based virtualization, e.g. HyPer4 [82] and HyperV [83] for the reference software target BMV2 [84]. HyPer4 was the first attempt exploit the reconfiguration capability of P4, in order to achieve data plane virtualization. A hypervisor-like program, based on P4 language, is used to provide partial virtualization of the data plane, enabling multiple P4 programs to run isolated on the same packet processing device. A table is used to dispatch the tenant network traffic between the P4 programs of the tenants. By updating certain table entries, the hypervisor can (de)activate the programs at run-time. HyPer4 design for supporting complex functionalities and on-the-fly program reconfigurability raises significant performance penalties. HyperV extended Hyper4 and further proposed novel techniques for providing full virtualization such as control flow sequencing for interpreting virtualized data planes into a uniform linear pattern and dynamic stage mapping to map arbitrary sequences of stages in a virtualized pipeline onto the hardware data plane with limited resources. The HyperV proposition is presented in more details in [85], including an implementation for the DPDK-target [86]. P4Visor [87] attempted to further reduce the resources needed to support virtualized programmable data planes compared to [82, 83], using a lightweight virtualization primitive for P4 programs through code merging. The P4Visor framework, operating between the P4 programs and the programmable data plane uses compiler optimizations and program analysis to achieve efficient source code merging; in essence it takes as input multiple P4 programs and produces a single P4 program while retaining their functionalities. A comparative description of all existing hypevisor-based or compiler-based approaches to date can be found in [88].

4.3.4 Wavelength Allocation

5G networks embrace multiple network segments (i.e. access, metro, and core) as well as technologies such as mobile, fixed, and optical to accommodate the heterogeneous service types required by the 5G verticals along with fulfilling their KPIs. In this context, the optical technologies are seen as essential to actually deal with some of these requirements and 5G service demands exploiting the huge transport capacity for interconnecting distributed cloud computing and storage centres (i.e. edge and core DCs). Specifically, optical technologies provide the leading solution for attaining effective network infrastructures coping with the expected 5G service requirements in terms of high-speed, low-latency connectivity, energy efficiency, etc. [89].

Required transport optical resources (i.e. optical spectrum, transmitter, receivers, etc.) accommodating 5G services (e.g. between remote DCs) are selected and programmed via a dedicated optical SDN controller, see Section 4.2.4. In general, technological SDN controllers are centrally managed by a higher-layer entity referred to as network (resource) orchestrator. An example of this is the IETF Application Based Network Orchestrator (ABNO) [90]. Typically, it relies on a hierarchical control architecture, where the *parent* controller ensures overarching control over a pool of multi-technology domains including packet and optical switching [91]. Additionally, to roll out 5G services and applications requiring storage, computing, and networking resources, the network orchestrator (e.g. ABNO) behaves as a VIM controller coordinated by a service orchestrator instance (i.e. MANO). In this joint IT/cloud and network orchestration, the dedicated element for handling the network interconnection is referred to as the Wide Area Network Information Manager (WIM).

In the framework of optical networks, it is well-known that traditional fixed grid DWDM networks need to be evolved toward the so-called flexi-grid DWDM networks. This is done to attain a more efficient use of the optical spectrum [92] fostering the tailored allocation of just enough optical spectrum to the service demands. To this end, flexi-grid DWDM networks exploit the flexibility provided by sliceable bandwidth variable transceivers (SBVTs). The introduction of flexi-grid DWDM networks with SBVT devices provides an effective transport infrastructure (e.g. interconnecting remote DC facilities) to fulfill stringent 5G service requirements with respect to both high throughput and low latency. To this end, the optical SDN governing the flexi-grid DWDM transport infrastructure takes over of selecting and configuring all the involved network elements such as optical switches, links' optical spectrum (i.e. central frequency and slot width), and endpoint SBVT parameters (e.g. modulation format) at the time of setting up targeted inter-DC optical flows. The SDN controller is required to (i) guarantee the 5G service data rate; (ii) deal with any physical transmission limitation imposed by the optical technology (e.g. maximum achievable data rate); and (iii) fulfill intrinsic technology restrictions such as spectrum continuity and contiguity [93]. The problem of selecting the spatial and spectral paths, resources, and SBVT parameters is typically addressed by the so-called Routing Spectrum and Modulation Assignment (RS(M)A) algorithms. In the last years, notable contributions on routing and spectrum assignment (RSA) algorithms have been produced. A complete survey tackling multiple RSA algorithm aspects can be found in [94].

In brief, the optical SDN controller receives the connection requirements from the network orchestrator. These requirements determine the pairs of source and destination DCs hosting the VNFs according to the resulting VNF Forwarding

Graph for a targeted 5G service, required bandwidth, maximum end-to-end latency, etc. [95]. These restrictions are then used as inputs to the SDN controller's RSA algorithm. The output of the RSA algorithm specifies the set of optical resources (i.e. traversed nodes, links, spectrum, SBVT parameters) to fulfill the connectivity demands. Then, these resources are allocated according to the corresponding control interface between the SDN controller and the agents handling each network element within the underlying optical network infrastructure, see Section 4.2.4.

4.3.5 Federation

Next generation mobile networks are expected to operate in highly heterogeneous environments. The network management of such scenarios, involving different technologies and network segments, requires of multi-domain orchestration. However, in next generation mobile networks, the term *domain* has to consider an additional and essential meaning, mainly imposed by the requirement of satisfying the diverse needs of different kinds of users. Some necessary functions to compose NSs or the allocation of infrastructure resources could be provided by different organizations, known as administrative domains (ADs), obeying to different criteria, like shortage of resources, simple availability of the service or capability to deploy the service satisfying different constraints, e.g. geographical or latency, hence requiring federation capabilities. Depending on how this process is done between ADs, one may distinguish between resource and service federation. Briefly, *resource federation* can be defined as the process by which a consumer AD requires the management of infrastructure resources to a provider AD to deploy an NS (or part of it). On the other hand, *service federation* is the process where the consumer AD requires the deployment of an NS to a provider domain, while the provider domain keeps the full management of its infrastructure resources.

In the literature, the tackling of the federation problem can be divided into two main groups. First, some work deals with the procedures and interfaces to make effective the federation process (in both forms) to allow the real instantiation of the NS among different ADs. Second, other works deal with the problem of distributing the different component parts of an NS among multiple ADs.

With respect to the definition of procedures and interfaces to perform federation operations, most of the work comes from Standard Development Organizations (SDOs), like the Open Networking Foundation (ONF) [96], the Metro Ethernet Forum (MEF) [97] and the ETSI NFV workgroups [98, 99].

With respect to the problem of distributing NSs across multiple domains, there are mainly two types of proposed approaches in the literature, namely *centralized* and *distributed*. In the *centralized* approach [100, 101], a central third party entity has full knowledge of the available resources and partitions the NS chain into "*sub-chains*," based mostly on ILP models. In the *distributed* approach [102, 103], ADs establish peer-to-peer relationships. The lack of topology information of the different ADs and its resource availability increases the difficulty of solving the multi-domain distribution problem and more messages are needed to be exchanged between adjacent ADs. However, it is this lack of exposure of essential information that network providers consider attractive.

AI/ML can be applied in various ways to assist the realization of the federation procedures. Last year, ETSI ZSM [104, 105] is focusing on the centralized federation approach and applying AI/ML as part of the self-planning, self-optimization, self-healing, and self-protecting processes. Opposite of applying the AI/ML models, for assisting the orchestration of federation procedures, it is applying federation to train AI/ML models through Federated ML [106]. In this case, the training of the AI/ML models occurs in different federated ADs, but the data is kept private and not shared to the federated ADs. In some cases, to increase the performances of the federated ML, a Distributed Ledger Technology (DLT), such as Blockchain, is applied [107].

4.4 Conclusions and Future Directions

We presented an overview of the main challenges that must be faced to successful develop 5G systems fulfilling the KPIs required by next-generation mobile services. In particular, we focused on such relevant aspects as radio access networks, optical networks, data plane management, network slicing, and service orchestration, and we discussed the most prominent solutions existing in the scientific literature as well as those proposed within relevant standards development organizations. While doing so, we highlighted two main approaches that will be required for the development of 5G-and-beyond systems: an autonomous, data-driven network management, and the federation among different ADs.

Given that 5G systems leverage technologies that are still in their early stage, a numerous of aspects need to be further investigated. In particular, under the scope of the 5Growth project [108], novel smart resource orchestration algorithms will be designed and implemented so as to meet the requirements in terms of reliability, throughput, and latency. In this context, it is critical to develop AI/ML approaches that can lead, not only to efficient solutions in terms of KPI fulfillment and resource usage but also to scalable and flexible mechanisms that can effectively deal with different services and greatly reduce the service deployment time. To this end, and drawing on the work carried out by previous 5G PPP projects such as 5G-TRANSFORMER [56], 5Growth [108] aims at designing an AI/ML-based architecture where the different aspects of service and resource orchestrations are tackled at different architectural layers.

First, upon receiving a vertical service request, an entity called vertical slicer, will take care of (i) assessing whether the service can be accommodated, given the amount of resources the vertical is entitled to use, (ii) which type of slice should be created, (iii) which, if any, sub-slice(s) can be reused for the service deployment. To accomplish these tasks, the vertical slicer will exploit data collected through a monitoring platform, query the underlying service orchestrator for information on the resource availability, and use these data to feed an AI/ML model to make sensible decisions.

Equipped with the above decisions, the service orchestrator can then perform the actual VNF placement and properly scale the resources to be allocated to existing sub-slices to meet traffic and workload demand. Importantly, the service orchestrator will have to ensure the smooth operation of the deployed slices, in spite of time-varying traffic loads and operational conditions. Again, by exploiting monitoring data, further scaling of the resources allocated to a slice may be needed, so as to meet the target KPIs. This may lead to a variation of the amount of resources assigned to the VMs and allocated on the virtual links to deal with a service traffic (scale up/down), or to the creation/deletion of a VNF replica (scale out/in). It is therefore essential to develop AI/ML solutions that can address these issues by making near-optimal decisions. On the one hand, reinforcement learning approaches will be adopted for real-time decisions, by defining reward functions that reflect the service requirements. On the other hand, a hybrid algorithm will be implemented, which combines genetic-based solutions with NNs, targeting to minimize end-to-end delay while ensuring that there will be no KPI violations. More precisely, genetic-based algorithms are examined and selected as solutions to the VNF placement problem due to their fast converge to the near-optimal solution. In this direction, NNs were selected so as to map traffic and application-aware metrics to VNF scaling decisions.

Furthermore, at the resources layer the 5Growth platform will introduce ML-based programmable SLA-aware traffic management algorithms at the data plane (e.g. scheduling, active queue management, etc.) focusing on potential bottlenecks (e.g. the RAN), assuring performance per slice via closed loop interactions. To this end appropriate slicing abstractions should be devised for programmable data planes, enabling also programmable traffic management. Coupling ML-based control with fully programmable data planes, will provide additional degrees of freedom for configuring and customizing slices to meet their performance requirements.

To realize the above vision, it is clear that pre-trained AI/ML models will be needed at the different architectural layers. It is therefore pivotal to develop a platform specifically designed to make off-line trained models available, as well as to provide the Vertical Slicer, the Service Orchestrator and Resource Layer with models that are trained on the spot, leveraging the data collected through the monitoring system. This is indeed the core of a closed-loop system as also envisioned by the ETSI ZSM [109]. Open interfaces will have to be developed, which allow service characteristics and target KPIs to be specified by the vertical, passed as hyper-parameters to the vertical slicer, and then down to the other layers of the network architecture. In this way, the AI/ML mechanisms can be tailored to the specific service and application requirements, leading to a fully-automated 5G system.

An efficient management of computing and storage resources, however, is not enough to meet the target KPIs required by verticals: RAN resources need to be smartly allocated as well. In particular, in the ultra-dense RAN ecosystems foreseen for 5G – and beyond – AI/ML approaches will require joint strategies that account for all types of resources. Distributed training, and as well distributed inference techniques are already being proposed in the literature presented earlier, in order to exploit the increased processing capabilities of diverse RAN elements, exploiting even end user device capabilities when applicable. As it can be inferred from the previous paradigms, AI/ML concepts are gradually becoming structural components of the network, introducing novel capabilities for intelligent RAN resource management, user and network device profiling, and spectrum allocation techniques.

Besides a joint resources management, one of the most crucial challenges is that up to now, AI/ML algorithms and network and communication protocols are designed separately. Joint RAN resource management and AI/ML algorithms design should be one of the high priorities toward a truly *AI-aware networking paradigm*, where coding and signal processing approaches are integrated with the AI framework. This will be realized by identifying common requirements and limitations that result from the two domains; for example, such joint approaches could be considering different dimensionality reduction or data encoding techniques for limited computing capabilities devices (such as IoT nodes), adaptive gradient aggregation for improved resilience, or joint channel coding and image compression techniques in noisy wireless environments.

In this line, the O-RAN Alliance specifies a series of use cases that bridge ML models with open and virtualized RANs. Relevant examples of these use cases are as follows:

- Flight path-based dynamic unmanned aerial vehicle (UAV) radio resources, where UAV steering and radio resource allocation are jointly optimized;
- RAN sharing, where virtualized (possibly tailored-made) instances of radio access points from different operators share common computing infrastructure at an edge cloud and/or radio spectrum; and
- Context-based dynamic handover management for V2X, where machine learning models assist in forecasting and classification tasks to customize handover sequences at UE granularity.

3GPP has already introduced a novel Network Data Analytics Function (NWDAF) in Rel.16 [110], which indicates the gradual adoption of AI/ML concepts in the core network architecture, also from the standardization perspective. Currently, this function has limited functionality and is deployed only as part of the Core Network (5GC) in order to facilitate operators' policy manipulation. Additionally, currently the data analytics is limited to only 3GPP-oriented information. Taking the aforementioned AI-aware networking paradigm as design guideline, the evolution of such an NF to a distributed, multi-domain, federated learning-based approach, exploiting also resource information from non-3GPP networks, could potentially boost the network AI capabilities toward seamless and more flexible RAN resource management at the network Edge. On top of that, the extraction of user and network behavioral patterns and their exploitation toward predictive RAN resource allocation, will offer considerable additional gains to the current RAN resource management approaches. Such an enabler will however require radical enhancements to the current architecture and NWDAF operation, enabling a distributed profile extraction approach exploiting edge nodes' – including even the UEs' – computing power.

Finally, a relevant aspect spanning across different domains is Federation. As previously mentioned, important gaps and possible solutions have been identified within the 5G-TRANSFORMER and 5Growth projects. In particular, the work therein extends the multi-AD service orchestration to cover the required orchestration operations and effectively interconnect the different NSs running in different ADs. In this context, an interesting research direction consists in designing architectural solutions that leverage DLTs to establish administrative relations between domains and Q-learning approaches for making better service split decisions among peering domains in which technical and business parameters may be involved, as proposed in [111, 112].

Bibliography

1 5GPPP (2020). 5G Network Support of Vertical Industries in the 5G Public-Private Partnership Ecosystem. *Technical report*. 5GPPP.

2 Technical Specification Group of Services and System Aspects (2020). System Architecture for the 5G System (5GS). *Technical Report 23.501*, 3GPP. version 16.4.0.

3 Felter, W., Ferreira, A., Rajamony, R., and Rubio, J. (2015). An updated performance comparison of virtual machines and Linux containers. *2015 IEEE International Symposium on Performance Analysis of Systems and Software (ISPASS)*, IEEE, pp. 171–172.

4 Cisco (2020). Annual Internet Report (2018-2023) White Paper. *Technical report*. Cisco.

5 O-RAN Alliance (2020). O-RAN-WG1-O-RAN Architecture Description - v01.00.00. Technical Specification, February 2020.

6 Ayala-Romero, J.A., Garcia-Saavedra, A., Gramaglia, M. et al. (2019). vrAIn: a deep learning approach tailoring computing and radio resources in virtualized RANs. *The 25th Annual International Conference on Mobile Computing and Networking*, pp. 1–16.

7 ETSI (2014). Network Functions Virtualisation (NFV); Management and Orchestration. Group Specification (GS) 001 v1.1.1, European Telecommunications Standards Institute (ETSI), 12 2014.

8 ETSI GS NFV (2014). ETSI GS NFV 002 Network Functions Virtualization (NFV), Architectural Framework. https://www.etsi.org/deliver/etsi_gs/NFV/001_099/002/01.01.01_60/gs_NFV002v010101p.pdf (accessed 15 April 2021).

9 OpenStack Cloud Operating System. https://www.openstack.org/ (accessed June 2020).

10 Open Source MANO Project. https://osm.etsi.org/ (accessed June 2020).

11 Open Network Automation Platform. https://www.onap.org/ (accessed June 2020).

12 Carella, G.A. and Magedanz, T. (2016). Open baton: a framework for virtual network function management and orchestration for emerging software-based 5G networks. *Newsletter*. https://sdn.ieee.org/newsletter/july-2016/open-baton.

13 Open Platform for NFV. https://www.opnfv.org/ (accessed June 2020).

14 P4 Language Consortium. https://p4.org/ (accessed June 2020).

15 Paolucci, F., Cugini, F., and Castoldi, P. (2018). P4-based multi-layer traffic engineering encompassing cyber security. *Optical Fiber Communication Conference*, Optical Society of America, pp. M4A–5.

16 Cugini, F., Gunning, P., Paolucci, F. et al. (2019). *Optical Fiber Communication Conference (OFC)*. P4 in-band telemetry (INT) for latency-aware VNF in metro networks. M3Z.6. https://doi.org/10.1364/OFC.2019.M3Z.6.

17 Yan, Y. (2019). P4-enabled smart NIC: architecture and technology enabling sliceable optical DCS. *European Conference on Optical Communications (ECOC)*, pp. 1–3.

18 Osi?ski, T., Kossakowski, M., Tarasiuk, H., and Picard, R. (2019). Offloading data plane functions to the multi-tenant cloud infrastructure using P4. *2019 ACM/IEEE Symposium on Architectures for Networking and Communications Systems (ANCS)*, pp. 1–6.

19 Paolucci, F., Civerchia, F., Sgambelluri, A. et al. (2019). P4 Edge node enabling stateful traffic engineering and cyber security. *IEEE/OSA Journal of Optical Communications and Networking* 11 (1): A84–A95.

20 Musumeci, F., Ionata, V., Paolucci, F. et al. (2020). Machine-learning-assisted DDoS attack detection with P4 language. *IEEE International Conference on Communications (ICC)*, pp. 1–6.

21 Kundel, R., Nobach, L., Blendin, J. et al. (2019). P4-BNG: central office network functions on programmable packet pipelines. *IEEE International Conference on Network and Server Management (CNSM)*, pp. 21–25.

22 OpenConfig. https://openconfig.net/ (accessed June 2020).

23 OpenROADM MSA. http://openroadm.org/ (accessed June 2020).

24 Telecom Infra Project. https://telecominfraproject.com/ (accessed June 2020).

25 White, T. (2012). *Hadoop: The Definitive Guide*. O'Reilly Media, Inc.

26 Apache Kafka. A distributed streaming platform. https://kafka.apache.org/ (accessed June 2020).

27 Apache Spark. Apache Spark - Unified Analytics Engine for Big Data. https://spark.apache.org/ (accessed June 2020).

28 Apache Flink. Stateful Computations over Data Streams. https://flink.apache.org/ (accessed June 2020).

29 Apache Drill. Schema-free SQL Query Engine for Hadoop, NoSQL and Cloud Storage. https://drill.apache.org/ (accessed June 2020).

30 Elasticsearch. A distributed, JSON based search and analytics engine. https://www.elastic.co/ (accessed June 2020).

31 Tableau. Business Intelligence and Analytics Software. https://www.tableau.com/ (accessed June 2020).

32 Grafana. The open observability platform. https://grafana.com/ (accessed June 2020).

33 Apache Superset. A modern, enterprise-ready business intelligence web application. https://superset.incubator.apache.org/ (accessed June 2020).

34 Papagiani, C., Mangues-Bafalluy, J., Bermúdez, P. et al. (2020). 5Growth: Ai-driven 5G for automation in vertical industries. *EUCNC*.

35 Apache Mesos. A distributed systems kernel. http://mesos.apache.org/ (accessed June 2020).

36 Apache AirFlow. A platform to programmatically author, schedule, and monitor workflow. https://airflow.apache.org/ (accessed June 2020).

37 Kubernetes (2014). Production-grade container orchestration. https://kubernetes.io/ (accessed June 2020).

38 Magdalinos, P., Barmpounakis, S., Spapis, P. et al. (2017). A context extraction and profiling engine for 5G network resource mapping. *Computer Communications* 109: 184–201.

39 Le, L., Sinh, D., Lin, B.P., and Tung, L. (2018). Applying big data, machine learning, and SDN/NFV to 5G traffic clustering, forecasting, and management. *2018 4th IEEE Conference on Network Softwarization and Workshops (NetSoft)*, pp. 168–176.

40 Valtorta, J.M., Martino, A., Cuomo, F., and Garlisi, D. (2019). A clustering approach for profiling LoRaWAN IoT devices. *Ambient Intelligence*, pp. 58–74.

41 Parwez, M.S., Rawat, D.B., and Garuba, M. (2017). Big data analytics for User-activity analysis and user-anomaly detection in mobile wireless network. *IEEE Transactions on Industrial Informatics* 13 (4): 2058–2065.

42 Karneyenka, U., Mohta, K., and Moh, M. (2017). Location and mobility aware resource management for 5G cloud radio access networks. *2017 International Conference on High Performance Computing Simulation (HPCS)*, pp. 168–175.

43 Xu, L. and Duan, R. (2018). Towards smart networking through context aware traffic identification kit (TriCK) in 5G. *2018 International Symposium on Networks, Computers and Communications (ISNCC)*, pp. 1–6.

44 Joud, M., García-Lozano, M., and Ruiz, S. (2018). User specific cell clustering to improve mobility robustness in 5G ultra-dense cellular networks. *2018 14th Annual Conference on Wireless On-demand Network Systems and Services (WONS)*, pp. 45–50.

45 Duan, X., Liu, Y., and Wang, X. (2017). SDN enabled 5G-vanet: adaptive vehicle clustering and beam formed transmission for aggregated traffic. *IEEE Communications Magazine* 55 (7): 120–127.

46 Lee, M., Yu, G., and Li, G. (2019). Learning to branch: accelerating resource allocation in wireless networks, 03 2019.

47 Chen, X., Zhifeng, Z., Wu, C. et al. (2019). Multi-tenant cross-slice resource orchestration: a deep reinforcement learning approach. *IEEE Journal on Selected Areas in Communications* 2377–2392.

48 Ahmed, K.I., Tabassum, H., and Hossain, E. (2019). Deep learning for radio resource allocation in multi-cell networks. *IEEE Network* 33 (6): 188–195.

49 Wei, Y., Yu, F.R., Song, M., and Han, Z. (2019). Joint optimization of caching, computing, and radio resources for fog-enabled IoT using natural actor-critic deep reinforcement learning. *IEEE Internet of Things Journal* 6 (2): 2061–2073.

50 Nasir, Y.S. and Guo, D. (2019). Multi-agent deep reinforcement learning for dynamic power allocation in wireless networks. *IEEE Journal on Selected Areas in Communications* 37 (10): 2239–2250.

51 Aihara, N., Adachi, K., Takyu, O. et al. (2019). Q-learning aided resource allocation and environment recognition in LoRaWAN with CSMA/CA. *IEEE Access* 7: 152126–152137.

52 Zhang, Y., Kang, C., Teng, Y. et al. (2019). Deep reinforcement learning framework for joint resource allocation in heterogeneous networks. *2019 IEEE 90th Vehicular Technology Conference (VTC2019-Fall)*, pp. 1–6.

53 Do, Q.V. and Koo, I. (2019). A transfer deep Q-learning framework for resource competition in virtual mobile networks with energy-harvesting base stations. *IEEE Systems Journal* 15 (1): 1–12.

54 Abiko, Y., Saito, T., Ikeda, D. et al. (2020). Flexible resource block allocation to multiple slices for radio access network slicing using deep reinforcement learning. *IEEE Access* 8: 68183–68198.

55 ETSI (2016). Network Functions Virtualisation (NFV); Management and Orchestration; Or-Vnfm reference point – Interface and Information Model Specification. Group Specification (GS) 007 v2.1.1, European Telecommunications Standards Institute (ETSI), 10 2016. https://www.etsi.org/deliver/etsi_gs/NFV-IFA/001_099/007/02.01.01_60/gs_NFV-IFA007v020101p.pdf (accessed 15 April 2021).

56 De la Oliva, A., Li, X., Costa-Perez, X. et al. (2018). 5G-TRANSFORMER: slicing and orchestrating transport networks for industry verticals. *IEEE Communications Magazine* 56 (8): 78–84.

57 Rost, P., Mannweiler, C., Michalopoulos, D.S. et al. (2017). Network slicing to enable scalability and flexibility in 5G mobile networks. *IEEE Communications Magazine* 55 (5): 72–79.

58 Sun, Q., Lu, P., Lu, W., and Zhu, Z. (2016). Forecast-assisted NFV service chain deployment based on affiliation-aware VNF placement. *2016 IEEE Global Communications Conference (GLOBECOM)*, pp. 1–6.

59 Harutyunyan, D., Shahriar, N., Boutaba, R., and Riggio, R. (2019). Latency-aware service function chain placement in 5G mobile networks. *2019 IEEE Conference on Network Softwarization (NetSoft)*, pp. 133–141.

60 Cziva, R., Anagnostopoulos, C., and Pezaros, D.P. (2018). Dynamic, latency-optimal VNF placement at the network edge. *IEEE INFOCOM 2018 - IEEE Conference on Computer Communications*, pp. 693–701.

61 Luizelli, M.C., Luis, W., Buriol, L.S., and Gaspary, L.P. (2017). A fix-and-optimize approach for efficient and large scale virtual network function placement and chaining. *Computer Communications* 102: 67–77.

62 Agarwal, S., Malandrino, F., Chiasserini, C.F., and De, S. (2019). VNF placement and resource allocation for the support of vertical services in 5G networks. *IEEE/ACM Transactions on Networking* 27 (1): 433–446.

63 Khoshkholghi, M.A., Taheri, J., Bhamare, D., and Kassler, A. (2019). Optimized service chain placement using genetic algorithm. *2019 IEEE Conference on Network Softwarization (NetSoft)*, pp. 472–479.

64 Manias, D.M., Jammal, M., Hawilo, H. et al. (2019). Machine learning for performance-aware virtual network function placement. *2019 IEEE Global Communications Conference (GLOBECOM)*, pp. 1–6.

65 Martinéz-Peréz, J., Malandrino, F., Chiasserini, C.F., and Bernardos, C.J. (2020). OKpi: all-KPI network slicing through efficient resource allocation. *IEEE INFOCOM 2020 - IEEE Conference on Computer Communications*.

66 Malandrino, F., Chiasserini, C.F., Einziger, G., and Scalosub, G. (2019). Reducing service deployment cost through VNF sharing. *IEEE/ACM Transactions on Networking* 27 (6): 2363–2376.

67 Pei, J., Hong, P., Pan, M. et al. (2020). Optimal VNF placement via deep reinforcement learning in SDN/NFV-enabled networks. *IEEE Journal on Selected Areas in Communications* 38 (2): 263–278.

68 Chai, H., Zhang, J., Wang, Z. et al. (2019). A parallel placement approach for service function chain using deep reinforcement learning. *2019 IEEE 5th International Conference on Computer and Communications (ICCC)*, pp. 2123–2128.

69 Khezri, H.R., Moghadam, P.A., Farshbafan, M.K. et al. (2019). Deep reinforcement learning for dynamic reliability aware NFV-based service provisioning, pp. 1–6.

70 Quang, P.T.A., Hadjadj-Aoul, Y., and Outtagarts, A. (2019). A deep reinforcement learning approach for VNF forwarding graph embedding. *IEEE Transactions on Network and Service Management* 16 (4): 1318–1331.

71 Houidi, O., Soualah, O., Louati, W. et al. (2017). An efficient algorithm for virtual network function scaling. *GLOBECOM 2017 - 2017 IEEE Global Communications Conference*, pp. 1–7.

72 Toosi, A.N., Son, J., Chi, Q., and Buyya, R. (2019). ElasticSFC: auto-scaling techniques for elastic service function chaining in network functions virtualization-based clouds. *Journal of Systems and Software* 152: 108–119.

73 Ren, Y., Phung-Duc, T., Liu, Y. et al. (2018). ASA: adaptive VNF scaling algorithm for 5G mobile networks. *2018 IEEE 7th International Conference on Cloud Networking (CloudNet)*, pp. 1–4.

74 Fei, X., Liu, F., Xu, H., and Jin, H. (2018). Adaptive VNF scaling and flow routing with proactive demand prediction. *IEEE INFOCOM 2018 - IEEE Conference on Computer Communications*, pp. 486–494.

75 Subramanya, T., Harutyunyan, D., and Riggio, R. (2020). Machine learning-driven service function chain placement and scaling in MEC-enabled 5G networks. *Computer Networks* 166: 106980.

76 Rahman, S., Ahmed, T., Huynh, M. et al. (2018). Auto-scaling VNFs using machine learning to improve QoS and reduce cost. *2018 IEEE International Conference on Communications (ICC)*, pp. 1–6.

77 Sivaraman, A., Winstein, K., Subramanian, S., and Balakrishnan, H. (2013). No silver bullet: extending SDN to the data plane. *Proceedings of the 12th ACM Workshop on Hot Topics in networks*, ACM, p. 19.

78 Kundel, R., Blendin, J., Viernickel, T. et al. (2018). P4-codel: active queue management in programmable data planes. *Proceedings of the IEEE 2018 Conference on Network Functions Virtualization and Software Defined Networks*, IEEE, pp. 27–29.

79 Papagianni, C. and De Schepper, K. (2019). PI2 for P4: an active queue management scheme for programmable data planes. *Proceedings of the 15th International Conference on Emerging Networking EXperiments and Technologies, CoNEXT '19*. New York, NY, USA: Association for Computing Machinery, pp. 84–86. ISBN 9781450370066. https://doi.org/10.1145/3360468.3368189.

80 De Schepper, K., Bondarenko, O., Tsang, I.-J., and Briscoe, B. (2016). PI2: A Linearized AQM for both Classic and Scalable TCP. *Proceedings of the 12th International on Conference on emerging Networking EXperiments and Technologies (CoNEXT '16)*. *Association for Computing Machinery*, New York, NY, USA, pp. 105–119. Doi: https://doi.org/10.1145/2999572.2999578.

81 Laki, S., Vörös, P., and Fejes, F. (2019). *Proceedings of the ACM SIGCOMM 2019 Conference Posters and Demos*, 2019 ACM/IEEE Symposium on Architectures for Networking and Communications Systems (ANCS), SIGCOMM Posters and Demos '19, New York, NY, USA: Association for Computing Machinery., pp. 148–150. ISBN 9781450368865. https://doi.org/10.1145/3342280.3342340.

82 Hancock, D. and Van der Merwe, J. (2016). HyPer4: using P4 to virtualize the programmable data plane. *Proceedings of the 12th International on Conference on emerging Networking EXperiments and Technologies*, ACM, pp. 35–49.

83 Zhang, C., Bi, J., Zhou, Y. et al. (2017). HyperV: a high performance hypervisor for virtualization of the programmable data plane. *2017 26th International Conference on Computer Communication and Networks (ICCCN)*, IEEE, pp. 1–9.

84 GitHub. P4lang/Behavioral-Model (BMv2). https://github.com/p4lang/behavioral-model (accessed 26 April 2021).

85 Zhang, C., Bi, J., Zhou, Y., and Wu, J. (2019). HyperVDP: high-performance virtualization of the programmable data plane. *IEEE Journal on Selected Areas in Communications* 37 (3): 556–569.

86 DPDK. The Data Plane Development Kit. https://www.dpdk.org/ (accessed 26 April 2021).

87 Zheng, P., Benson, T., and Hu, C. (2018). P4Visor: lightweight virtualization and composition primitives for building and testing modular programs. *Proceedings of the 14th International Conference on Emerging Networking EXperiments and Technologies*, ACM, pp. 98–111.

88 Han, S., Jang, S., Choi, H. et al. (2020). Virtualization in programmable data plane: a survey and open challenges. *IEEE Open Journal of the Communications Society* 1: 527–534.

89 Thyagaturu, A.S., Mercian, A., McGarry, M.P. et al. (2016). Software defined optical networks (SDONs): a comprehensive survey. *IEEE Communication Surveys and Tutorials* 18: 2738–2786.

90 Kinf, D. and Farell, A. (2015). IETF RFC 7491, A PCE-Based Architecture for Application-Based Network Operations. https://tools.ietf.org/html/rfc7491 (accessed 15 April 2021).

91 Casellas, R., Martinez, R., Vilalta, R., and Munoz, R. (2018). Control, management, and orchestration of optical networks: evolution, trends, and challenges. *IEEE/OSA Journal of Lightwave Technology* 36: 1390–1402.

92 Gerstel, O., Jinno, M., Lord, A., and Yoo, S.J.B. (2012). Elastic optical networking:a new dawn for the optical layer? *IEEE Communications Magazine* 50: s12–s20.

93 Wang, Y., Lu, P., Lu, W., and Zhu, Z. (2017). Cost-efficient virtual network function graph (vNFG) provisioning in multidomain elastic optical networks. *IEEE/OSA Journal of Lightwave Technology* 35: 2712–2723.

94 Chatterjee, B.C., Sarma, N., and Oki, E. (2015). Routing and spectrum allocation in elastic optical networks: a tutorial. *IEEE Communication Surveys and Tutorials* 17: 1776–1800.

95 Fichera, S., Martinez, R., Martini, B. et al. (2019). Latency-aware resource orchestration in SDN-based packet over optical flexi-grid transport networks. *IEEE/OSA Journal of Optical Communications and Networks* 11: B83–B96.

96 ONF (2016). SDN Architecture TR-521. *Technical Report 1.1*. Open Networking Foundation (ONF). https://www.opennetworking.org/images/stories/ downloads/sdn-resources/technical-reports/TR-521_SDN_Architecture_issue_ 1.1.pdf (accessed 15 April 2021).

97 MEF (2016). Lifecycle Service Orchestration (LSO): Reference Architecture and Framework. Service Operations Specification MEF 55, Metro Ethernet Forum (MEF), 3 2016. https://www.mef.net/Assets/Technical_Specifications/ PDF/MEF_55.pdf (accessed 15 April 2021).

98 ETSI (2018). Network Functions Virtualisation (NFV) Release 3; Management and Orchestration; Report on architecture options to support multiple administrative domains. Group Report (GR) 028 v3.1.1, European Telecommunications Standards Institute (ETSI), 01 2018. https://www.etsi.org/deliver/ etsi_gr/NFV-IFA/001_099/028/03.01.01_60/gr_NFV-IFA028v030101p.pdf (accessed 15 April 2021).

99 ETSI (2019). Network Functions Virtualisation (NFV) Release 3; Management and Orchestration; Multiple Administrative Domain Aspect Interfaces Specification. Group Report (GR) 030 v3.2.2, 05 2019. https://www.etsi.org/ deliver/etsi_gs/NFV-IFA/001_099/030/03.01.01_60/gs_NFV-IFA030v030101p .pdf (accessed 15 April 2021).

100 Dietrich, D., Abujoda, A., Rizk, A., and Papadimitriou, P. (2017). Multi-provider service chain embedding with nestor. *IEEE Transactions on Network and Service Management* 14 (1): 91–105. https://doi.org/10.1109/TNSM.2017. 2654681.

101 Chang, V., Sun, G., and Li, Y. (2018). Service function chain orchestration across multiple domains: a full mesh aggregation approach. *IEEE Transactions on Network and Service Management* 15 (3): 1175–1191. https://doi.org/10.1109/TNSM.2018.2861717.

102 Zhang, Q., Wang, X., Kim, I. et al. (2016). Vertex-centric computation of service function chains in multi-domain networks. *IEEE Network Softwarization Conference and Workshops (NetSoft)*, IEEE, June 2016, pp. 211–218.

103 Abujoda, A. and Papadimitriou, P. (2016). DistNSE: distributed network service embedding across multiple providers. *8th International Conference on Communication Systems and Networks (COMSNETS)*, January 2016, IEEE, pp. 1–8.

104 ETSI (2019). Zero-touch network and Service Management (ZSM); Reference Architecture. Group Specification (GS) 002 v1.1.1, European Telecommunications Standards Institute (ETSI), 08 2019. https://www.etsi.org/deliver/etsi_gs/ZSM/001_099/002/01.01.01_60/gs_ZSM002v010101p.pdf (accessed 15 April 2021).

105 Benzaid, C. and Taleb, T. (2020). Ai-driven zero touch network and service management in 5G and beyond: challenges and research directions. *IEEE Network* 34 (2): 186–194. https://doi.org/10.1109/MNET.001.1900252.

106 Konecný, J., McMahan, H.B., Yu, F.X. et al. (2016). Federated learning: strategies for improving communication efficiency. *CoRR*, abs/1610.05492. http://arxiv.org/abs/1610.05492 (accessed June 2020).

107 Ramanan, P. and Nakayama, K. (2020). BAFFLE: blockchain based aggregator free federated learning. *2020 IEEE International Conference on Blockchain (Blockchain 2020)*, November 2020, IEEE, pp. 72–81.

108 H2020 project 5Growth, 5G-enabled Growth in Vertical Industries. http://5growth.eu/ (accessed June 2020).

109 ETSI. Zero-touch network & Service Management. https://www.etsi.org/technologies/zero-touch-network-service-management (accessed June 2020).

110 3GPP (2019). Network Data Analytics Services, Release 16, December 2019. Technical specification (TS) 29.520 v16.2.0, 3rd Generation Partnership Project (3GPP), December 2019.

111 Antevski, K. and Bernardos, C.J. (2020) Federation of 5G services using distributed ledger technologies. *Wiley Internet Technology Letters* 1–6. https://doi.org/10.1002/itl2.193.

112 Antevski, K., Martín-Pérez, J., Garcia-Saavedra, A. et al. (2020). A Q-learning strategy for federation of 5G services. *2020 IEEE International Conference on Communications (ICC 2020)*, June 2020, IEEE, pp. 1–6.

5

AI in 5G Networks: Challenges and Use Cases

Stanislav Lange[1], Susanna Schwarzmann[2], Marija Gajić[1], Thomas Zinner[1], and Frank A. Kraemer[1]

[1]Department of Information Security and Communication Technology, Norwegian University of Science and Technology, Trondheim, Norway
[2]Department of Telecommunication Systems, TU Berlin, Berlin, Germany

5.1 Introduction

Network softwarization paradigms such as software defined networking (SDN) and network functions virtualization (NFV) alongside techniques for network slicing pave the way for flexible and programmable 5G networks [1]. At the same time, the heterogeneity of services and devices that need to be supported call for a high degree of automation to enable quick adaptation to changing network conditions and to maintain resource efficiency and service quality.

Due to numerous advances in the fields of Machine Learning (ML) and Artificial Intelligence (AI) as well as their successful application in a wide range of domains, current research work discusses utilizing ML/AI techniques to manage communication networks and services [2]. To enable a seamless integration and to maximize automation benefits, AI-based mechanisms have to be embedded in the 5G architecture and connected to the management and orchestration (MANO) systems as well as to functions that provide network- and service-related monitoring data.

Figure 5.1 displays the key components of a 5G integrated network architecture that features end-to-end (E2E) network slicing. Using the slicing paradigm, operators can subdivide their physical networks into several virtual and isolated logical networks that are tailored to specific requirements and use them to achieve differentiated treatment of network traffic. In the following, we describe the contents of the figure layer by layer, starting with the E2E slices at the highest layer. These E2E slices are virtual networks with Service Level Objective (SLO) requirements and are used to achieve differentiated treatment of traffic based on the requested service type and the subscription type of the device making the request. For instance,

Communication Networks and Service Management in the Era of Artificial Intelligence and Machine Learning,
First Edition. Edited by Nur Zincir-Heywood, Marco Mellia, and Yixin Diao.

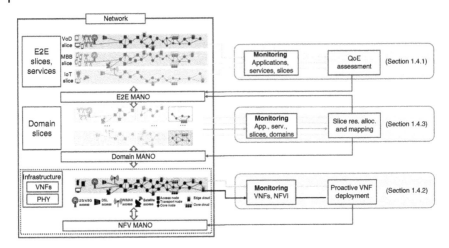

Figure 5.1 Overview of AI/ML use cases in a service-aware 5G integrated network. Extended view, Source: Based on Ordonez-Lucena et al. [3].

different E2E slices could be used to provide bandwidth-intensive video stream-ing services and delay-critical healthcare services. These E2E slices cover multiple domains such as access, transport, and core, which come with their individual characteristic components, standards [4, 5], and challenges. Hence, E2E slices are mapped to domain slices (shown in the middle layer of Figure 5.1) that cover the corresponding part of the network and feature their own MANO components. Dif-ferentiated treatment at this layer is based on Service Level Agreements (SLAs) regarding Quality of Service (QoS) metrics.

To meet the SLO requirements of E2E slices, appropriate domain slices need to be chosen to handle the corresponding traffic. This mapping between E2E and domain slices is also an opportunity for operators to benefit from economies of scale since a single domain slice can serve multiple E2E slices. For instance, two E2E slices could share their transport and core slices while having separate access slices since the latter tends to be a bottleneck. Finally, a mapping between domain slices and resources from the infrastructure layer is necessary to make sure that the SLA guarantees are maintained. The infrastructure layer is composed of physical links, nodes, as well as physical and virtual network functions (NFs). At this layer, operators can once again benefit from economies of scale by sharing infrastructure resources between domain slices while avoiding SLA violations.

For readability, we omit some details in the figure, but would like to point out two additional aspects: domains can be broken down in a recursive fashion before being mapped to physical resources [6]. Furthermore, different slices might be maintained by different operators, which introduces additional challenges regard-ing communication and management.

AI-based techniques can help achieving autonomous management in the context of such complex systems that are additionally subject to temporal dynamics. Exemplary use cases encompass quality of experience (QoE)-based application monitoring and control, autoscaling of virtualized network functions (VNFs), and automated slice resource allocation.

Before applying such AI-based mechanisms, several challenges need to be taken into consideration. These include general challenges like data acquisition and preparation as well as algorithm and feature selection. Other challenges are more context-specific. For instance, the impact on the networking ecosystem itself – including reliable and robust operation – has to be taken into account.

In this chapter, we discuss three application areas for AI within the 5G ecosystem. We present results from case studies to highlight both general and specific challenges that arise in this context and discuss viable ways to overcome them.

The remainder of this chapter is structured as follows. We discuss related work regarding the general application of ML techniques in the networking domain as well as ML-based management of softwarized 5G networks and QoE assessment in Section 5.2. In Section 5.3, we present case studies that cover QoE management, VNF deployment, and slice management. We identify the main challenges, illustrate how they can be addressed, and extract guidelines for operators and practitioners. Finally, we conclude the chapter with a summary of key findings in Section 5.4.

5.2 Background

In this section, we cover related work regarding three main directions. First, we discuss the application of ML methods to networking problems in general. Furthermore, ML applications in the specific context of virtualized networks and QoE management are highlighted.

5.2.1 ML in the Networking Context

Current and future communication networks need to deal with heterogeneity regarding numerous aspects such as use cases, applications, devices, communication paradigms, and deployment options for NFV-based solutions [7]. All these factors contribute to a steady growth of the parameter space in which optimizations need to be carried out and therefore limit or prevent the applicability of traditional methods such as those based on Integer Linear Programming (ILP) or simulations. To address the goals w.r.t. flexibility, adaptability, and automation [8], recent publications in the networking domain propose ML-based approaches [2, 9]. Such approaches cover not only general networking problems such as heavy hitter detection [10] but also include problems that are specific to softwarized networks [11] like the SDN controller placement problem [12]. Furthermore, the interest in 5G networks has led to contributions regarding traffic forecasting [13], prediction of user quantities [14], and cognitive networking [15]. Depending on various factors such as the particular use case and deployment environment, mechanisms based on all the four principal ML paradigms are applied, i.e. supervised learning (SL), semi-SL, unsupervised learning, and reinforcement learning.

5.2.2 ML in Virtualized Networks

In virtualized networks, ML approaches have been proposed in the form of neural network-based admission control in Virtual Network Embedding (VNE) scenarios [16] as well as resource assignment [17] based on reinforcement learning. Additionally, ML techniques have been applied to several demand, resource, and performance prediction tasks in NFV-based networks. While examples for predicted metrics include the CPU usage of VNF instances [18–20] and performance metrics in cloud environments [21], the ML mechanisms under consideration include neural networks, random forest regression, and support vector regression. In addition to their immediate use, the outputs of the prediction models could also be integrated as new features to improve the performance of ML-based decision-making mechanisms for tasks like VNF autoscaling [22, 23].

5.2.3 ML for QoE Assessment and Management

ML is widely applied for estimating QoE from network-level key performance indicators (KPIs). In this context, many works focus on video streaming, as this is one of the most prominent applications in today's networks. Due to the ongoing trend toward traffic encryption, ML is often applied on traffic meta-data to estimate QoE-relevant video streaming metrics, such as resolution or bitrate [24–28], as well as to predict [29] or classify [30, 31] the QoE in terms of Mean Opinion Score (MOS).

As mobile video streaming is getting more and more popular and due to additional KPIs such as channel quality and factors such as the movement characteristics of clients, QoE assessment in mobile environments needs dedicated evaluations. Therefore, Lin et al. [32] focus on mobile networks when studying the performance of various classifiers and the impact of the used features retrieved from network- and application-related data. However, it is desirable to estimate the QoE solely using network-related features, as such data is typically available to a mobile network operator (MNO) and easier to obtain at scale than detailed application-level information. Such a solution is presented in [33], focusing on long term evolution (LTE) networks. New opportunities and challenges for ML-based QoE assessment arise with the introduction of 5G. While [34] focus on the challenges and propose a data-driven architecture for QoE management, Schwarzmann et al. [35] discuss how the newly introduced NFs Application Function (AF) and Network Data Analytics Function (NWDAF) can support the QoE estimation process. The authors of [36] use the Least Absolute Shrinkage and Selection Operator (LASSO) to perform a deeper investigation of the relevant statistics at the NWDAF in order to achieve a certain QoE estimation accuracy. The work in [37] goes a step beyond the QoE assessment by presenting a ML-based resource allocation for 5G networks. The proposed system determines the network performance level via clustering, predicts network KPIs by means of regression, and dynamically provisions resources in a proactive way. One of the new key features of 5G, network slicing, is exploited in conjunction with deep learning [38] and reinforcement learning [39] to achieve an optimized resource utilization.

5.3 Case Studies

In this section, we discuss three use cases in the 5G context that benefit from AI-based techniques. In addition to providing an overview of the underlying technical problems and resulting challenges both for traditional as well as ML-based approaches, we present the methodology for addressing these challenges alongside evaluation results and guidelines for tackling similar tasks.

First, we cover the topic of deriving QoE estimates and management actions in environments that contain multiple applications and only provide access to QoS metrics. Second, we demonstrate how ML can assist in proactively deploying VNF instances to optimize both resource efficiency and request admission time in the context of Service Function Chaining (SFC). Finally, we discuss the potential of leveraging insights from both areas to devise strategies for QoE-aware slice resource allocation and management.

Table 5.1 illustrates how our specific 5G-related use cases fit into the ML landscape by categorizing them w.r.t. the problem type, applied algorithms, and the process from data acquisition to preprocessing. We model the first two use cases as SL tasks where the goal consists of learning the relationship between monitoring data and continuous QoE levels or a set of discrete management actions, respectively. In both cases, we carefully design simulations that integrate relevant aspects related to standards, communication interfaces, and temporal dynamics. Subsequently, we utilize established models to enrich the data with ground truth labels for model training. To assess the general feasibility of ML-based approaches for these tasks, we apply existing algorithms to the training data and evaluate their performance in terms of accuracy.

Table 5.1 Categorization of covered use cases.

Use case	QoE estimation (Section 5.3.1)	VNF deployment (Section 5.3.2)	Slice mgmt. (Section 5.3.3)
Problem type	Regression-based prediction	Classification-based decision-making	Estimation, prediction, decision making
Algorithm choice	SVM, LASSO	XGBoost, GBM, neural networks	Unsupervised learning (k-means clustering), reinforcement learning
Data collection, analysis, and preparation	Omnet simulation, ITU-T model(s) for ground truth	Simulation, ground truth labeling based on ILP solutions	Simulation, testbed

In the case of slice management, we identify two different tasks that can be approached with different classes of ML algorithms. To determine which services can be aggregated into one slice, clustering approaches can be applied to representative traffic characteristics. On the other hand, strategies based on reinforcement learning can be used to dynamically map and allocate resources between layers, e.g. between domain slices and physical resources.

5.3.1 QoE Estimation and Management

Driven by business incentives, providing a good QoE to customers is important for network providers and application or content providers (CPs), such as YouTube or Netflix. Delivering a good QoE to each user depends on two key factors: (i) a holistic view of the current QoE in the system, e.g. all users connected to one base station in a mobile network and (ii) the capability to accordingly adapt the system, e.g. certain configurations of the base station. The inherent dilemma is, however, that only the MNO is capable of performing QoE-aware network management, while the information regarding the current system QoE is only available to the CP. This issue can be overcome with newly introduced NFs in 5G. In particular, the AF is fundamental for the information exchange between MNOs and CPs. It is a standardized interface which allows third-party tenants, e.g. the CP, to communicate information, such as application quality metrics or QoE, to the MNO. As a consequence, this enables the MNO to perform a QoE-aware resource control.

But despite having the AF, an operator faces two challenges: (i) the MNO depends on the information provided via the third party AF, i.e. the MNO has no control regarding the frequency and amount of information that is communicated.

(ii) The information is transmitted via the control plane and hence increases the amount of costly control plane traffic. To solve these challenges, a second 5G NF can fundamentally change the network management: The NWDAF [40]. It collects and provides analytics from and to other NFs in the 5G architecture and can therefore potentially enable data-driven QoE monitoring and management: The QoE obtained via the third party AF can be correlated with network monitoring statistics, i.e. throughput (TP), packet loss, or channel quality indicator (CQI). ML models exploiting these correlations are then capable of estimating the QoE from these network-level statistics, which are available to the MNO. This allows gathering information about the system QoE, even in the absence of QoE information provided by the third party AF.

5.3.1.1 Main Challenges

Realizing such an ML-driven, QoE-aware resource control introduces a wide range of challenges, which are summarized in the following.

- *Identification of relevant features*: ML algorithms estimate the QoE based on different statistics, so-called features. While some of these statistics are highly relevant for the estimation process, others only have negligible impact on the model output. The task of *feature engineering*, i.e. extracting the most relevant features and transforming them into a format that is compatible with the chosen ML approach, poses significant challenges as it requires both domain knowledge and knowledge about the ML approach. Examples for aspects to consider include the applied ML technique, the environment (mobile vs. wired access), and the service type.
- *Identification of appropriate models*: Out of the wide range of available ML-based techniques, we need to examine which one is suitable for the given problem. The applicability of a specific technique not only depends on its performance in terms of estimation accuracy, but also on factors such as the complexity for training and applying the model as well as its adaptability and comprehensibility. For instance, while the behavior of basic regression models can be understandable to humans, the decisions of complex neural networks can be hard or impossible to understand, as they act like a black box.
- *Limitations in terms of data quality and granularity*: An MNO's capability for network monitoring has practical limits, which means that important metrics can only be collected for a fraction of the active user equipments (UEs). Furthermore, the time intervals between two measurements, e.g. the TP at a base station, cannot be arbitrarily short. This means that the collected data is temporally or spatially aggregated, or only snap shot data is available.
- *Quantification of the costs for deploying ML*: Deploying an ML-based QoE estimation within the 5G architecture is expensive. It introduces, among others, costs

for collecting, transmitting, processing, and storing all relevant data. There is the need to quantify such costs and to examine how different factors, such as the monitoring granularity or the relevant features, influence these costs.

- *Trade-off between accuracy and costs*: The used features, ML algorithms, and monitoring not only affect the costs, but also the accuracy that can be achieved. In general, a finer monitoring granularity and a larger number of features result in a better estimation accuracy. However, they increase the costs. Analyzing the cost vs. accuracy trade-off is a crucial challenge that needs to be tackled when it comes to ML for 5G.
- *Integration into the 5G architecture*: It needs to be examined, how the ML-based QoE estimation can be integrated in compliance with the 5G networking architecture specifications. This includes for example the tasks of the different stakeholders and NFs involved. It needs to be specified whether the NWDAF trains the model itself or if it is equipped with a final, externally trained model. Does the third party AF communicate the QoE or only relevant metrics, so that the NWDAF derives the QoE using a standardized QoE model?
- *Implementation of feedback control loops*: The available QoE estimation can now be used to trigger automated network control actions, e.g. if the QoE is below a predefined threshold for a certain amount of time. To optimize the system for autonomous QoE management, the effects of the control actions need to be monitored and adapted if needed.

5.3.1.2 Methodology

Our proposed integration of ML-based QoE estimation in the 5G architecture is illustrated in Figure 5.2. It considers three phases.

Phase 1 – data collection: The third party AF communicates application performance data to the NWDAF, enriching it by ground truth QoE ①.

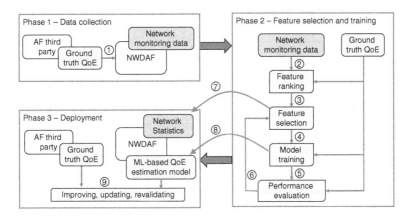

Figure 5.2 Possible framework for ML integration in 5G.

Phase 2 – feature selection and model training: This phase covers the typical ML pipeline of training and testing a suitable model. In particular, a vast number of network features is generated by statistical processing of the network monitoring data. These features are ranked according to their significance in terms of estimating the QoE ②. With a subset of significant features ③ and with the ground truth QoE values, the ML-based models for QoE estimation ④ are trained. Next, the model performance is evaluated ⑤. This process can be repeated for different feature sets and different models ⑥ until a desired estimation accuracy is obtained. The identified feature set dictates the necessary network statistics the NWDAF has to provide ⑦ for a reliable QoE estimation during the deployment phase based on the trained model ⑧.

Phase 3 – deployment: Although the MNO can now estimate the QoE without the need of application metrics provided by the AF, the CP can still communicate such information to facilitate updates, verification, and improvements of the trained model ⑨.

With this procedure, we tackle the challenge regarding how to integrate an ML-based QoE estimation in the 5G architecture. To evaluate the relevant features and to study the accuracy vs. cost trade-off in terms of features used, we applied the ML approach *LASSO* on data generated from simulations using the discrete event network simulator *OMNeT++* [41]. LASSO is a regression analysis method which performs feature selection and trains a model to predict the outcome based on the selected features [42]. Its regularization parameter allows to tune the number of regression coefficients set to zero, which reduces the number of features. For any feature set size, it selects the most appropriate features out of all features that are available. As a result, LASSO is an appropriate way to study the impact of an increased number of features on the estimation accuracy.

We studied a video streaming use case and simulated clients in a mobile cell with different movement characteristics. Network-relevant information is monitored as time series within the OMNeT++ simulator: TP of UEs, TP of the base station, CQI, and round-trip time (RTT). From these time series, we derive features by applying typical statistics, such as mean, standard deviation, quartiles, minimum, and maximum. Furthermore, we collect all QoE-relevant video streaming metrics, such as video stallings and quality, to compute a user's QoE using the standardized ITU-T P.1203 model [43]. We apply LASSO to this data and use different values of the regularization parameter to obtain feature sets of different sizes [36].

5.3.1.3 Results and Guidelines

When estimating the QoE on MOS scale based on 33 different network-related features, LASSO is capable to achieve a mean squared error (MSE) of roughly 0.15. Given the fact that MOS ranges from 1 to 5, LASSO can achieve a high accuracy, despite being a basic regression method. In general, we can observe a tendency

Table 5.2 Impact of including different monitoring types on QoE estimation accuracy.

Monitoring type	UE DL TP	CQI UL	CQI DL	UE UL TP	AN UL TP	RTT	AN DL TP
MSE	0.318	0.236	0.215	0.193	0.189	0.153	0.151

of increased accuracy with an increasing feature set size. These findings can be helpful for an MNO to derive which monitoring points are crucial and what needs to be prioritized in terms of data collection. Table 5.2 shows the MSE that can be achieved when considering additional monitoring points to generate features from. For instance, we obtain an MSE of about 0.32 if only features generated from UE downlink (DL) TP are used. Having additionally features that are generated from the uplink (UL) CQI, the MSE falls below 0.25. When also taking the DL CQI-related features into account, the MSE can be reduced to about 0.21. However, only minor performance gains can be achieved by additionally including features related to the access node (AN) TP or RTT.

We did not only study the feature relevance itself, but additionally evaluated if the clients' movement characteristics influence the QoE estimation process. Indeed, we could observe that mobility has a significant impact on the QoE, and as a consequence, on the features selected by LASSO. The features for static clients are mostly generated from statistics such as average, median, or different percentiles. However, for moving clients, the majority of features are generated from statistics that express variance. For example, standard deviation, covariance, or skewness. As a result, an MNO needs to monitor with a higher granularity if it aims a reliable QoE estimation for mobile clients. Otherwise, the variations in the time series, e.g. in CQI or DL TP, cannot be captured accurate enough. As an implication, the costs for QoE estimation are higher in mobile scenarios.

5.3.2 Proactive VNF Deployment

Softwarization paradigms like SDN and NFV provide network operators with benefits regarding flexibility, scalability, and cost efficiency. Furthermore, they are key enablers for the concept of SFC that allows linking together different NFs to form service chains and also dynamically changing their structure and size to adapt to network events. Since modern communication networks need to support numerous heterogeneous services that run on the same physical substrate and whose requirements change dynamically during their lifetime, efficient management and operation of these networks requires a high degree of automation paired with proactive decision-making. This ensures that resource efficiency is maintained without affecting service quality.

A particularly important step consists of determining the optimal number of VNF instances required to accommodate current and upcoming service requests. This directly affects resource efficiency and constitutes the foundation for subsequent decisions such as placement and chaining of VNFs. In our recent work [23, 44], we devise a fast and proactive decision-making scheme based on ML that uses monitoring data to predict whether to adapt the current number of VNF instances. By making these decisions ahead of time, arriving requests can be admitted directly upon arrival without having to wait for new VNFs to be instantiated. To address the lack of realistic data sets, we additionally present a methodology for generating labeled training data that integrates temporal dynamics and heterogeneous demands of real-world networks. Using two different network topologies that represent a wide area network (WAN) and a multi-access edge computing (MEC) scenario, we demonstrate the applicability of the proposed methodology and provide insights into the ability of models to generalize. Furthermore, we provide guidelines for network operators regarding feature relevance, amount of required training data, and the accuracy trade-offs that result from long-term predictions.

5.3.2.1 Problem Statement and Main Challenges

We use a system model that is similar to that of the VNF placement problem [45, 46]. The underlying physical network is represented by a undirected graph whose nodes and edges have resources such as CPU cores and bandwidth capacities and impose delays on traffic that passes them. The characteristics of VNFs are stored in a catalog that includes their resource requirements and their packet processing capacity. Service function chain requests are characterized by their time of arrival, their duration, the nodes between which traffic is exchanged, as well as constraints regarding the minimum bandwidth and maximum latency of the corresponding flow. Each SFC request also contains an ordered list of VNFs that need to be traversed by each packet. Finally, we assume that a monitoring system continuously collects various network statistics like the number of active requests in the network, the arrival rate of SFC requests, the amount bandwidth that is required by each request, as well as the number of active instances per VNF type.

With this system model, the prediction task for VNF deployment actions is defined as follows: at the current time t_{cur}, we are given monitoring data from a window of width a as well as a prediction horizon p. The goal of the prediction task is to determine whether we should increase, decrease, or keep the current number of VNFs to be able to accommodate the requests that will be active after time p. We treat this a classification problem where a feature vector is extracted from the monitoring data and mapped to a decision that corresponds to one of the three deployment actions. A graphical representation of the prediction process is provided in Figure 5.3.

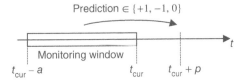

Figure 5.3 Overview of the prediction process.

To solve the outlined prediction task, several challenges need to be addressed. First, there is a lack of realistic and publicly available data sets that can be used to train and evaluate algorithms. The issues regarding data quality include a high degree of homogeneity in terms of VNFs and services, a limited amount of temporal dynamics, and the usage of synthetic network topologies that are not necessarily representative of real-world networks. Additionally, determining the optimal configuration and with that the number of VNF instances at each point in time is an non-deterministic polynomial-time (NP)-hard problem due to its relationship to the VNF placement problem [45]. Although such optimizations can be performed with ILP-based solutions, they only provide results for a given configuration and cannot be used in the context of predictions that involve uncertainties regarding upcoming arrival and departure events of service requests.

Network operators who would like to apply ML-based prediction algorithms to the deployment prediction task also face operational questions. These include the choice, relevance, and availability of metrics that can be monitored at a reasonable cost, the amount of training data required to train an ML model, and the extent to which the model generalizes to changes in network topology and traffic patterns.

5.3.2.2 Methodology

To overcome the above challenges and allow other researchers to reproduce and extend our work, we propose the following workflow that covers the entire process including request trace generation, generation of labeled training data, as well as model training and evaluation.[1]

1. **General configuration** of parameters such as the network topology, the traffic matrix which determines communication patterns among nodes, and the VNF catalog with their characteristics. Additional parameters for tuning the overall system load include the maximum request arrival rate as well as their required bandwidth and duration.

2. **SFC request generation** integrates realistic temporal dynamics by using data from real-world traffic matrices to modulate the request arrival process. This

---

1 An implementation of the entire procedure is available at https://github.com/dpnm-ni/2019-ni-deployment-prediction.

way, both inter- and intra-day phenomena regarding the number of active requests are reflected in the resulting trace.

3. **Optimal placement calculation** is performed for the system configuration at each arrival event and uses an ILP-based algorithm [45]. In particular, the output contains the optimal number of VNF instances which serves as ground-truth for model training.

4. **Training data generation** encompasses the construction of labeled feature vectors. To this end, various features are extracted from a monitoring window that is constructed around each arrival event, while labels (deployment actions) are determined by comparing the ILP outputs for the current and future point in time.

5. **Model training and evaluation** is performed by first splitting the data set into training and test sets and feeding the former to several state-of-the-art SL algorithms, e.g. XGBoost, Gradient Boosting Machine (GBM), and neural networks. In addition to their competitive performance on various SL tasks, the boosting-based algorithms allow assessing feature importance in a straightforward manner – a crucial capability for deriving operational guidelines. Finally, trained models are evaluated using the test set.

5.3.2.3 Evaluation Results and Guidelines

We use the proposed methodology to evaluate the performance of ML-based approaches to address the VNF deployment prediction task in the context of two different scenarios: the WAN scenario covers the perspective of a service provider who can instantiate VNFs at different locations of a regional or global cloud provider, whereas the MEC scenario is representative of an operator who can choose to place VNFs at capacity-constrained central offices that are close to end-users or at a central data center that comes with a latency trade-off. In the following, we summarize our main results and translate them into guidelines for network operators planning to deploy such mechanisms.

- *Feature importance*: We observe that accurate predictions require features from two categories. The first category comprises VNF-specific information such as the total remaining capacity of active instances and their utilization. The second category contains contextual global features such as the time since the last request arrival or the arrival rate during the monitoring interval. Furthermore, by quantifying the features' relative contribution to prediction accuracy, operators can make more informed decisions regarding whether to invest the monitoring and communication overhead for collecting the corresponding data.

- *Training set size*: By providing different amounts of training data to the ML algorithms and applying the resulting models to the test set, we analyze the impact of the training set size on performance. Additionally, we perform this experiment with request arrival processes that have different degrees of variability: one

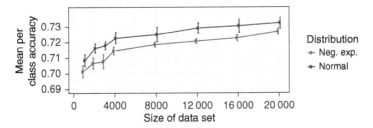

Figure 5.4 Mean per class accuracy on the testing set when using different amounts of training data as well as different distributions for the request interarrival time.

with normally distributed interarrival times and a coefficient of variation equal to 0.25 and one that uses a negative exponential distribution with a coefficient of variation equal to 1. Figure 5.4 shows numerical results from [44]. Two effects can be observed. First, convergence is faster and therefore less data is required in the case of the less variable and therefore more predictable arrival process. Second, even with as few as 1000 labeled examples, accuracy levels of around 70% can be achieved while the accuracy stops improving significantly for training set sizes beyond 20 000. Hence, operators need to be aware of the particular traffic characteristics in their networks.

- *Impact of prediction horizon*: While long-term predictions leave more time to execute deployment actions and find the optimal location for placing new VNFs, they also come with trade-offs in terms of accuracy. We quantify this trade-off in a parameter study regarding the prediction horizon p and find an almost linear decrease of prediction accuracy when varying p in a range from 10 to 100 seconds. Again, parameters of the ML model need to be aligned with characteristics of the technical system at hand.
- *Generalizability*: Finally, the WAN and MEC scenarios alongside their different network topologies serve the purpose of analyzing the generalizability of models by training them on one scenario and testing them on the other. We show that models pick up enough scenario-independent knowledge to generalize well across scenarios, but that the direction matters. In particular, special properties of the MEC scenario are not learned during training on the WAN scenario and therefore result in a higher accuracy penalty. A similar behavior is expected when using models that are trained on simulation-based data to make predictions in a physical deployment that introduces additional complexity.

While this and the Section 5.3.1 deal with monitoring and management tasks on a single layer of the architecture shown in Figure 5.1, the tasks covered in Section 5.3.3 are concerned with mapping resources and requirements between layers.

5.3.3 Multi-service, Multi-domain Interconnect

Network services have different requirements in terms of resources, TP, latency, and jitter. While the currently applied best-effort operational mode satisfies basic data access, it might fail to fulfill SLAs for services with strict requirements. For example, real-time video streaming requires high TP and minimal latency which is often not available through best-effort operational mode. One solution is to make more network resources available, but that is not desirable by network operators since overprovisioning is expensive. Instead, network operators aim for solutions that represent a good compromise between these two scenarios, that means, providing services with adequate quality levels while reducing the overprovisioning.

To keep customers satisfied and avoid SLA violations and overprovisioning, network operators can transform their QoS-based networks to QoE-based user- and application-aware networks. To achieve this, valid QoE assessment based on QoS parameters becomes crucial. As shown in Section 5.3.1, ML is a feasible way to instrument the mapping. QoE-based networks require a complex QoE management framework to manage the heterogeneous and co-existing services. In the most flexible case, the framework processes every packet flow and assigns required resources to it. This approach is presented in [47] and it demonstrates that QoE-aware resource management outperforms QoS-based approaches. However, taking a growing number of users and services with strict requirements into account, this per-flow processing approach is not scalable. A novel solution is to classify services into a specific number of classes based on their requirements. For example, as suggested in [48], such classification can be done as follows: *Basic Quality* (best-effort), *Background Quality* (least-effort), *Improved Quality*, and *Assured Quality*. Another classification approach could be based on service types such as *Video streaming, Web browsing, Voice-over-IP, Email*, and *File download*. The classes can be further differentiated in different quality sub-classes, e.g. *Premium Quality Video* and *Basic Quality Video* streaming.

Different network services are assigned to these classes based on their requirements for identified key service features such as bandwidth, latency, and jitter allowance. SLAs for each class are based on E2E service requirements. The number of classes and the clustering process have to be carefully selected and designed. This is an optimization problem for trade-off between scalability (control plane overhead) and performance gain. For instance, having five classes is scalable from a management and control plane overhead perspective, but the granularity in service differentiation is low and some services might be under- or over-provisioned. The number of classes, clustering, and acceptance criteria for the classes is an optimization problem suitable for a ML-based solution. Unsupervised learning, and more specifically the k-means clustering algorithm can be used to cluster services into a specified number of k classes. However, the service clusters have to be evaluated in terms of their performance and cost, for every value of k that is of interest. After evaluation (possibly via simulation or implementation to physical testbed), the cluster with best performance and cost metric is chosen as optimal.

As mentioned earlier in this section, there is a need for a complex framework governing the service classes and resource allocations. Components of the framework are shown in Figure 5.5. The figure shows coexistence of different applications sharing the same physical resources. Within the E2E MANO, the QoE modeler takes QoS data from the network as an input and predicts the QoE level for each slice. The approach presented in Section 5.3.1 can be used for the QoE prediction. Once the QoE level is estimated, the mapper/allocator checks this information, compares it with subscription data for each user and performs actions such as reallocation of the domain slices or route changes to make sure that SLAs are satisfied. Domain MANO implements similar, recursive logic for allocation of physical resources to domain slices.

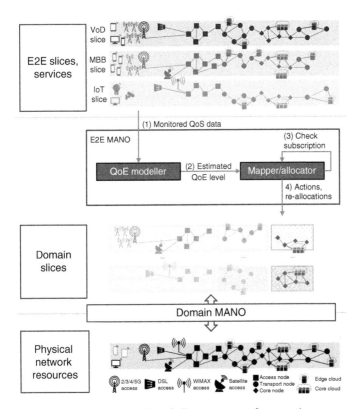

Figure 5.5 Components of the QoE management framework.

On the application level, each service class corresponds to one E2E slice. As mentioned previously, such slices cover multiple domain slices including access, transport, and core. Operators can benefit from economies of scale since coexisting applications can share the same domain slices. However, the slices have to be shared in a way which does not harm the guaranteed SLAs for each application. Optimizing the mapping between domain slices and E2E slices, as well as mapping between domain slices and physical infrastructure resources is a future work challenge.

Since the optimal allocation depends on the current monitored state-of-the-network, the mapping from E2E slices to domain slices should be dynamic and in real time. Key network features have to be identified in order to define trigger points when the current allocation becomes no longer optimal. Some key features examples are number of active users, utilization of the links, and router queue size. The first challenge is to determine how often to monitor the data. Furthermore, trigger point threshold values have to be carefully chosen. Reallocation of domain slices imposes changes in the resources allocated to each service. Doing this too often is not preferred because it may result in reordering of packets, jitter, and delay. On the other hand, doing the reallocations seldom may result in overlooking some significant network state changes, which further leads to nonoptimal QoE levels and thus user dissatisfaction.

The number of users and services is so huge that manual management and reallocation is not feasible. Ideally, an automated agent that monitors the features and performs an optimal number of reallocations is needed. The agent is part of the QoE Manager (Figure 5.5). Since this is a future vs. current state cost and benefit trade-off, the agent can take advantage of reinforcement learning. Based on a set of key features, the agent can determine the current network state, monitor the changes in the state, and elaborate on what was done the last time when the system was in the same or a similar state. It would be trained in the feedback loop where actions with positive reward would improve the SLAs, while the negative would worsen or break the SLAs. In this context, actions refer to the rearrangement of the domain slices allocated to the E2E slice representing one service class. The challenges identified in this section represent future work in the area of applying ML to automated QoE-based and user-centric networks.

5.4 Conclusions and Future Directions

In the 5G context, paradigms like SDN, NFV, and network slicing contribute to a complex architecture which aims at providing a high degree of programmability and automation to accommodate a wide range of services that are sensitive to different network characteristics and have different temporal dynamics. Additionally, operators want to maximize resource efficiency and service-awareness to optimize costs while maintaining user satisfaction.

The architecture is composed of several layers that feature MANO entities within and between each other. These components provide network operators and service providers with the opportunity to optimize the entire stack, ranging from service quality assessment via QoE estimation, to resource efficiency maximization via optimized resource allocation and resource mapping between layers.

While the high degree of complexity often makes traditional optimization mechanisms inapplicable due to a rapidly growing state space, recent advances regarding AI and ML methods have demonstrated their suitability for approaching network-related problems. In this chapter, we highlight how AI and ML can help with the aforementioned tasks by providing insights from our recent studies. First, we discuss the general workflow and point out important design considerations and challenges when approaching the above tasks with ML. These challenges include the acquisition of realistic data sets, appropriate system models, and the required combination of ML-related and domain-specific expertise. In addition to demonstrating the applicability of the ML-based mechanisms, we provide insights into operational aspects such as feature importance, sensitivity to different parameters, and generalizability. Such investigations help operators in assessing the feasibility of the proposed approaches in the context of their particular network and traffic conditions. For instance, we find that a small subset of features is already sufficient for making reliable QoE estimations, allowing operators to minimize the overhead for collecting and processing monitoring data. This example also highlights the important roles that domain knowledge and careful feature engineering play. In the case of VNF deployment decisions, we demonstrate that models generalize well across network topologies, i.e. they can be used to make decisions in new environments without explicit retraining.

Finally, we give an outlook on new directions that have a high potential regarding efficiency improvements and ways to approach them with ML-based techniques that integrate and extend existing mechanisms. In particular, we focus on identifying appropriate granularity levels for service differentiation and devising service-aware slice management strategies that target optimal mapping and sharing of resources.

Bibliography

1 Afolabi, I., Taleb, T., Samdanis, K. et al. (2018). Network slicing and softwarization: a survey on principles, enabling technologies, and solutions. *IEEE Communication Surveys and Tutorials* 20 (3): 2429–2453.

2 Boutaba, R., Salahuddin, M.A., Limam, N. et al. (2018). A comprehensive survey on machine learning for networking: evolution, applications and research opportunities. *Journal of Internet Services and Applications* 9 (1): 16.

3 Ordonez-Lucena, J., Ameigeiras, P., Lopez, D. et al. (2017). Network slicing for 5G with SDN/NFV: concepts, architectures, and challenges. *IEEE Communications Magazine* 55 (5): 80–87.

4 3GPP (2020). Technical Specification TS 23.501 - System architecture for the 5G System (5GS), Rev. 16.4.0. https://www.3gpp.org/DynaReport/23501.htm (accessed 16 April 2021).

5 3GPP (2019). Technical Specification TS 28.530 - 5G; Management and orchestration; Concepts, use cases and requirements, Rev. 16.1.0. https://www.3gpp .org/DynaReport/28530.htm (accessed 16 April 2021).

6 ETSI (2019). Group Specification GS ZSM 002 - Zero-touch network and Service Management (ZSM); Reference Architecture, Rev. 1.1.1. https://www .etsi.org/deliver/etsi_gs/ZSM/001_099/002/01.01.01_60/gs_zsm002v010101p.pdf (accessed 16 April 2021).

7 Linguaglossa, L., Lange, S., Pontarelli, S. et al. (2019). Survey of performance acceleration techniques for network function virtualization. *Proceedings of the IEEE* 107 (4): 746–764.

8 Kellerer, W., Kalmbach, P., Blenk, A. et al. (2019). Adaptable and data-driven softwarized networks: review, opportunities, and challenges. *Proceedings of the IEEE* 107 (4): 711–731.

9 Wang, M., Cui, Y., Wang, X. et al. (2018). Machine learning for networking: workflow, advances and opportunities. *IEEE Network* 32: 92–99.

10 Sivaraman, V., Narayana, S., Rottenstreich, O. et al. (2017). Heavy-hitter detection entirely in the data plane. *Symposium on SDN Research*, Santa Clara, CA, USA, April 2017, pp. 164–176.

11 Xie, J., Yu, F.R., Huang, T. et al. (2018). A survey of machine learning techniques applied to software defined networking (SDN): research issues and challenges. *IEEE Communication Surveys and Tutorials* 21 (1): 393–430.

12 He, M., Kalmbach, P., Blenk, A. et al. (2017). Algorithm-data driven optimization of adaptive communication networks. *International Conference on Network Protocols*, Toronto, Canada, October 2017, pp. 1–6.

13 Alawe, I., Ksentini, A., Hadjadj-Aoul, Y., and Bertin, P. (2018). Improving traffic forecasting for 5G core network scalability: a machine learning approach. *IEEE Network* 32 (6): 42–49.

14 Polese, M., Jana, R., Kounev, V. et al. (2018). Machine learning at the edge: a data-driven architecture with applications to 5G cellular networks. *arXiv preprint arXiv:1808.07647*.

15 Bega, D., Gramaglia, M., Fiore, M. et al. (2019). DeepCog: cognitive network management in sliced 5G networks with deep learning. *IEEE Conference* *on Computer Communications (IEEE INFOCOM)*, Paris, France, May 2019, pp. 280–288.

16 Blenk, A., Kalmbach, P., Van Der Smagt, P. et al. (2016). Boost online virtual network embedding: using neural networks for admission control. *International Conference on Network and Service Management*, Montreal, Canada, October 2016, pp. 10–18.

17 Mijumbi, R., Gorricho, J.-L., Serrat, J. et al. (2014). Design and evaluation of learning algorithms for dynamic resource management in virtual networks. *IEEE Network Operations and Management Symposium (NOMS)*, Krakow, Poland, May 2014, pp. 1–9.

18 Jmila, H., Khedher, M.I., and El Yacoubi, M.A. (2017). Estimating VNF resource requirements using machine learning techniques. *International Conference on Neural Information Processing*, Guangzhou, China, November 2017, pp. 883–892.

19 Mestres, A., Rodriguez-Natal, A., Carner, J. et al. (2017). Knowledge-defined networking. *ACM SIGCOMM Computer Communication Review* 47: 2–10.

20 Kim, H., Lee, D., Jeong, S. et al. (2019). Machine learning-based method for prediction of virtual network function resource demands. *2019 IEEE Conference on Network Softwarization (NetSoft)*, IEEE, pp. 405–413.

21 Moradi, F., Stadler, R., and Johnsson, A. (2019). Performance prediction in dynamic clouds using transfer learning. *2019 IFIP/IEEE Symposium on Integrated Network and Service Management (IM)*, IEEE, pp. 242–250.

22 Mijumbi, R., Hasija, S., Davy, S. et al. (2017). Topology-aware prediction of virtual network function resource requirements. *IEEE Transactions on Network and Service Management* 14 (1): 106–120.

23 Lange, S., Kim, H.-G., Jeong, S.-Y. et al. (2019). Predicting VNF deployment decisions under dynamically changing network conditions. *International Conference on Network and Service Management*, Halifax, Canada, October 2019, pp. 1–9.

24 Wassermann, S., Seufert, M., Casas, P. et al. (2019). Let me decrypt your beauty: real-time prediction of video resolution and bitrate for encrypted video streaming. *2019 Network Traffic Measurement and Analysis Conference (TMA)*, Paris, France, June 2019, IEEE, pp. 199–200.

25 Mangla, T., Halepovic, E., Ammar, M., and Zegura, E. (2018). eMIMIC: Estimating HTTP-based video QoE metrics from encrypted network traffic. In *IEEE TMA*, Vienna, Austria, June 2018, pp. 1–8.

26 Seufert, M., Casas, P., Wehner, N. et al. (2019). Stream-based machine learning for real-time QoE analysis of encrypted video streaming traffic. *IEEE ICIN*, Paris, France, February 2019, pp. 76–81.

27 Dimopoulos, G., Leontiadis, I., Barlet-Ros, P., and Papagiannaki, K. (2016). Measuring video QoE from encrypted traffic. *ACM IMC*, Santa Monica, California, USA, November 2016, pp. 513–526.

28 Wassermann, S., Seufert, M., Casas, P. et al. (2019). I see what you see: real time prediction of video quality from encrypted streaming traffic. *ACM Internet-QoE*, Los Cabos, Mexico, October 2019, pp. 1–6.

29 Khokhar, M., Ehlinger, T., and Barakat, C. (2019). From network traffic measurements to QoE for internet video. *IFIP Networking Conference*, Warsaw, Poland, May 2019, pp. 1–9.

30 Orsolic, I., Pevec, D., Suznjevic, M., and Skorin-Kapov, L. (2017). A machine learning approach to classifying YouTube QoE based on encrypted network traffic. *Multimedia Tools and Applications* 76 (21): 22267–22301.

31 Oršolić, I., Rebernjak, P., Sužnjević, M., and Skorin-Kapov, L. (2018). In-network QoE and KPI monitoring of mobile YouTube traffic: insights for encrypted iOS flows. *IEEE CNSM*, Rome, Italy, November 2018, pp. 233–239.

32 Lin, Y., Oliveira, E., Jemaa, S., and Elayoubi, S. (2017). Machine learning for predicting QoE of video streaming in mobile networks. *IEEE ICC*, Paris, France, May 2017, pp. 1–6.

33 Begluk, T., Baraković Husić, J., and Baraković, S. (2018). Machine learning-based QoE prediction for video streaming over LTE network. *IEEE INFOTEH*, East Sarajevo, Bosnia and Herzegovina, March 2018, pp. 1–5.

34 Wang, Y., Li, P., Jiao, L. et al. (2016). A data-driven architecture for personalized QoE management in 5G wireless networks. *IEEE Wireless Communications* 24 (1): 102–110.

35 Schwarzmann, S., Marquezan, C.C., Bosk, M. et al. (2019). Estimating video streaming QoE in the 5G architecture using machine learning. *Internet-QoE Workshop on QoE-Based Analysis and Management of Data Communication Networks*, Los Cabos, Mexico, October 2019, pp. 7–12.

36 Schwarzmann, S., Marquezan, C.C., Trivisonno, R. et al. (2020). Accuracy vs. cost trade-off for machine learning based QoE estimation in 5G networks. *IEEE International Conference on Communications: Next-Generation Networking and Internet Symposium (IEEE ICC NGNI Symposium)*, Dublin, Ireland, June 2020, pp. 1–6.

37 Martin, A., Ega na, J., Flórez, J. et al. (2018). Network resource allocation system for QoE-aware delivery of media services in 5G networks. *IEEE Transactions on Broadcasting* 64 (2): 561–574.

38 Yan, M., Feng, G., Zhou, J. et al. (2019). Intelligent resource scheduling for 5G radio access network slicing. *IEEE Transactions on Vehicular Technology* 68 (8): 7691–7703.

39 Li, R., Zhao, Z., Sun, Q. et al. (2018). Deep reinforcement learning for resource management in network slicing. *IEEE Access* 6: 74429–74441.

40 3GPP (2019). Technical Specification TS 23.288 - Architecture Enhancements for 5G Systems (5GS) to Support Network Data Analytics Services, Rev. 16.1.0. https://www.3gpp.org/DynaReport/23288.htm (accessed 16 April 2021).

41 Varga, A. (2010). OMNeT++. In: *Modeling and Tools for Network Simulation*, (K. Wehrle, M. Güneş, J. Gross) 35–59. Springer.

42 Tibshirani, R. (1996). Regression shrinkage and selection via the lasso. *Journal of the Royal Statistical Society: Series B (Methodological)* 58 (1): 267–288.

43 Raake, A., Garcia, M., Robitza, W. et al. (2017). A bitstream-based, scalable video-quality model for HTTP adaptive streaming: ITU-T P. 1203.1. *IEEE QoMEX*, Erfurt, Germany, June 2017, pp. 1–6.

44 Lange, S., Kim, H.-G., Jeong, S.-Y. et al. (2019). Machine learning-based prediction of VNF deployment decisions in dynamic networks. *Asia-Pacific Network Operations and Management Symposium (APNOMS)*, Matsue, Japan, September 2019, pp. 1–6.

45 Bari, Md.F., Chowdhury, S.R., Ahmed, R. et al. (2015). On orchestrating virtual network functions. *International Conference on Network and Service Management*, Barcelona, Spain, November 2015, pp. 50–56.

46 Lange, S., Grigorjew, A., Zinner, T. et al. (2017). A multi-objective heuristic for the optimization of virtual network function chain placement. *International Teletraffic Congress*, Genoa, Italy, September 2017, pp. 152–160.

47 Sieber, C., Schwarzmann, S., Blenk, A. et al. (2020). Scalable application- and user-aware resource allocation in enterprise networks using end-host pacing. *ACM Transactions on Modeling and Performance Evaluation of Computing Systems* 5 (3): 1–41.

48 Lønsethagen, H., Vazquez-Castro, A., Gil-Casti neira, F. et al. (2016). Service Level Awareness and open multi-service internetworking. *NetWorld 2020 White Paper*.

6

Machine Learning for Resource Allocation in Mobile Broadband Networks

Sadeq B. Melhem, Arjun Kaushik, Hina Tabassum, and Uyen T. Nguyen

Department of Electrical Engineering and Computer Science, York University, Toronto Ontario, Canada

6.1 Introduction

The demand for extremely high data rates, ultra-reliability, and low latency in wireless networks is increasing due to the proliferation of smart phones, video streaming, social networks (e.g. Facebook), massive machine type communications (mMTCs), vehicular connectivity, and so on. To date, communication system designers have mainly relied on conventional non-data driven optimization to manage network resource allocations. However, due to the complexity of channel propagation and blockages in large-scale multi-band wireless networks,[1] the traditional network optimization and radio resource management (RRM) solutions are generally not applicable [1]. The reason is that the traditional solutions are not scalable and do not consider the unique channel propagation issues in various frequency bands [2, 3]. Subsequently, integrating artificial intelligence (AI) in conventional RRM algorithms becomes a necessity.

AI can be defined as a study that enables a machine to solve a problem efficiently by itself. In other words, AI can be described as any mechanism that observes from the perceived environment and exploits the observation results to solve a specific problem. As such, AI can help to create a smart network that will be able to learn, realize, and proactively allocate resources. Machine learning (ML) is among the most potent AI techniques that can automate the next generation wireless networks by enabling autonomous connections to a variety of spectral bands such as millimeter-wave and terahertz frequencies, autonomous regulation of transmission powers via energy learning [4, 5], autonomous transmission

---

1 The term "multi-band" refers to a combination of conventional low frequency (sub 6 GHz) and high frequency (millimeter-wave, terahertz, and visible light) communication networks.

Communication Networks and Service Management in the Era of Artificial Intelligence and Machine Learning,
First Edition. Edited by Nur Zincir-Heywood, Marco Mellia, and Yixin Diao.

protocols with quality of service (QoS) learning, and so on. ML can offer a smart decision via learning the characteristics of data traffic, management, controls, and other characteristics automatically and master knowledge in network operations and maintenance. The potential of ML increases further with the massive information and big data sets that can be collected from the communication nodes and access points (APs).

To this end, this chapter delivers a comprehensive overview of existing ML techniques that have been applied to date to wireless networks and discuss their benefits, shortcomings, and application scenarios. We then provide an in-depth survey of existing ML techniques in the context of wireless channel and power allocation, user scheduling, and user association. Finally, we list the key performance metrics of emerging 6G wireless networks and discuss potential ML techniques in 6G such as transfer learning, imitation learning, federated-edge learning, and quantum ML.

6.2 ML in Wireless Networks

This section will discuss different ML techniques and their applications to wireless network problems. Figure 6.1 shows a categorization of ML techniques.

6.2.1 Supervised ML

Supervised learning is an ML process that learns the features of output from input, provided the data set is labeled (i.e. the solution to the problem is known beforehand). Supervised learning implies the features of a labeled training data

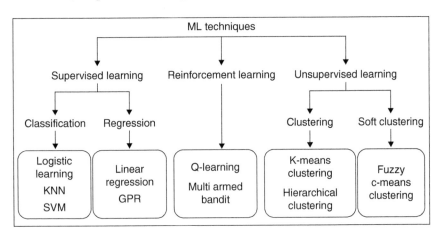

Figure 6.1 Brief overview of ML techniques.

set. It has been used for error detection, channel coding, and decoding in wireless networks [6]. Supervised learning can be divided into classification and regression based on the reliability of the output.

6.2.1.1 Classification Techniques

The goal of a classification algorithm is to predict which group or class label an object identifies in. Classification algorithms are mostly discrete and are used to group data into categories. Common classifier algorithms include logistic regression, K-nearest neighbors (KNN), and support vector machine (SVM).

- **Logistic regression** is a method that can predict the probability of binary classification (1/0), and it can be regularized to prevent over-fitting. It can be updated to fit new data. However, it does not perform well when presented with nonlinear data. In [7], the authors presented an application of logistic regression to wireless communications. Their model consisted of a logistic regression technique used in the physical layer to detect spoofing attacks within the network. Without knowing the channel model, the proposed logistic regression method has shown to have a high detection accuracy and low computational complexity.
- **KNN** is a classifier ML technique that groups data into clusters. Clusters are made such that the data points in a cluster are similar, and the similarity factor is determined by using Euclidean distances to find the nearest neighbor. A weighted KNN model is proposed to predict the characteristics of network traffic and categorize different temporal and spatial patterns of radio resource margins [8]. This model of KNN could capture the dynamics of traffic loads to enable load balancing and increasing efficiency of radio resource utilization.
- **SVM** can classify data by finding a linear boundary that separates data points from one class to another. SVM is suitable for nonlinearly separable data; it is simple to interpret, and has high accuracy. Nevertheless, SVM requires long processing time and training times. SVM is applied to classify base stations according to their traffic loads [9]. The authors used three different classes, class-1 (always loaded), class-2 (morning peak loaded), and class-3 (evening peak loaded) to classify the base stations. Call data records (CDR), which contain useful information such as user ID, cell ID, and geographical position of cell towers that a user was attached to were used as training data for SVM with prediction accuracy of 87.14%.

6.2.1.2 Regression Techniques

Regression algorithms can fit mathematical models to describe a set of data for approximation or interpolation. Common regression algorithms include linear regression and Gaussian process regression (GPR). These regression techniques can estimate relationships between variables and are powerful statistical tools for predicting and forecasting a continuous variable [10]. A regression model has been used to build a relationship between signal-to-interference-plus-noise ratio (SINR) and packet reception ratio [11]. With this model, an analytical framework was constructed to display the trade-offs between overhead and accuracy of interference measurements.

- **Linear regression** is a statistical modeling technique used to describe a continuous response as a function of one or more predictor variables. This technique is simple to interpret but may not capture complex patterns.
- **GPR** models can predict the value of a continuous response variable, and it is nonparametric. This algorithm is accurate for interpolating spatial data in the presence of uncertainty.

6.2.2 Unsupervised ML

Unsupervised ML is a process that explores a hidden structure from unlabeled data. It does not have a controller or supervisor which is labeled data, so the machine trains itself based on unlabeled input data. In addition, unsupervised learning uses cluster analysis on data sets in which the data is partitioned into categories based on similar characteristics. However, there are many challenges faced when using unsupervised learning as the data given to the machine does not contain any labels; as a result, the implementer may not know what the correct output should be. Therefore, evaluation of unsupervised learning algorithms must be analyzed manually. There are two types of cluster algorithms: hard clustering and soft clustering.

6.2.2.1 Clustering Techniques

Clustering algorithms group data points into one cluster such that no clusters overlap. Some popular clustering algorithms include *K*-means and hierarchical clustering.

- *K***-means clustering** divides data into *K* partitions of mutually exclusive clusters, hence the name. *K*-means clustering is a simple algorithm which clusters data points such that Euclidean distances from one data point to another is minimized. The algorithm continues this process to find a centeroid which is a data point having the minimal distance to its neighboring data points to create clusters. In [12], the contributors proposed a *K*-means algorithm for a hot spot clustering problem to maximize spectrum utilization. The *K*-means algorithm suggested clustering users based on locations and distances. The user with minimum distance from the centroid is selected as the hot spot for the cluster.

- **Hierarchical clustering** groups data into a binary hierarchical tree where clusters are formed to deduce similarities between the data points. This algorithm provides a visualization guide for cluster selection. However, it has cubic time complexity, causing it to be slower than other clustering algorithms. In [13], the researchers proposed a combination of hierarchical clustering and K-means clustering for user-activity analysis and user-anomaly detection. They were able to verify the genuine identity of users with dynamic spatiotemporal activities accurately.

6.2.2.2 Soft Clustering Techniques

Soft clustering unsupervised learning allows data points to coincide within different clusters/groups. The most common algorithm for soft clustering is fuzzy c-means. Fuzzy c-means partitions data into clusters where the data points may belong to one or more clusters. The data points closer to the centroid have a high membership value in that cluster and farther data points have a lower degree. Fuzzy c-means is beneficial for pattern recognition and for data points that share more than one characteristics. A fuzzy c-means algorithm is proposed for achieving energy efficiency in wireless sensor networks [14]. Fuzzy c-means is used to form an optimum number of clusters. These clusters are then used to elect cluster heads and eliminate redundant data generation and transmission to avoid loss of energy by turning off unnecessary sensors.

6.2.3 Reinforcement Learning

Reinforcement learning (RL) is the most popular technique in the context of resource allocation in wireless networks. Reinforcement learning relies on iterative learning and decision-making process. This technique was applied to coverage and capacity optimization and resource allocation [15]. The distinction between RL and supervised/unsupervised learning is that RL is based on a feedback system where a benefit is received if a correct decision is taken, or else a warning is given. The RL method will thus evolve constantly, while supervised/ unsupervised learning usually has static solutions. Reinforcement learning algorithms are generally of two types: Q-learning and multi-armed bandit techniques.

- **Q-Learning** is a commonly used reinforcement technique that interacts with the environment/problem to learn "Q" values. The Q values are considered to be the reward, which are tuples consisting of a state and action taken by the machine. Once the machine is trained, it has the ability to compute the actions which maximize Q values as a reward needs to be maximized. Q-learning is applied to allocate channel resources autonomously [16]. The Q-learning system consists of three main components, the reward R, channel state S, and channel action A. A is chosen so that free channels are selected for allocation. However, all actions have to be exhausted and updated. S indicates the channel state, which also contains information about the quality and usage of the channel, and the idleness in each time before the action chooses the channel. The machine is then trained with these constraints to build a table with the reward maximizing actions. Drawbacks of this method include scalability into larger networks with more channels and time complexity of taking all actions.

- **Stateless Q-learning**, another application of Q-learning is used, where the authors apply a stateless variation of Q-learning to improve spatial reuse in a wireless network [17]. The authors focused on allowing the network to modify both transmission power and channel resources based on experienced throughput. Stateless Q-learning is applied to a decentralized scenario where all information about wireless node (WN) is available, and each WNs is considered to be an agent running the stateless Q-learning algorithm. A greedy action selection strategy is used such that the actions correspond to all combinations, which can be chosen with respect to the channel and transmission power [17]. This method of Q-learning allows performance enhancements by utilizing reward maximizing actions. The algorithm is able to exploit optimal channel and power allocation.

- **Multi-armed bandit** takes the path of exploring and exploiting. This algorithm is used when there are multiple paths to select, and for each selection, a reward is given. Like Q-learning, the priority is to maximize reward, but there are trade-offs between taking the current path and gathering information to receive a larger reward. Multi-armed bandit was applied to dynamically select channels within a four-channel wireless local area networks (WLANs) [18]. Communication quality such as throughput is considered; the reward value and multi-armed bandit are used to maximize the throughput by either exploiting actions taken or exploring new ones. However, in depth exploration can result in losses in maximizing reward while quick decisions of exploiting actions may cause missing the best choice. Therefore, the authors proposed an ultra-fast multi-armed bandit solution that utilizes chaotically oscillating waveforms to achieve a latency of 1 ns. This system was implemented in a four channel WLAN, which successfully demonstrated adaptive and autonomous channel selection to improve throughput.

6.2.4 Deep Learning

Deep learning (DL) algorithms are based on artificial neural networks (ANNs) that are capable of modeling and processing nonlinear relationships. DL depends on large amounts of data and processing abilities, which in turn provides multilayer models that learn efficient representations of data from unstructured sources. Also, DL algorithms can handle large amounts of data, which can lead to better prediction accuracy. DL is able to learn features at its different levels and allow the target system to learn complex functions. However, the biggest drawback of DL is that it requires large amounts of data for training and testing, which may not be available and difficult to generate (especially when the data refers to "labels" in supervised learning).

DL can be used to find optimal solutions for wireless resource allocation problems [19]. DL was combined with a supervised learning technique's outputs to train a deep neural network (DNN). This has been shown to produce optimal solutions with low computational complexity [20]. The authors used this form of a DL network (DNN) to solve a power allocation problem in an interference channel, with performance being very close to weighted minimum mean-squared error (WMMSE).

Another application of DL for wireless resource allocation presented in [21]. DL is combined with reinforcement learning to produce a deep Q-learning (DQL) based spectrum allocation algorithm, which is applied to a wireless IoT network consisting of primary and secondary users. Primary users were given fixed power control, whereas the secondary users could adjust their power by autonomously learning about the shared common spectrum using the proposed DQL algorithm.

6.2.5 Summary

ML techniques are summarized in Table 6.1. Since learning accuracy is strongly dependent on the representation of the training data, certain supervised learning methods might be computationally intensive and not scalable. Moreover, the output quality derived from the outcomes of supervised learning is limited by the quality of the labels. Reinforced Q-learning is a good choice to optimize resource allocation via the action, state, and reward methodology. However, the reinforcement learning becomes computationally intensive with the increase in the number of actions and states. A combination of supervised learning and other ML techniques (RL or DL) would thus be beneficial such as the algorithm proposed in [21] for cellular IoT networks. Unsupervised ML is typically beneficial in classifying in–out data into various groups based on data delivery and used for resource management, resource allocation, and smart caching in wireless communication [22].

6.3 ML-Enabled Resource Allocation

The dynamic nature of wireless networks necessitates regular re-execution of resource allocation algorithms. However, the quality of RRM algorithms may

Table 6.1 Summary of ML techniques.

References	Objective	Methodology	Main conclusion	Scenario
[12]	Improve the maximization of spectrum utilization	K-Means clustering	K-Means proposed to cluster users based on locations and distances Provided optimal centroid selection for possible hot spot	Cellular network
[10]	Building relationship between SINR and packet reception ratio	Regression	Analytical framework created to display trade-offs between overhead and accuracy of interference measurements	Sensor networks
[8]	Predict characteristics of network traffic while categorizing different temporal and spatial patterns of radio resource margins	KNN	Proposed a weighted KNN that can capture the dynamics of traffic loads which enables load balancing	Wireless network
[17]	Improve spatial reuse in wireless networks using stateless Q-learning	Stateless Q-learning	The method addressed performance enhancements by utilizing reward maximizing actions. Method exploits optimal channel and power allocations	Wireless network
[18]	To dynamically select channels within a WLAN with low latency	Multi-armed bandit	ML technique achieved latency of 1 ns and adaptive channel selection to improve throughput	WLAN

degrade when they are deployed in real-life situations as they are generally based on mathematical models not the real-world data. Conventional resource management can be enhanced by capturing valuable user- and network-related knowledge and incorporating them in ML algorithms. In this section, we provide a review of ML algorithms used for power control, user scheduling, user association, and spectrum allocation.

6.3.1 Power Control

6.3.1.1 Overview
Power control in wireless networks refers to regulating a transmitter's power to achieve improved signal transmission or minimize interference. Efficient power control can decrease inter-user interference throughout the bandwidth, boost network reliability while retaining the QoS constraints in real-time settings, enhance the capacity in cellular networks, and increase device throughput. Power control is essential in the cellular network for interference and energy management since the battery capacity of devices is limited. Also, power control is needed for connectivity management because of wireless channel variations. The receiver must be able to retain a constant degree of transmitted signal so that it can remain connected to the transmitter and predict the channel state. Thus, power control allows preserving logical compatibility for the transmission of a particular signal [33, 34].

6.3.1.2 State-of-the-Art
Deep reinforcement learning has been used in wireless networks for power control due to its robust features, efficiency, and sufficient time processing. The authors in [23] devised a non-cooperative power allocation game in device-to-device (D2D) communication. They introduced three separate deep reinforcement learning algorithms. These algorithms consisted of DQL, dueling of DQL, and double DQL for multi-agent learning. Another deep-reinforcement learning approach was proposed to solve the dynamic problem of resource allocation within the non-orthogonal multi-carrier access framework [24].

A model-free distributed power allocation algorithm has been proposed [35]. A deep RL-based power allocation system has been built that can control the evolution of networks and perform with minimal dissemination of data in a distributed way. The approach assumed a central training system, where each transmitter functions as a learning agent. It investigated the unfamiliar environment with a greedy strategy. It then transferred this expertise to a control center to optimize the weighted efficiency sum-rate function, which could be defined to achieve the full sum-rate. In [25], the researchers provided a method to address a resource allocation problem in heterogeneous wireless networks. It manages the multi-agent network infrastructure to maximize the sum capacity of the network, thus reducing complexity and providing QoS to users. An ML technique combined with both SVM and deep neural network algorithms was implemented for joint power control and transmission scheduling in cellular networks [26].

6.3.1.3 Lessons Learnt

Table 6.2 summarizes power control algorithms based on ML techniques. Overall, DQL is recognized to be an effective ML technique applied in power control scenarios [23]. Here, Q-values could be expressed in a table format or by a network of neurons with specific overheads for memory and computation. Using deep reinforcement learning to estimate conventional high-complexity power algorithms is a useful approach to perform real-time power allocation and to deal with continuous action issues. Finally, multi-agent learning methods for non-cooperative issues and advanced DL techniques are promising approaches to improve learning accuracy, reduce the trial error, and reduce the complexity of the algorithm.

6.3.2 Scheduling

6.3.2.1 Overview

Scheduling is a critical aspect of wireless networks, which enables deciding which users are going to transmit using a specified time/frequency/antenna resource. The scheduling of radio resources is a mechanism where frames of resources are allocated to users. Numerous studies have been published in the literature for wireless scheduling algorithms. Also, with the massive connectivity requirements, wireless networks would have to support not only individual users but possibly a large number of low-cost machines and sensors as well. These machines would have traffic features distinct from ordinary human applications. For instance, a sensor may wake up, relay its measurements across a wireless network, and go back to sleep. In this context, AI techniques have the potential to achieve better efficiency over traditional methods.

6.3.2.2 State-of-the-Art

In [36], the authors introduced a control gradient-based scheduler to automate traffic flow to the cellular network. Explicitly, they set the scheduling issue as a Markov decision process (MDP) and indicate that the process will respond dynamically to traffic differences, allowing mobile networks to accommodate more data traffic, an increase of 14.7% compared with existing schedulers, while outperforming by a factor of 2. Likewise, the study conducted in [37] used deep Q learning for roadside network scheduling. Relations between vehicle populations are structured as an MDP, including the sequence of acts, findings, and reward signals.

Table 6.2 Machine learning-based power control.

References	Objective	Methodology	Main conclusion	Scenario
[23]	Improve network efficiency while maintaining the QoS limitations in real-time environments	DRL	Three separate deep reinforcement learning algorithms was introduced to maximize energy-efficient resource allocation	D2D
[24]	Optimizing the efficiency of the non-orthogonal multiple access multi-carrier framework	DRL	A deep reinforcement learning system with optimum power weights using an attention-based neural network is considered to resolve the channel assignment problem	non-orthogonal multiple access (NOMA) system
[35]	To optimize the weighted efficiency sum-rate function that could be defined to achieve the full sum-rate	Multi-agent DL	Proposed a model-free interference power allocation algorithm that achieves comparable efficiency by implementing enhanced learning techniques to different complex wireless networks	Wireless network
[25]	Manage the multi-agent network infrastructure and the total difficulty of the distribution of resources	Q-Learning	The method addressed the optimization problem in dense heterogeneous wireless networks as a distributed solution, thus reducing complexity	HetNets
[26]	Performance optimization power management in cellular networks	SVM	ML technique combined with both SVM and deep belief neural network algorithms to address a nonlinear problem	Wireless network
[27]	Femtocell trains itself to change its transmitting capacity to support its consumer service while preserving the macrocell consumer	Q-Learning	Solve the issue of power optimization in compact heterogeneous networks thus substantially reducing resource consumption	HetNets

By approximating the Q value function, the agent discovers a scheduling strategy that results in lower latency and compared to conventional scheduling approaches.

In [38], the contributors addressed simultaneously the issues of user scheduling and caching of content. In particular, they train a deep reinforcement learning (DRL) agent to determine which base station will service such content and whether the content should be cached. Besides, an observer is appointed to measure the value function and supply the performer with input. Simulations over a base station cluster demonstrate that the agent can provide a low delay in transmission. In [26], the researchers studied link scheduling and power management in wireless networks for performance optimization. They decomposed the initial problem into a linear and nonlinear system and introduced an ML algorithm combined with both SVM and deep-seated neural network to solve the nonlinear sub-problem.

A deep-Q learning method to minimize energy usage in real-time networks was presented in [28]. An auto encoder (AE) is utilized in their proposal to approximate the Q function, and the system conducts experience replay to optimize the testing cycle and speed up convergence. Two neural network-based algorithms were implemented to prove that free slots in a multiple frequency time division multiple access (MF-TDMA) network can be accurately predicted using an approximation approach [30]. Scheduling algorithms based on ML are summarized in Table 6.3.

6.3.2.3 Lessons Learnt
DL techniques such as DQL are useful to solve scheduling problems. DQL can be used to schedule frequency scaling allowing an adequate amount of energy [28]. ML has lately been implemented for user scheduling and shown to offer significant efficiency over traditional techniques. Collaborative scheduling, which transfers information to automate user scheduling, is also useful in eliminating the interference. Nevertheless, certain nonideal connections can face very long latency, which is a problem for coordinated scheduling in large-scale networks. Moreover, the mobility can create channel prediction errors, which is another challenge for producing robust scheduling algorithms.

6.3.3 User Association

6.3.3.1 Overview
The association of users, i.e. to associate a user with a specific serving base station, impacts the efficiency of the network. Cellular networks are becoming heterogeneous mainly through the addition of small cells. Because of the different transmit powers of various base stations, user association measurements such as SINR

Table 6.3 blackMachine-learning based scheduling.

References	Objective	Methodology	Main conclusion	Scenario
[36]	Enables cellular networks to accommodate more data traffic, although outperforming better	DRL	Develop a regulation gradient-based scheduler to maximize traffic flow to the mobile network. They placed the scheduling concern for forecasting system performance	HetNets
[37]	Reduces the energy consumption	Q-Learning	Discovers a scheduling strategy that results in reduced latency and busy power consumption relative to conventional scheduling approaches	V2V
[38]	The operator can produce low delay in transmission	DRL	The RL method is introduced to improve the predictive policy with a view to reducing the average latency in transmission	Cellular networks
[26]	Minimizing ultra-reliable time of low latency contact	RL	A machine-based learning algorithm introduced combined with SVM and deep-belief network approaches to tackle the nonlinear sub-problem	Wireless networks
[28]	Reduce the use of power in real time networks	Deep Q-learning	Adopt a deep Q learning scheduling method for frequency scaling, to reduce power efficiency in real time applications	Real time systems
[29]	Reduce the overall delay costs and energy usage for all user	Deep Q-learning	Uses deep Q learning to simultaneously automate offloading decisions and the utilization of computing resources	Mobile edge computing
[30]	The number of collisions with other networks are reduced by half	Imitation learning	Presenting two neural network-based algorithms to show that an approximation method can reliably predict free slots in the MF-TDMA network	HetNets

can lead to significant load disparities and underused small cells [3]. A crucial missing component in current association metrics seems to be the load that gives a vision of resource allocation, thereby impacting the long-term levels. In general, because of the unknown binary variables, identifying an appropriate load-aware user association solution is an issue involving the exponential complexity of nonlinear optimization. As such, AI can serve as a practical and less complex technique to obtain optimal global solutions. User association algorithms based on ML are summarized in Table 6.4.

6.3.3.2 State-of-the-Art
A RL-based algorithm has been introduced for the downlink of heterogeneous cellular networks to achieve optimum long-term total network efficiency while ensuring consumer QoS specifications [31]. A method of distributed optimization is being designed based on multi-agent RL to jointly optimize user association and resource allocation. There have been numerous solutions to the user association problem in millimeter-wave networks to optimize the use of cell services [42–45].

In [39], the authors designed a novel user-based ML technique to enable multi-connectivity in mmwave networks, where the user association challenge was defined as a multi-label classification task. Through implementing effective algorithms for different classifiers, the initial problem transformed into a set of the single-label classification tasks. Using spatiotemporal traffic flow in vehicular networks helps to create an online association framework that captures the load variation of different cells [40].

6.3.3.3 Lessons Learnt
Traditional solutions to the user association problem include (i) maximum SINR, which allows a user to select a base station that maximizes its SINR and (ii) gradient descent and dual decomposition methods. Li et al. [40] shows that reinforcement learning provides solutions that can respond more adaptively to the dynamics of wireless networks compared with the traditional solutions. A promising solution is to turn the user association problem into a sequential decision-making problem which can be solved using sequence-to-sequence learning. Another solution is to apply an RL strategy that will eliminate the reliance on the training samples, hence saving time and resources. Unsupervised learning algorithms for user association will also be of immediate relevance.

6.3.4 Spectrum Allocation

6.3.4.1 Overview
Spectrum allocation is the process of allocating resources to clients in a manner that achieves some goal (e.g. total data rate). That process is performed in

Table 6.4 Machine learning-based user association.

References	Objective	Methodology	Main conclusion	Scenario
[31]	Maximize the long-term downlink potential while maintaining the QoS specifications of the UEs	Multi-agent Q-learning	The multi-agent distributed DRL was proposed to achieve a mutually optimized user association for the HetNets	HetNets
[39]	To support multi-connectivity in millimetre wave (mmWave) networks	Multi-label classification	Proposed a machine-based computing method to address user-association problems in mmWave communication networks	mmWave networks
[32]	Identify user associations which can reduce the occurrence of breaks in presence	Federated learning	Develop federated learning algorithm based on echo state networks to identify the association of users	Wireless virtual reality networks
[40]	Conduct load balancing between the base stations	RL	Online RL algorithm considered to deliver reasonable service level for vehicles	Vehicular networks
[41]	Enhance association between users and base stations taking into account various factors that influence QoS	Collaborative filtering	Collaborative filtering based network was adopted that indeed balance the traffic load	HetNets

mobile networks by collecting the users' spectrum status information and then determining the optimum response. Exposure to the spectrum would be a crucial concern for future wireless networks. The model-dependent approaches cannot precisely accommodate real systems. As such, ML-based dynamic spectrum access approaches can be applied in a centralized manner to support a large number of users consuming spectrum [46]. ML-based spectrum allocation is summarized in Table 6.5.

6.3.4.2 State-of-the-Art

In [47], the study introduced a centralized, self-organizing approach for efficient spectrum control in a Cognitive Radio Vehicular Adhoc Network (CRAVENET) system. The method consists of five scale user demands for connectivity. Every user provides a decision-making input to the self-organizing director. The framework then automatically auto-organizes the spectrum allocation as per customer needs using a reinforcement learning technique. It helps to establish the self-organizing efficient management of spectrum, along with improved QoS in the CRAVENET system for significantly lower latency. Another study has addressed the complex spectral control problem in cognitive radio networks using DRL-based methods [48]. The authors have suggested a DRL and reservoir computing approach for shared spectrum access. The secondary user is capable of sensing all streams in real time and considering the sensing failure and conflict interaction with other secondary users. They showed that DRL can provide accurate channel status and robust predictions for complex spectrum allocation.

Deep Q-networks is presented to optimize secondary network performance [49]. Moreover, the use of deep Q-network has been shown to produce faster deployment and greater adaptability to complex wireless network architectures. In [50], the researchers implemented a proactive resource allocation algorithm. They applied long short-term memory cells at the unmanned aerial vehicle level, allowing for efficient distribution of radio resources and enabling small wireless networks to perform channel estimation and marginal spectrum allocation promptly.

6.3.4.3 Lessons Learnt

Since base stations are allowed to control spectrum resources, ML can be a viable technique for base stations to efficiently and effectively allocate spectrum resources. Also, algorithms such as Q-learning can be a feasible candidate as it has the ability to learn from its environment. It will allow base stations to predict reliable spectrum resource allocations. Spectrum sharing is an essential technique to increase the spectrum of resources in future wireless networks [52].

Table 6.5 Machine learning-based spectrum allocation.

References	Objective	Methodology	Main conclusion	Scenario
[47]	Helps to ensure economic and social justice in CRAVENET	RL	Methodology for controlling spectrum was built for QoS in CRAVENET system with small average latency	CRAVENET
[48]	DRL is used to learn about the availability of spectrum resources for test the performance	DRL	Deep Q-network approach is shown to converge faster with better performance comparing Q-learning	Cognitive radio networks
[49]	Maximize secondary network efficiency when fulfilling the main link interruption limit	Deep Q-networks	Using deep Q-network resulted in greater integration and improved the adaptability of cognitive radio networks	Cognitive radio networks
[50]	Estimate spectrum allocation with low signal latency using DRL	DRL	Using unlicensed spectrum for small long term evolution (LTE) cells thus maintaining compatibility for existing Wi-Fi networks and other LTE providers	HetNets
[51]	Boost the number of V2I connections and the packet transmission rate for V2V connections	Multi-agent RL	Enhanced spectrum utilization by upgrading Q-networks using the feedback obtained	Vehicular networks

6.4 Conclusion and Future Directions

This chapter provides a comprehensive overview of existing ML techniques that are crucial for wireless resource allocation, especially in the context of wireless channel and power allocation, user scheduling, and user association. We highlight the application scenarios where different ML techniques can be applied. We conclude that ML-based algorithms will enable NP-hard resource allocation problems to be solved in a scalable and cost-effective manner with low computational complexity. Block-chain-based spectrum sharing is an interesting solution to improving standard spectrum sharing efficiency and security. Another promising approach is quantum communications that can enhance computing efficiency and security. Significant learning and processing efficiency would be required to ensure cooperation across highly integrated 6G networks with multi-access edge computing, vehicular communication, massive-IoT, drones, integrated aerial-terrestrial and satellite networks, wireless body sensors in the human body.

ML techniques can fully automate physical layer architectures, decision-making, resource management, and resource utilization in 6G. Advanced ML techniques such as transfer learning, imitation learning, federated edge learning, and quantum learning will be the key highlights for resource allocation in 6G. In the following, we discuss potential ML techniques for 6G, such as transfer learning, imitation learning, federated-edge learning, and quantum ML.

6.4.1 Transfer Learning

Resource allocation is crucial in wireless networks for performance optimization. Nevertheless, due to the complexity of these optimization problems, they tend to become mixed integer nonlinear programming problems (MINLP), especially spectrum allocation, subchannel allocation, user association, and scheduling problems. Transfer learning via self-imitation can be applied to these problems, which significantly improves performance but suffer from task mismatch issues that occur when network parameters change [53]. Transfer learning is an ML technique where a model created from a previous task is reused as a starting point for a model on other tasks. Therefore, applying transfer learning eliminates the task of creating a model from scratch. Overall, this is a unique method of optimizing resource allocation as it is able to adapt to different network parameters with a significantly minimal number of training sets, thereby reducing training time.

6.4.2 Imitation Learning

Resource allocation in wireless networks is usually formulated using MINLP and branch and bound (BB) is a commonly used algorithm to solve such resource allocation problems. However, it converges slowly and has very high time complexities. The use of imitation learning improves the performance of BB by using a pruning policy that autonomously discards nonoptimal nodes [54]. The main goal of the algorithm is to find an optimal solution by comparing all solutions to one another. This method of imitation learning achieves good optimality and reduces computational time, but it relies on the result from the BB algorithm, which can be either good or poor. In addition, the proposed imitation learning algorithm speeds up the BB algorithm, but at times the speed-up rates are not high enough to see significant changes [54].

Imitation learning techniques aim to copy human behaviors in a given task. The machine (agent) is trained to complete tasks from previous demonstrations by learning a mapping between observations and actions. Imitation learning is also a combination of supervised learning and reinforced learning with no explicit labels. However, examples are provided on how to reach an object if needed. Imitation learning is more suitable for applications such as connected and autonomous vehicles.

6.4.3 Federated-Edge Learning

Federated learning is a distributed ML technique that allows training on a large amount of data residing on devices like mobile phones. It consists of two main components, the data owners (participants) and the model owner (federated learning server). Applications of federated ML include edge computing and caching. The primary role of cache is to improve processing time and computational time typically on computers and electronic devices. Frequently used data are stored in a cache and it can be accessed instantaneously. The primary constraint on a cache is its size limit. Edge computing, on the other hand, is a technology that allows information processing to be done close to the edge of a network, where objects and people produce or consume information.

Federated learning is considered to reduce backhaul traffic in wireless networks with caching and edge computing [55]. Federated learning trains a global model with local user data to determine the "popular information" to be stored in the cache. A federated model, after being trained, can make future predictions about what users tend to browse and enable applications to be accessed frequently. Furthermore, it allows applications that have strict delay and bandwidth to meet those requirements.

The biggest drawback in implementing distributed learning in a large population is privacy concerns of sharing user interactions with a network, as described by Niknam et al. [55]. Security and privacy challenges can be alleviated by modifying federated learning to include secure aggregation, which gives users a private space that is not revealed to other users. This provides privacy to the user (learner) rather than a single data packet. However, with added algorithms, performance and computational resources are sacrificed.

Overall, federated learning is a technique that can be applied in wireless network resource allocation, which can address energy, bandwidth, delay, and privacy concerns. Simulations of federated learning were carried out in a cache enabled network with augmented reality applications [55]. It will be interesting to see federated learning applied in wireless networks with different types of users as there are many constraints to take into consideration. These constraints include privacy concerns of individual data interacting with the network, overall convergence time as it would depend on training rates on user individual devices and the quality of the channel. In addition, as federated learning depends on decentralized data, the number of users willing to train a global model will also affect the performance.

6.4.4 Quantum Machine Learning

The 6G vision is a heavily linked promising paradigm expected to respond rapidly to user' requests over real-time network context learning as defined by the edge of the network (e.g. base station [BS] ranges and cache functionality), radio controller (e.g. communication spectrum and transmission link), and device capabilities (e.g. battery capacity and destinations). The multidimensional complexity of the entire network, which needs real-time information, can be provided as an issue of quantum uncertainty. The evolving concepts of ML, quantum computing (QC), and Q-learning techniques and their strategies with communication systems can be recognized as necessary in 6G. Quantum ML will play a significant role in utilizing available resources and large-scale data to allow smart interactions in 6G wireless communications. It will assist in all processes ranging from proactive caching to estimating a massive number of channels in cell-free massive-multiple input multiple output (MIMO).

Bibliography

1 Khalili, A., Akhlaghi, S., Tabassum, H., and Ng, D.W.K. (2020). Joint user association and resource allocation in the uplink of heterogeneous networks. *IEEE Wireless Communications Letters* 9 (6): 804–808.

2 Ibrahim, H., Tabassum, H., and Nguyen, U.T. (2020). The meta distributions of the SIR/SNR and data rate in coexisting Sub-6GHz and millimeter-wave cellular networks. *IEEE Open Journal of the Communications Society* 1: 1213–1229.

3 Sayehvand, J. and Tabassum, H. (2020). Interference and coverage analysis in coexisting RF and dense terahertz wireless networks. *IEEE Wireless Communications Letters* 9 (10): 1738–1742.

4 Thuc, T.K., Hossain, E., and Tabassum, H. (2015). Downlink power control in two-tier cellular networks with energy-harvesting small cells as stochastic games. *IEEE Transactions on Communications* 63 (12): 5267–5282.

5 Tabassum, H., Hossain, E., Hossain, Md.J., and Kim, D.I. (2015). On the spectral efficiency of multiuser scheduling in RF-powered uplink cellular networks. *IEEE Transactions on Wireless Communications* 14 (7): 3586–3600.

6 Singh, A., Thakur, N., and Sharma, A. (2016). A review of supervised machine learning algorithms. *2016 3rd International Conference on Computing for Sustainable Global Development (INDIACom)*, pp. 1310–1315.

7 Xiao, L., Wan, X., and Han, Z. (2018). PHY-layer authentication with multiple landmarks with reduced overhead. *IEEE Transactions on Wireless Communications* 17 1676–1687.

8 Feng, Z., Li, X., Zhang, Q., and Li, W. (2017). Proactive radio resource optimization with margin prediction: a data mining approach. *IEEE Transactions on Vehicular Technology* 66 1–2.

9 Hammami, S., Afifi, H., Marot, M., and Gautheir, V. (2016). Network planning tool based on network classification and load predicition, pp. 1–6 *Arxiv*.

10 Wang, J., Li, X., Jiang, C. et al. (2019). Thirty years of machine learning: the road to pareto-optimal wireless networks. *IEEE Communication Surveys and Tutorials* 1–46.

11 Chang, X., Huang, J., Liu, S. et al. (2016). Accuracy-aware interference modeling and measurements in wireless sensor networks. *IEEE Transactions on Mobile Computing* 270–290.

12 Sun, Y., Peng, M., Zhou, Y. et al. (2019). Application of machine learning in wireless networks: key techniques and open issues, pp. 5–6.

13 Parwez, S., Rawat, D., and Garuba, M. (2017). Big data analytics for user-activity analysis and user-anomaly detection in mobile wireless network. *IEEE Transactions on Industrial Informatics* 13 20581–2065.

14 Jain, A. and Goel, A. (2018). Energy efficient algorithm for wireless sensor network using fuzzy C-means clustering. *International Journal of Advanced Computer Science and Applications* 9 1–8.

15 Wu, C., Yoshinaga, T., Chen, X. et al. (2018). Cluster-based content distribution integrating LTE and IEEE 802.11p with fuzzy logic and Q-learning. *IEEE Computational Intelligence Magazine* 13 (1): 41–50.

16 Guan, M., Wu, Z., Cui, Y. et al. (2019). An intelligent wireless channel allocation in HAPS 5G communication system based on reinforcement learning. *EURASIP Journal on Wireless Communications and Networking* 138.

17 Wilhelmi, F., Bellata, B., Cano, C., and Jonsson, A. (2017). Implications of decentralized Q-learning resource allocation in wireless networks, pp. 1–5 *Arxiv.org*.

18 Takeuchi, S., Hasegawa, M., Kanno, K. et al. (2020). Dynamic channel selection in wireless communication via a multi-armed bandit algorithm using laser chaos time series. *Scientific Reports* 1574.

19 Ahmed, K.I., Tabassum, H., and Hossain, E. (2019). Deep learning for radio resource allocation in multi-cell networks. *IEEE Network* 33 (6): 188–195.

20 Liang, L., Ye, H., Yu, G., and Li, G. (2019). Deep learning based wireless resource allocation with application to vehicular networks, pp. 1–14 *Arxiv.org*.

21 Hussain, F., Hussain, R., Hassan, S., and Hossain, E. (2020). Machine learning for resource management in cellular and IoT networks: potentials, current solutions and open challenges. *IEEE Communication Surveys and Tutorials* 22 1–26.

22 Jiang, C., Zhang, H., Ren, Y. et al. (2017). Machine learning paradigms for next-generation wireless networks. *IEEE Wireless Communications* 24 (2): 98–105.

23 Nguyen, K.K., Duong, T.Q., Vien, N.A. et al. (2019). Non-cooperative energy efficient power allocation game in D2D communication: a multi-agent deep reinforcement learning approach. *IEEE Access* 7: 100480–100490.

24 He, C., Hu, Y., Chen, Y., and Zeng, B. (2019). Joint power allocation and channel assignment for NOMA with deep reinforcement learning. *IEEE Journal on Selected Areas in Communications* 37 (10): 2200–2210.

25 Amiri, R., Mehrpouyan, H., Fridman, L. et al. (2018). A machine learning approach for power allocation in HetNets considering QoS. *2018 IEEE International Conference on Communications (ICC)*, pp. 1–7.

26 Cao, X., Ma, R., Liu, L. et al. (2018). A machine learning-based algorithm for joint scheduling and power control in wireless networks. *IEEE Internet of Things Journal* 5 (6): 4308–4318.

27 Amiri, R., Almasi, M.A., Andrews, J.G., and Mehrpouyan, H. (2019). Reinforcement learning for self organization and power control of two-tier heterogeneous networks. *IEEE Transactions on Wireless Communications* 18 (8): 3933–3947.

28 Zhang, Q., Lin, M., Yang, L.T. et al. (2019). Energy-efficient scheduling for real-time systems based on deep Q-learning model. *IEEE Transactions on Sustainable Computing* 4 (1): 132–141.

29 Li, J., Gao, H., Lv, T., and Lu, Y. (2018). Deep reinforcement learning based computation offloading and resource allocation for MEC. *2018 IEEE Wireless Communications and Networking Conference (WCNC)*, pp. 1–6.

30 Mennes, R., Camelo, M., Claeys, M., and Latré, S. (2018). A neural-network-based MF-TDMA MAC scheduler for collaborative wireless networks. *2018 IEEE Wireless Communications and Networking Conference (WCNC)*, pp. 1–6.

31 Zhao, N., Liang, Y., Niyato, D. et al. (2019). Deep reinforcement learning for user association and resource allocation in heterogeneous cellular networks. *IEEE Transactions on Wireless Communications* 18 (11): 5141–5152.

32 Chen, M., Semiari, O., Saad, W. et al. (2020). Federated echo state learning for minimizing breaks in presence in wireless virtual reality networks. *IEEE Transactions on Wireless Communications* 19 (1): 177–191.

33 Chiang, M., Hande, P., Lan, T., and Tan, C.W. (2008). Power control in wireless cellular networks. *Foundations and Trends® in Networking* 2 (4): 381–533. http://dx.doi.org/10.1561/1300000009.

34 Umoren, I.A., Shakir, M.Z., and Tabassum, H. (2020). Resource efficient vehicle-to-grid (V2G) communication systems for electric vehicle enabled microgrids. *IEEE Transactions on Intelligent Transportation Systems*.

35 Nasir, Y.S. and Guo, D. (2019). Multi-agent deep reinforcement learning for dynamic power allocation in wireless networks. *IEEE Journal on Selected Areas in Communications* 37 (10): 2239–2250.

36 Chinchali, S., Hu, P., Chu, T. et al. (2018). Cellular network traffic scheduling with deep reinforcement learning. *AAAI*.

37 Atallah, R., Assi, C., and Khabbaz, M. (2017). Deep reinforcement learning-based scheduling for roadside communication networks. *2017 15th International Symposium on Modeling and Optimization in Mobile, Ad Hoc, and Wireless Networks (WiOpt)*, pp. 1–8.

38 Wei, Y., Zhang, Z., Yu, F.R., and Han, Z. (2018). Joint user scheduling and content caching strategy for mobile edge networks using deep reinforcement learning. *2018 IEEE International Conference on Communications Workshops (ICC Workshops)*, pp. 1–6.

39 Liu, R., Lee, M., Yu, G., and Li, G.Y. (2020). User association for millimeter-wave networks: a machine learning approach. *IEEE Transactions on Communications* 68 1.

40 Li, Z., Wang, C., and Jiang, C. (2017). User association for load balancing in vehicular networks: an online reinforcement learning approach. *IEEE Transactions on Intelligent Transportation Systems* 18 (8): 2217–2228.

41 Meng, Y., Jiang, C., Xu, L. et al. (2016). User association in heterogeneous networks: a social interaction approach. *IEEE Transactions on Vehicular Technology* 65 (12): 9982–9993.

42 Kwon, G. and Park, H. (2019). Joint user association and beamforming design for millimeter wave UDN with wireless backhaul. *IEEE Journal on Selected Areas in Communications* 37 (12): 2653–2668.

43 Shen, L., Chen, Y., and Feng, K. (2020). Design and analysis of multi-user association and beam training schemes for millimeter wave based WLANs. *IEEE Transactions on Vehicular Technology* 69 1.

44 Mesodiakaki, A., Adelantado, F., Alonso, L. et al. (2017). Energy- and spectrum-efficient user association in millimeter-wave backhaul small-cell networks. *IEEE Transactions on Vehicular Technology* 66 (2): 1810–1821.

45 Su, Z., Ai, B., Lin, Y. et al. (2018). User association and wireless backhaul bandwidth allocation for 5G heterogeneous networks in the millimeter-wave band. *China Communications* 15 (4): 1–13.

46 Shafin, R., Liu, L., Chandrasekhar, V. et al. (2020). Artificial intelligence-enabled cellular networks: a critical path to beyond-5G and 6G. *IEEE Wireless Communications* 27 (2): 212–217.

47 Ghanshala, K.K., Sharma, S., Mohan, S. et al. (2018). Self-organizing sustainable spectrum management methodology in cognitive radio vehicular adhoc network (CRAVENET) environment: a reinforcement learning approach. *2018 1st International Conference on Secure Cyber Computing and Communication (ICSCCC)*, pp. 168–172.

48 Chang, H., Song, H., Yi, Y. et al. (2019). Distributive dynamic spectrum access through deep reinforcement learning: a reservoir computing-based approach. *IEEE Internet of Things Journal* 6 (2): 1938–1948.

49 Shah-Mohammadi, F. and Kwasinski, A. (2018). Deep reinforcement learning approach to QoE-driven resource allocation for spectrum underlay in cognitive radio networks. *2018 IEEE International Conference on Communications Workshops (ICC Workshops)*, pp. 1–6.

50 Challita, U., Dong, L., and Saad, W. (2018). Proactive resource management for LTE in unlicensed spectrum: a deep learning perspective. *IEEE Transactions on Wireless Communications* 17 (7): 4674–4689.

51 Liang, L., Ye, H., and Li, G.Y. (2019). Spectrum sharing in vehicular networks based on multi-agent reinforcement learning. *IEEE Journal on Selected Areas in Communications* 37 (10): 2282–2292.

52 Monemi, M. and Tabassum, H. (2020). Performance of UAV-assisted D2D networks in the finite block-length regime. *IEEE Transactions on Communications* 68.

53 Shen, Y., Shi, Y., Zhan, J., and Letaief, K. (2018). Transfer learning for mixed integer resource allocation problems in wireless networks, pp. 1–7 *Arxiv.org*.

54 Lee, M., Yu, G., and Li, G. (2019). Learning to branch: accelerating resource allocation in wireless networks, pp. 1–13 *Arxiv.org*.

55 Niknam, S., Dhillon, H., and Reed, J. (2020). Federated learning for wireless communications: motivations, opportunities and challenges, pp. 1–7 *Arxiv.org*.

7

Reinforcement Learning for Service Function Chain Allocation in Fog Computing

José Santos, Tim Wauters, Bruno Volckaert, and Filip De Turck

Department of Information Technology, Ghent University – imec, IDLab, Gent, Technologiepark-Zwijnaarde, Oost-vlaanderen, Belgium

7.1 Introduction

With the advent of the Internet of Things (IoT), distributed cloud architectures have become a potential business opportunity for most cloud providers [1]. Low-latency and high mobility constraints are among the strictest requirements imposed by IoT services, making centralized cloud solutions impractical. In response, cloud computing evolved toward a novel paradigm called Fog Computing (FC) [2], where a distributed cloud infrastructure is set up to provide services close to end users. Furthermore, micro-services are currently revolutionizing the way developers build their software applications [3]. An application is decomposed in a set of self-contained containers deployed across a large number of servers instead of the traditional single monolithic application. In fact, containers are the most promising alternative to the conventional Virtual Machines (VMs), due to their low overhead and high portability. Nevertheless, several challenges in terms of resource provisioning and service scheduling persist which prevent service providers and end users from fully benefiting from micro-service patterns. One key challenge that remains is called Service Function Chaining (SFC) [4], where services are connected in a specific order, forming a service chain that each request needs to traverse to access a particular Network Service (NS). For instance, a service chain can be composed of an Application Programming Interface (API), a database, and a Machine Learning (ML) service. Sensors access the API to send their data to the infrastructure while users access the database service to retrieve the sensor's collected data. This data may have already been filtered and modified by an ML service. In fog–cloud environments, the interactions between fog locations and cloud are crucial to ensure that services operate properly due to the hierarchical architecture. For example, the database

Communication Networks and Service Management in the Era of Artificial Intelligence and Machine Learning,
First Edition. Edited by Nur Zincir-Heywood, Marco Mellia, and Yixin Diao.

service must be allocated close to the users in a fog location, but the ML service could be instantiated in the cloud where more computing resources are available. We need proper provisioning strategies to ensure both services are allocated close enough so that users do not experience latency in accessing the inferred results. These chain requirements (e.g. service location, low-latency, minimum available bandwidth) must be guaranteed during SFC allocation in FC, which are currently not being studied since SFC concepts are still mostly unexplored in FC environments.

Although the theoretical foundations of FC have already been established, the adoption of its concepts is still in its early stages, and practical implementations are still scarce. Furthermore, current studies on resource allocation are mainly focused on theoretical modeling and heuristic-based solutions, which in most cases cannot cope with the dynamic behavior of the network and leads to poor resource utilization and scalability issues. In fact, resource allocation is a difficult online decision-making problem where appropriate actions depend on fully understanding the network environment. Thus, in this chapter, we explore a subset of ML called Reinforcement Learning (RL) [5] to provide a suitable solution for SFC allocation in FC. The SFC allocation problem has been translated into an RL problem where the best resource allocation decisions (i.e. actions) are learned depending on the current status of the network infrastructure (i.e. environment). Based on a previously presented Mixed-Integer Linear Programming (MILP) formulation, an environment has been developed where agents learn to allocate service chains in FC directly from interacting with the environment without any knowledge or information at the beginning. Our results show that RL techniques perform comparably to state-of-the-art Integer Linear Programming (ILP)-based implementations but provide more scalable solutions.

In summary, FC is one of the most challenging topics in modern cloud computing, along with resource allocation and service chaining concepts. The rest of the chapter is organized as follows: Section 7.2 provides a brief overview of the technical background. Section 7.3 discusses the current state-of-the-art on resource allocation for FC. Section 7.4 presents the proposed RL approach for SFC allocation in FC, which is followed by the evaluation setup in Section 7.5. Next, in Section 7.6, results are shown. Finally, future research directions and open challenges are discussed in Section 7.7, concluding the chapter.

7.2 Technology Overview

This section provides a brief overview of the FC paradigm. Then, the fundamental concepts related to resource allocation and SFC are discussed. Finally, the main concepts of RL are introduced.

7.2.1 Fog Computing (FC)

The FC paradigm is an extension of cloud computing to provide resources on the edges of the network to deal with the exponential growth of connected devices [7]. Figure 7.1 presents a high-level view of the FC environment. In contrast to a centralized cloud, fog nodes are distributed across the network to act as an intermediate layer between end devices and the cloud. These so-called fog nodes, edge locations or even Cloudlets [8] are essentially small cloud entities, which bring processing power, storage procedures, and memory capacity closer to devices and end users to enable local operations. Cloud nodes are the traditional cloud servers where a high amount of resources is available.

7.2.2 Resource Provisioning

Resource provisioning or also known as resource allocation has been studied for years in the network management domain [9–11]. Resource provisioning is related to the allocation of computing, network and storage resources needed to instantiate services requested by users and devices over the Internet. In recent years, cloud providers and users have been working together toward an efficient manner of dealing with computing resources. On one hand, users want to receive the best Quality of Service (QoS) for the minimum cost while cloud providers want to increase their revenue. Users want to maximize their service plan without increasing their costs while cloud providers want to respect the agreed QoS level by using a minimum percentage of their infrastructure. Thus, energy efficiency is essential for cloud providers while low-latency is crucial for users. Reducing costs by

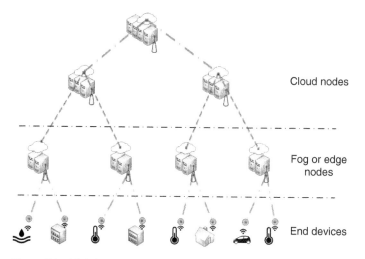

Cloud nodes

Fog or edge nodes

End devices

Figure 7.1 High-level view of a fog computing environment [6].

using the minimum amount of hardware while guaranteeing users QoS level, or increase the number of active nodes to reduce latency between the deployed service and the user to a minimum. Efficient allocation strategies are crucial for both cloud providers and users. Different provisioning policies can be applied depending on the status of the network infrastructure or the current user demand.

In addition, with the advent of FC, resource allocation has become an even more important research topic. FC has been introduced as an answer to the inherent provisioning challenges introduced by IoT services. For example, IoT services are highly challenging in terms of latency. Delay-sensitive services (e.g. connected vehicles, interactive video applications) require latencies in the order of milliseconds. If the latency increases, surpassing the communication threshold, the user connection can become unstable and the user control over the service is potentially lost. Also, since vehicles and users are continuously moving in the network area, mobility is another important factor to consider. Allocation strategies must consider service reallocations in case user connectivity is lost to ensure proper service operation at all times. Centralized infrastructures cannot fully satisfy the dynamic demands of these types of services. Therefore, FC is essential to rapidly modify the allocation of services according to highly variable demand patterns.

7.2.3 Service Function Chaining (SFC)

SFC placement [12, 13] has been studied in the network management domain during the last few years. SFC is related to the services' proper ordering ensuring that each user has to traverse the given service chain to access a particular NS as shown in Figure 7.2. The circles represent different service functions while the arrows show how the traffic is steered in the network. User requests are routed through the service chain according to a service graph, which aims to optimize resource allocation to further improve application performance. SFC enables cloud providers to dynamically reconfigure softwarized NSs without having to implement changes at the hardware level. SFC provides a flexible and reliable alternative to today's static network environment.

7.2.4 Micro-service Architecture

Recently, micro-service patterns [15] gained tremendous attention. An application is decomposed in a set of loosely coupled services that can be developed, deployed, and maintained independently. Each service is responsible for a single task and communicates with the other services through lightweight protocols. These services can then be developed in different programming languages and even using different technologies. Nowadays, containers are the most promising alternative to the traditional monolithic application paradigm, where almost everything is centralized and code-heavy.

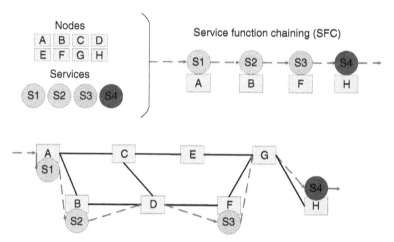

Figure 7.2 An example of a service function chaining deployment [14].

7.2.5 Reinforcement Learning (RL)

In recent years, RL methods have become an important area in ML research [16–18]. The typical scenario in RL is represented in Figure 7.3. In most cases, RL techniques are used to solve sequential decision-making problems. An RL agent learns to make better decisions directly from experience interacting with an environment. The environment represents the problem to solve. In the beginning, the agent knows nothing about the problem at hand and learns by performing actions in an environment. For each action taken, the agent receives a reward and a new observation that describes the new state of the environment. Depending on the goal and how well the agent is performing on the given task, the reward can be positive or negative. The agent learns to be successful by repeated interaction with the environment, by determining the inherent synergies between states, actions, and subsequent rewards. Ultimately, RL algorithms try to maximize the total reward an agent would collect by experiencing multiple problem rounds. For instance, let us consider an agent allocating resources in a cloud infrastructure.

Figure 7.3 The representation schema of most RL scenarios.

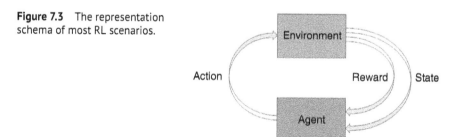

The agent would receive a reward for each action applied in the system. If the action translates into an appropriate allocation scheme, the agent will receive a positive reward. Otherwise, if the agent performs a bad action (e.g. terminate a service needed in the network), which produces an inappropriate allocation scheme, the reward would be negative. To maximize the reward, the agent would need to apply actions that translate into proper allocation schemes at all times. The ultimate goal in this scenario would be to train an agent able to learn good allocation actions to maximize performance and minimize costs.

7.3   State-of-the-Art

With the advent of FC concepts, efficient resource provisioning is needed in modern cloud infrastructures. This section provides a summary of the relevant previous studies concerning specifications and implementations of resource allocation strategies for fog–cloud infrastructures. First, research on resource allocation for FC is introduced, which is followed by recent advances in resource provisioning provided by ML methods. Finally, recent works on RL for resource allocation are highlighted.

7.3.1   Resource Allocation for Fog Computing

Recently, a handful of research efforts has been performed in the context of resource provisioning in FC environments. In [19], the authors proposed an allocation scheme to support crowdsensing applications in the context of IoT. Their approach has been formulated as a MILP model which takes cost-efficient provisioning and task distribution into account. Results confirmed that their proposal could outperform traditional cloud infrastructures. In [20], an optimization formulation for the QoS-aware deployment of IoT applications over fog infrastructures has been proposed and implemented as a prototype called FogTorch. Their approach focused not only on hardware and software demands but also on QoS requirements, such as network latency and bandwidth. Results showed that their algorithm ensures optimal service deployment while decreasing hardware capacity and increasing resource demands. Additionally, in [21], the IoT resource allocation problem in FC has been modeled as an ILP formulation, where QoS metrics and deadlines for the deployment of each application have been considered. Results proved that their formulation can prevent QoS violations and reduce costs when compared to a traditional cloud approach. Furthermore, in [22], a particle swarm optimization algorithm has been proposed for the resource allocation problem in fog–cloud infrastructures specifically focused on smart buildings. Results showed that their approach can reduce the response time and the cost of VM allocations. In [23], an ILP model for the fog resource provisioning problem has been formulated followed by a heuristic-based algorithm able to find suboptimal solutions, albeit achieving better time efficiency. In their work, the authors studied the trade-off between maximizing the reliability and minimizing the overall system cost. Moreover, in [24], service placement strategies for FC based on matching game algorithms have been introduced. On one hand, the first approach is based on SFC concepts since the ordered sequence of services requested by each application is considered. On the other hand, the second one formulates the problem while overlooking the chain structure to lower the computation complexity without compromising performance. Also, in [25], an edge container orchestrator for low powered devices called FLEDGE has been presented. Their results showed that FLEDGE minimizes resource costs when compared with other platforms. Recently, their work has been extended in [26], where the scalability and volatility of a fog–cloud infrastructure have been studied. The authors proposed a scheduling algorithm to allocate services in a large-scale fog deployment capable of adapting to network changes.

Although most of the cited research has dealt with allocation issues in FC, none of the aforementioned studies considered realistic QoS requirements or any kind of constraints coming from the high demands imposed by IoT (e.g. latency thresholds, service location, container-based services). Furthermore, most research is focused on theoretical modeling and simulation studies which limit their practical implementation.

7.3.2 ML Techniques for Resource Allocation

Due to recent advances in ML, studies have been carried out to apply ML methods to resource allocation problems. In [27], the authors proposed supervised learning techniques to predict future Network Function Virtualization (NFV) requests. Their goal is to proactively allocate resources based on previously observed patterns. Results showed that their proposal can proactively satisfy NFV requests. In [28], neural-network models have been proposed to address the Virtual Network Function (VNF) auto-scaling problem in 5G Networks. Their goal is to predict the required number of VNFs at a given moment based on previous traffic demands. Furthermore, an ILP formulation has been presented to solve the SFC allocation problem. Results proved that the average end-to-end (E2E) latency reduces significantly when service chains are allocated at the edge. In [29], an ML model has been employed to predict VNF resource demands with high accuracy (e.g. CPU). Their approach can be applied to SFC allocation problems, such as auto-scaling and optimal placement. Additionally, in [30], the fog infrastructure has been modeled as a distributed intelligent processing system called SmartFog by using ML techniques and graph theory. Their approach provides low-latency decision-making and adaptive resource management through a nature-inspired fog architecture.

In summary, supervised and unsupervised ML techniques have been implemented in the literature to improve decision-making in cloud infrastructures. Most cited research deals with allocation and auto-scaling problems that traditional methods (e.g. theoretical modeling, heuristic algorithms) have not been able to fully resolve due to the dynamic behavior of the network.

7.3.3 RL Methods for Resource Allocation

Recently, RL methods have been given significant attention in the field of resource allocation. In [31], a deep RL technique called DeepRM has been presented to solve the task placement problem in a cloud management system. Their initial results showed that DeepRM performs comparably to heuristics-based solutions and that it can learn different strategies depending on the network status. In [32], the IoT service allocation problem is addressed by employing an RL mechanism to calculate satisfactory levels of Quality of Experience (QoE). Their evaluations proved the efficiency of the applied methods. Furthermore, in [33], an RL-based optimization framework has been presented to tackle the resource allocation problem in wireless multi-access edge computing (MEC). Their objective is to minimize costs while optimizing resources. Simulation results have been presented where their proposed methods achieved significant cost reductions. In [34], RL techniques have been studied for their applicability to the SFC allocation problem in NFV–SDN (Software Defined Networking) enabled metro-core optical networks. Their results demonstrated the advantages of using RL-based optimizations over rule-based methods. Additionally, in [35], a fog resource scheduling mechanism based on deep RL has been presented. Their approach is focused on vehicular FC use cases aiming to minimize the time consumption of safety-related applications. Results showed that their proposed schemes can reduce time consumption when compared to traditional approaches.

In summary, RL methods have proven their potential applicability to resource allocation issues during the past years. However, the performance of RL techniques is deeply interconnected with the way the environment and the reward system are set up. Depending on the assumptions made in the system, RL methods could deliver completely different results. To the best of our knowledge, RL methods have not yet been applied to SFC allocation where fog–cloud infrastructures and container-based services have been assumed. Also, the dynamic behavior of the network and different scheduling strategies (e.g. low-latency, energy efficiency) have not been entirely addressed. However, RL techniques have proven that learning directly from experience could work in a practical deployment and offer a real alternative to current heuristic-based approaches. Thus, in Section 7.4, a novel RL approach for SFC allocation in FC is proposed.

7.4 A RL Approach for SFC Allocation in Fog Computing

This section introduces the RL approach for SFC allocation in FC. First, the IoT allocation problem formulation is presented. Then, the observation and action spaces from our RL environment are described. Finally, the reward function and the agent are introduced.

7.4.1 Problem Formulation

As mentioned, the IoT allocation problem has been modeled as an MILP formulation previously presented in [36]. The model considers a fog–cloud infrastructure where containerized service chains can be allocated. An IoT application is decomposed in a set of micro-services, which have a particular replication factor for load balancing or redundancy. Multiple users are expected to access these micro-services. The fog–cloud infrastructure manages a set of nodes, in which micro-service instances must be allocated based on its requirements and subject to multiple constraints. For example, nodes have limited capacities (e.g. CPU and memory) and all micro-services composing a given application must be allocated in the network so that the application can be considered deployed. The MILP formulation has been translated into an RL environment called gym-fog[1] where actions can be performed and at each time step, a new observation is given which describes the new state of the environment. It should be noted that only the cloud formulations have been considered for the designing of the RL approach and wireless aspects available in the model have not been used.

For this work, we designed a new objective for the MILP model: the minimization of the overall cost of the system, which translates into increased energy efficiency. By using the nomenclature of the MILP formulation presented in Table 7.1, this objective can be expressed as shown in (Eq. (7.1)). The agent will try to learn how to minimize the overall system cost as a MILP formulation by interacting with the gym-fog environment.

$$\sum_{a \, \epsilon \, A} \sum_{id \, \epsilon \, ID} \sum_{s \, \epsilon \, S} \sum_{\beta_i \, \epsilon \, \beta} \sum_{n \, \epsilon \, N} P_{s,\beta_i}^{a,id}(n) \times \varpi_n \times \omega_s \times \gamma_s \times \delta_s \tag{7.1}$$

---

1 https://github.com/jpedro1992/gym-fog

Table 7.1 Variables used for the minimization of the overall system cost.

Symbol	Description
$P_{s,\beta_l}^{a,id}(n)$	The placement matrix. If $P_{s,\beta_l}^{a,id}(n) = 1$, the replica β_l of micro-service s is executed on node n for the application a with the SFC identifier id
ϖ_n	The associated weight to node n
ω_s	The CPU requirement (in cpu) of the micro-service s
γ_s	The memory requirement (in GB) of the micro-service s
δ_s	The bandwidth requirement (in Mbit/s) of the micro-service s

7.4.2 Observation Space

The observation space corresponds to the state representing the environment at a given step. For instance, consider an agent playing a chess game, the observation will be the board status of a particular game. In the implemented gym-fog, the observation space has been designed as shown in Table 7.2. For an easier understanding of our methodology, let us consider a small infrastructure where all user requests coming to our system are made based on an IoT application decomposed in two individual micro-services. The observation space will be constituted by five metrics. Two metrics (RS1 and RS2) represent the ratio between the allocation scheme proposed by the agent and the MILP model for each micro-service. Then, two metrics (RL and MILP) represent the overall system cost given by the agent and the MILP model, respectively. Finally, the last metric (UR) is about the exact number of user requests made in the network at that particular step. Thus, the

Table 7.2 A sample fraction of the observation space of the gym-fog environment.

Metric name	Description
Ratio $S1$ (RS1)	The relation of allocated micro-service 1 instances between the MILP model and the agent
Ratio $S2$ (RS2)	The relation of allocated micro-service 2 instances between the MILP model and the agent
Cost RL (RL)	The agent allocation scheme cost
Cost MILP (MILP)	The MILP allocation scheme cost
User requests (UR)	The number of user requests at the given moment

observation space increases linearly with the number of applications available in the MILP model and their corresponding micro-services.

7.4.3 Action Space

The action space corresponds to all actions that the agent could apply in the environment. Considering the same chess analogy as before, the action space in a chess game would be selecting each piece and move it to a certain board position. In a fog–cloud infrastructure, the action space must include the allocation and termination of all micro-services available in the system. The action space of the gym-fog environment has been designed as shown in Table 7.3 assuming the same IoT application constituted by two micro-services and that the fog–cloud infrastructure is represented by only one node. The action space is composed of five discrete actions. The action space also increases linearly with the number of micro-services and the number of nodes available in the infrastructure. The first action is called as *DoNothing* since when applied by the agent, no allocation or termination will be performed in the network. Thus, the agent should only select this action when the current allocation scheme meets the current network demand. The second set of actions corresponds to the allocation of micro-service instances (Deploy-Si-Ni). The agent can choose which micro-service instance should be allocated and on which node it should be executed. The action space has been designed in this manner to make sure that the agent can find better allocation decisions by choosing a particular micro-service instance to be deployed on a certain node. The agent can apply a given action from this set several times if more instances of the same micro-service are needed in the network to support all user requests. Finally, the last set of actions (Stop-Si-Ni) corresponds to the termination of micro-service instances. As in allocation actions, the agent chooses which micro-service should be terminated and which instance should it be since

Table 7.3 A sample fraction of the action space of the gym-fog environment.

Action label	Description
DoNothing	The agent does nothing
Deploy-S1-N1	Allocate a micro-service 1 instance in node 1
Deploy-S2-N1	Allocate a micro-service 2 instance in node 1
Stop-S1-N1	Terminate the micro-service 1 instance in node 1
Stop-S2-N1	Terminate the micro-service 2 instance in node 1

the node where the micro-service is deployed is also given. Our goal is to teach the agent that a certain number of micro-service instances needs to be allocated for the proper chain operation and to support all user requests.

7.4.4 Reward Function

The purpose of the reward function is to teach the agent how to maximize the accumulated reward by selecting appropriate actions depending on the observation provided by the environment. A certain reward is obtained for each action the agent selects. This reward can be positive or negative. Thus, the agent can learn if its chosen action was appropriate based on the received reward. The design of an appropriate reward function through the manual tuning of ML parameters is needed to ensure the agent learns what it is supposed to. The reward function implemented in the gym-fog environment is shown in Algorithm 7.1. The agent's purpose is to learn how to allocate micro-services in a fog–cloud infrastructure according to the MILP formulation. The MILP model provisions services in the network area by minimizing the overall system cost, as shown previously. Therefore, the closer the agent is to achieve the MILP's solution, the higher the reward it receives. First, rewards are calculated based on constraints included in the MILP model. For instance, a constraint has been added to limit the allocation of one instance of the same micro-service per node. Thus, if the agent selects an action that would revoke this constraint, the agent would receive a negative reward (i.e. −1). Then, individual rewards are calculated for each micro-service ratio as shown in Algorithm 7.2. First, if the number of allocated micro-service instances by the agent is equal to zero, a reward of −5 is retrieved because the agent is not allocating a single instance of this micro-service, which prevents the service chain from proper operation. Second, if the number of allocated micro-service instances by the agent is equal to the ones allocated by the MILP model, a reward of 5 is returned. Finally, a reward of −1 is retrieved in case the agent is allocating a higher number of replicas that are not required (i.e. over-provisioning) or if the agent is allocating fewer instances than needed (i.e. under-provisioning).

After ratio reward calculation, a cost reward function is performed as shown in Algorithm 7.3. First, if the agent's cost is lower than the MILP one, a negative reward is returned because the agent cannot have a lower cost since the MILP solution is optimal. Thus, the agent is probably violating several constraints of the IoT service problem. Then, if the agent's cost is equal or up to 10% higher than the MILP one, 10 is returned since the agent is performing similar to the MILP model. Then, depending on how higher the agent cost is compared to the MILP one, a decreasing reward is returned, meaning that the agent is being taught that the closer it stays to the MILP cost, the higher reward it receives. Nevertheless, the ultimate goal is to achieve similar costs to the MILP model and allocate all necessary

---

Algorithm 7.1 Reward Function of the gym-fog environment

Input: Observation state after action step in
Output: Reward out

1:   *// Return the reward for the given state*
2:   **getReward(obs):**
3:     *reward* = 0
4:     *ratioS1* = *obs.get*(1)
5:     *ratioS2* = *obs.get*(2)
6:     *costRL* = *obs.get*(3)
7:     *costMILP* = *obs.get*(4)
8:
9:     *// Reward based on Keywords for MILP constraints*
10:    *// Constraint: MAX micro-services on a single Node*
11:    *// Constraint: Terminate micro-service without deployment first*
12:    *// Constraint: MAX micro-service instances reached*
13:    **if** *constraintMaxServicesOnNode* == *True* **then**
14:      **return** −1
15:    **if** *constraintTerminateServiceFirst* == *True* **then**
16:      **return** −1
17:    **if** *constraintMAXServiceInstances* == *True* **then**
18:      **return** −1
19:
20:    *// Micro-service ratio Reward calculation*
21:    *reward* = *reward* + *getRatioReward*(*ratioS1*)
22:    *reward* = *reward* + *getRatioReward*(*ratioS2*)
23:
24:    *// Cost Reward Calculation*
25:    *reward* = *reward* + *getCostReward*(*costRL*, *costMILP*)
26:
27:    *// Ultimate Goal calculation*
28:    **if** *ratioS1* == 1 **and** *ratioS2* == 1 **then**
29:      **if** *costRL* > *costMILP* **then** *// High Reward*
30:        *reward* = *reward* + 10
31:      **if** *costRL* == *costMILP* **then** *// MAX Reward*
32:        *reward* = *reward* + 100
33:
34:    **return** *reward*

Algorithm 7.2 Micro-service Ratio Reward Calculation

Input: Micro-service Ratio observation state in
Output: Ratio reward out
1: // *Return the reward for the given micro-service ratio*
2: **getRatioReward(ratio):**
3:      **if** *ratio* == 0 **then** // *No service deployed - Bad solution*
4:          **return** −5
5:      **else if** *ratio* == 1 **then** // *Equal to the MILP - Good solution*
6:          **return** 5
7:      **else then** // *Under / Over-provisioning scheme*
8:          **return** −1

Algorithm 7.3 Cost Reward Calculation

Input: CostRL, CostMILP in
Output: Cost reward out
1: // *Return the reward for the relation between the CostRL and CostMILP*
2: **getCostReward(costRL, costMILP):**
3:      **if** costRL < costMILP **then** // *Lower than MILP - Bad solution*
4:          **return** −10
5:      **else if** costMILP ≤ costRL ≤ 1.10×costMILP **then** // *Best Solution*
6:          **return** 10
7:      **else if** 1.10×costMILP < costRL ≤ 1.25×costMILP **then**
8:          **return** −2
9:      **else if** 1.25×costMILP < costRL ≤ 1.75×costMILP **then**
10:         **return** −4
11:     **else if** 1.75×costMILP < costRL ≤ 2.0×costMILP **then**
12:         **return** −6
13:     **else if** 2.0×costMILP < costRL ≤ 3.0×costMILP **then**
14:         **return** −8
15:     **else if** 3.0×costMILP < costRL ≤ 4.0×costMILP **then**
16:         **return** −10
17:     **else then** // *> 4×costMILP*
18:         **return** −20

---

micro-service instances for the acceptance of all user requests. Thus, two bonus rewards can be given to the agent if all micro-service ratios are equal to 1. First, if the agent's cost is higher than the MILP cost, a bonus reward of 10 is given since the agent allocated all micro-service instances needed in the network, despite the higher cost. Second, if the agent's cost matches the MILP cost, a bonus reward of 100 is retrieved because the agent accomplished exactly what it was supposed to.

The agent learned how to allocate micro-services in a fog–cloud infrastructure as a MILP formulation.

7.4.5 Agent

This section introduces the Q-learning agent used in the evaluation of the gym-fog environment. Q-Learning [37] is a classical RL algorithm that learns the best action to select at a given state by experiencing each state–action pair $Q(s, a)$. Q-Learning is an off-policy RL method since the agent learns the optimal policy (π) independently of the applied actions based on a two-step process. The first process is called **exploitation** where a Q-table is calculated as a baseline for all possible actions for a given state. Then, the action with a higher value (i.e. maximum reward) would be applied. The second operation is called **exploration** since the agent instead of selecting actions based on the maximum future reward, the agent selects an action at random which allows the exploration and discovery of new states that otherwise could have not been explored due to the exploitation process. Exploration and exploitation rates can be settled at run time, thus complete control over the algorithm is provided.

The main issue with Q-learning agents is that it requires to see all action-state pairs for a given environment to be able to apply actions that would maximize reward. As the problem size grows, representing all state–action pairs in memory becomes prohibitive. For instance, increasing the complexity of the gym-fog environment (e.g. adding nodes to the infrastructure, adding extra services to the service chain), has a serious impact on memory and execution time because it is directly linked with the size of the action and state space. Thus, to reduce the space complexity, the observation space has been discretized as shown in Table 7.4, where a specific range for each observation metric has been attributed reducing significantly the number of states that the Q-learning agent needs to consider. Assuming the previous fog–cloud infrastructure, the observation space would have been reduced into 288 discrete states. First, the observation metrics regarding micro-service allocations have been reduced into three spaces. For instance, the *Ratio S1* can only be equal to 0, equal to 1, or anything else (i.e. all other possibilities are grouped). These three states are the only states that the Q-learning agent needs to consider to find good actions regarding the *Ratio S1* metric. Additionally, the two cost observation metrics (costRL and costMILP) have been combined into a new metric called cost where the difference between these two is used to formulate eight states based on the previously shown cost reward function. Finally, user requests are also aggregated into four states based on the solutions provided by the MILP model, which vary depending on the service chains to be allocated and on the considered fog–cloud infrastructure.

Table 7.4 The reduced observation space of the gym-fog environment.

Metric name	Number of states
RatioS1	Three states (ratio calculation): $[RS1 = 0, RS1 = 1, \text{else}]$
RatioS2	Three states (ratio calculation): $[RS2 = 0, RS2 = 1, \text{else}]$
Cost	Eight states (cost calculation): [RL < MILP, MILP $\leq$ RL $\leq$ 1.10×MILP, ..., RL > 4.0×MILP]
UserRequests	Four states: [UR <= 20, UR <= 32, UR <= 40, UR <= 50]
Total	288 states $(3 \times 3 \times 8 \times 4)$

7.5 Evaluation Setup

This section describes the fog–cloud infrastructure used for the evaluation of the gym-fog environment. Then, the environment implementation is detailed followed by the respective configuration applied in the evaluation.

7.5.1 Fog–Cloud Infrastructure

The fog–cloud infrastructure illustrated in Figure 7.4 has been represented in the gym-fog environment. A total area of 324 km² has been considered. The fog–cloud infrastructure is deployed on five locations L, where the micro-service allocation is possible. Each location manages a set of three nodes. The hardware configurations of each node are shown in Table 7.5. Each node has a given computing capacity (i.e. CPU, RAM, and Bandwidth) and a certain weight, which are the necessary information to calculate the overall system cost based on the MILP formulation.

7.5.2 Environment Implementation

The gym-fog environment was developed based on the OpenAi gym [38]. OpenAi gym is an open-source toolkit for RL research written in Python. It includes a collection of benchmark problems that expose a standardized interface comparing RL algorithms in terms of performance. The MILP formulation initially developed in Java has been rewritten in Python to ease the interaction between the MILP model and the OpenAi gym. The gym-fog environment was built based on the OpenAi gym structure as shown in Figure 7.5. To begin the experiment, the initialize function is triggered. Then, during the training, at each iteration, the agent selects an

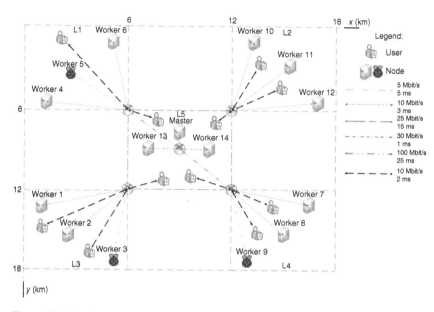

Figure 7.4 The fog–cloud infrastructure for the gym-fog environment evaluation.

Table 7.5 The hardware configuration of each node.

Node	CPU (cpu)	RAM (Mi)	Band. (Mbit/s)	Weight
Worker 1	2.0	4.0	10.0	2.0
Worker 2	2.0	4.0	10.0	2.0
Worker 3	1.0	2.0	5.0	1.0
Worker 4	2.0	4.0	10.0	2.0
Worker 5	1.0	2.0	5.0	1.0
Worker 6	2.0	4.0	10.0	2.0
Worker 7	2.0	4.0	10.0	2.0
Worker 8	2.0	4.0	10.0	2.0
Worker 9	1.0	2.0	5.0	1.0
Worker 10	2.0	4.0	10.0	2.0
Worker 11	2.0	4.0	10.0	2.0
Worker 12	2.0	4.0	10.0	2.0
Worker 13	6.0	16.0	30.0	3.0
Worker 14	6.0	16.0	30.0	3.0
Master	8.0	24.0	30.0	3.0

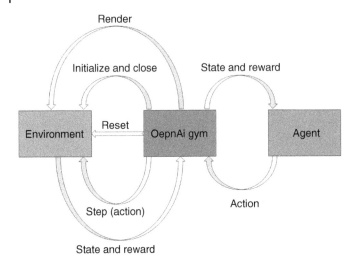

Figure 7.5 The OpenAi gym environment structure.

action which then is passed to the environment by OpenAi gym through a step function, where a new state and the respective reward are returned. Also, a reset function is used at the beginning or when each episode is finished so that the initial state of the environment is reinstated. Furthermore, a render function can be used to render the environment after each step. Finally, a close function is called when the learning process is completed to properly terminate the environment. The implemented gym-fog uses the mentioned functions to interact with the MILP model. Essentially, OpenAi gym acts as a bridge between our MILP model and the agent.

7.5.3 Environment Configuration

The gym-fog environment configuration is shown in Table 7.6 based on the described fog–cloud infrastructure. One application is available, which is decomposed in three micro-services composing a service chain. The maximum replication factor corresponds to 5, meaning that the MILP model or the agent can only deploy five micro-service instances of the same type of micro-service. The action space is constituted by 91 actions (3 micro-services, 15 nodes), while the observation space is constituted by 6 observation metrics which have been reduced into 864 discrete states. Each episode duration is constituted of 100 steps. The agent's explore rate and learning rate have been settled to 0.01 and 0.001, respectively. For the evaluation, the agent and the MILP calculations have been executed on a 6-core Intel i7-9850H CPU @ 2.6GHz processor with 16GB of memory.

Table 7.6 The gym-fog environment configuration.

Name	Description
Number applications	1
Number of micro-services	3
Number of locations	5
Number of nodes	15
The SFC structure	$a_1 : s_1 \rightarrow s_2 \rightarrow s_3$
Max. replication factor	5
Action space	91 actions
Observation space	6 observation Metrics
Reduced observation space	864 states
Episode duration	100 steps
Agent explore/learning rate	0.01/0.001

7.6 Results

This section presents the obtained results. First, a static scenario has been evaluated where the number of user requests is kept constant throughout the evaluation. Then, a dynamic use case is assessed where the network demand is constantly changing since users join and leave randomly.

7.6.1 Static Scenario

As a first evaluation, the Q-learning agent has been trained during 10 000 episodes by considering a static use case where the number of user requests has been kept constant. Thus, dynamic changes in terms of user requests are not expected in this experiment. The Q-learning agent should be able to learn significantly faster adequate actions in this use case than in a dynamic scenario since the number of requests is constant throughout all training. In Figure 7.6, both the reward accumulated and the cost difference between the Q-learning agent and the MILP model are illustrated. A smoothing window of 100 has been applied to reduce spikes in both curves. As shown, the agent can reduce the overall system cost reaching solutions 5% worse than the MILP model. Additionally, accumulated rewards of 1200 have been obtained in a single episode meaning that the agent is receiving on average a reward of 12 per step, which based on our implemented reward function means that the agent is allocating all the required micro-service instances in the infrastructure though it is not able to fully optimize the overall system cost as

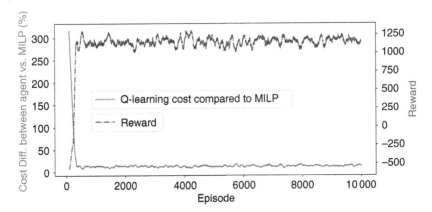

Figure 7.6 The accumulated reward and the cost difference for the static use case.

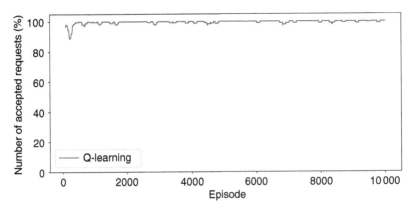

Figure 7.7 The percentage of accepted requests for the static scenario.

the MILP model. Also, another important factor to consider is the percentage of accepted requests in each episode shown in Figure 7.7 since the agent can reach low costs without allocating all the necessary micro-services which would translate into unaccepted user requests. The first 10 steps have been disregarded regarding the acceptance of requests as a warming up period, ensuring that the agent has enough steps to properly select actions. As shown, the Q-learning agent can accept all user requests (i.e. 100%) consistently after 500 episodes. Finally, in Figure 7.8, the execution time of each episode is presented. The Q-learning agent solves a single episode in on average 0.15 and 0.20 seconds, which compared to ILP-based calculations is significantly faster because an ILP formulation needs to calculate the optimal allocation scheme on each episode step. Results prove that the Q-learning agent is not only able to learn allocation schemes with low costs but also accept all user requests for a static scenario.

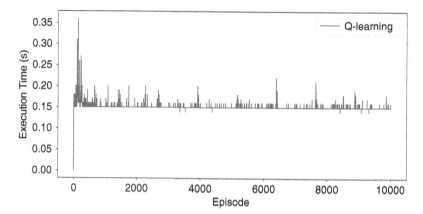

Figure 7.8 The execution time of each episode run.

7.6.2 Dynamic Scenario

In the dynamic scenario, the network demand is constantly changing during the episode. The number of user requests may decrease or increase and the agent must adapt its allocation scheme according to the network demand. The number of user requests has been changed every 5 steps between 1 and 50 based on specific probabilities (increase: 50%, equal: 35%, decrease: 15%). The total increase or decrease is random, which makes this dynamic scenario more challenging than the previous static case since no pattern is given to the agent throughout the experiment since several patterns occur in different episodes. In Table 7.7, the MILP execution time for each configuration is shown. For instance, for user requests higher than 25, the MILP model requires at least five seconds to obtain the optimal allocation scheme. For even higher values of user requests, the MILP model takes on average at least 10 seconds. These calculations represent a single step on an episode, which proves the drawback of ILP-based solutions because every change on the network, would require a new calculation making these solutions impractical. In Figure 7.9, both the accumulated reward and the average cost difference between the Q-learning agent and the MILP model are illustrated. The Q-learning agent can reduce the overall system cost reaching solutions 50% worse than the MILP model. Additionally, the agent only accumulates rewards of 300 in a single episode, meaning that

Table 7.7 The MILP model execution time.

	Number user requests							
	1	5	10	20	25	30	40	50
Execution time (s)	0.05	0.12	0.20	0.27	5.09	5.95	9.33	48.83

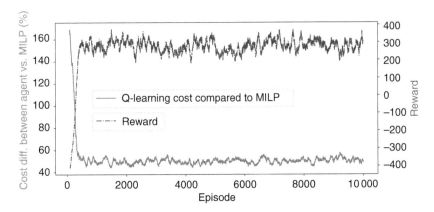

Figure 7.9 The accumulated reward and the cost difference for the dynamic case.

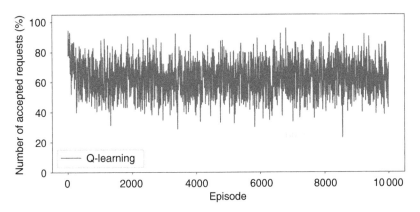

Figure 7.10 The percentage of accepted requests for the dynamic scenario.

the agent is receiving on average a reward of 3 per step. Based on our implemented reward function, this means that the agent is not able to allocate all the required micro-service instances in the infrastructure, affecting the percentage of accepted requests as shown in Figure 7.10. Due to the dynamic demand, the agent needs to constantly adapt the allocation scheme in the infrastructure, which translates into under-provisioning and over-provisioning schemes during several steps in a single episode. Thus, the acceptance of requests oscillates between 40% and 90% during the 10 000 episode training when a smoothing window of 10 is applied. The agent is constantly reacting to demand changes, which makes this scenario notably more challenging than the previous static use case. Results prove that efficient solutions for dynamic resource allocation are still missing due to the problem complexity. It is hard to find practical approaches that meet user demands while decreasing the overall system cost. Nevertheless, our early results show that RL can be applied to SFC allocation in FC and should be further explored in future research. The extension of the observation space is left for future work as it could improve these cost and acceptance results.

7.7 Conclusion and Future Direction

Over the past years, ML techniques have become an interesting research field in the networking domain. Several efforts have been made to adapt ML methods to common network problems. This chapter focuses on RL agents to provide an efficient solution for SFC allocation in fog–cloud infrastructures. Resource provisioning has been studied for years in the network management field. However, networks and services are continuously evolving, with new protocols and technologies introduced to address current problems and improve the overall QoS. Recent examples include the adoption of SFC, micro-service paradigms, and FC. Services are connected in a specific order to improve flexibility and resource allocation performance. Also, the micro-service pattern revolutionized the way developers are currently building their software applications. An application evolved from a single monolithic into a set of small containers, which may be deployed across several servers. Thus, traditional centralized clouds evolved into small distributed fog locations to distribute computing resources across the network area. And when all these concepts come into place, resource allocation is a quite complex online task. Resource provisioning research in fog–cloud infrastructures is still in its early stages. Distributing the infrastructure has increased operational costs for service providers, and energy consumption has become a growing concern. We addressed this challenge in this chapter by employing RL agents to find proper allocation decisions, focused on reducing the overall system cost. An environment called gym-fog has been developed to bridge the gap between ILP-based solutions with RL algorithms. Observation and action spaces have been designed for the resource allocation problem to teach RL agents how to allocate services in FC. A reward system has been set up to incentivize RL agents to select appropriate actions for SFC allocation focused on reducing the overall system cost, translating into higher energy efficiency. Results proved that our developed agent can obtain comparable performance to state-of-the-art ILP formulations for static use cases, where 100% of requests have been accepted with overall costs 5% worse than our MILP model. In contrast, dynamic use cases also proved their complexity by showing that practical solutions able to reduce the overall cost and accept all user-requests are still missing. Our agent can reduce costs up to 50% and accept on average 60% of the requests.

Developing RL systems able to learn directly from experience without any prior knowledge and capable of reallocating services in the infrastructure by reacting to sudden network changes will be the next main topic in this research field. RL methods have already proven their potential applicability to the resource provisioning domain. However, the performance of these techniques is deeply interconnected with the way the RL system is set up. The environment is the key to the problem. The interactions between the agent and the environment affect greatly the performance of these algorithms. Further, the state and action space of the problem can grow exponentially depending on the infrastructure size (i.e. the number of nodes, the number of services) used in the environment, which could lead to an unsolvable problem. Finally, the importance of the reward system should not be neglected. The agent will only learn to properly allocate services if it is compensated with positive rewards during the learning process, even if it was not able to reach desirable solutions. The key is to give the agent higher rewards the closer it is to reach the ultimate goal, otherwise, it is quite challenging for the agent to learn proper actions. In contrast, ILP-based methods are difficult to implement in practice due to their resolution time. Also, they require a lot of initial information to be fed to the algorithm so that optimal allocation schemes can be found. These methods can take hours or even days to find the optimal service allocation and when network changes occur, service reallocations should be made as fast as possible. Another challenge is the lack of expertise in both RL and resource allocation fields. Few experts have significant knowledge in both domains, which makes it difficult to implement RL solutions adapted for resource allocation problems. Most RL methods used in networking have been created for other types of applications (e.g. video games).

In summary, several challenges persist in the resource allocation domain. Nevertheless, given the dynamic behavior of the network and the need for efficient scheduling strategies (e.g. energy efficiency, low-latency), RL methods have proven that with enough training they can be an adequate solution for resource provisioning in fog–cloud infrastructures. Furthermore, these methods have shown their potential in practical scenarios where current ILP-based solutions have several drawbacks, especially in terms of scalability. As future work, we will extend our gym-fog environment by designing more complex reward functions capable of fully addressing the challenges of dynamic use cases, as well as experiment with different RL agents as deep queue networks and actor-critic methods.

Bibliography

1 Biswas, A.R. and Giaffreda, R. (2014). IoT and cloud convergence: opportunities and challenges. *2014 IEEE World Forum on Internet of Things (WF-IoT)*, IEEE, pp. 375–376.

2 Naha, R.K., Garg, S., Georgakopoulos, D. et al. (2018). Fog computing: survey of trends, architectures, requirements, and research directions. *IEEE Access* 6: 47980–48009.

3 Newman, S. (2015). *Building Microservices: Designing Fine-Grained Systems.* O'Reilly Media, Inc.

4 Bhamare, D., Jain, R., Samaka, M., and Erbad, A. (2016). A survey on service function chaining. *Journal of Network and Computer Applications* 75: 138–155.

5 Sutton, R.S. and Barto, A.G. (2018). *Reinforcement Learning: An Introduction.* MIT Press.

6 Santos, J., Wauters, T., Volckaert, B., and De Turck, F. (2018). Fog computing: enabling the management and orchestration of smart city applications in 5G networks. *Multidisciplinary Digital Publishing Institute* 20: 4.

7 Bonomi, F., Milito, R., Zhu, J., and Addepalli, S. (2012). Fog computing and its role in the internet of things. *Proceedings of the 1st Edition of the MCC Workshop on Mobile Cloud Computing*, pp. 13–16.

8 Verbelen, T., Simoens, P., De Turck, F., and Dhoedt, B. (2012). Cloudlets: bringing the cloud to the mobile user. *Proceedings of the 3rd ACM Workshop on Mobile Cloud Computing and Services*, pp. 29–36.

9 Famaey, J., Latré, S., Strassner, J., and De Turck, F. (2010). A hierarchical approach to autonomic network management. *2010 IEEE/IFIP Network Operations and Management Symposium Workshops*, IEEE, pp. 225–232.

10 Deboosere, L., Vankeirsbilck, B., Simoens, P. et al. (2012). Efficient resource management for virtual desktop cloud computing. *The Journal of Supercomputing* 62 (2): 741–767.

11 Pradhan, P., Behera, P.K., and Ray, B.N.B. (2016). Modified round robin algorithm for resource allocation in cloud computing. *Procedia Computer Science* 85: 878–890.

12 Moens, H. and De Turck, F. (2014). VNF-P: A model for efficient placement of virtualized network functions. *10th International Conference on Network and Service Management (CNSM) and Workshop*, IEEE, pp. 418–423.

13 Bhamare, D., Samaka, M., Erbad, A. et al. (2017). Optimal virtual network function placement in multi-cloud service function chaining architecture. *Computer Communications* 102: 1–16.

14 Santos, J., Wauters, T., Volckaert, B., and De Turck, F. (2020). Towards delay-aware container-based Service Function Chaining in Fog Computing. *NOMS 2020-2020 IEEE/IFIP Network Operations and Management Symposium.* Accepted for publication.

15 Nadareishvili, I., Mitra, R., McLarty, M., and Amundsen, M. (2016). *Microservice Architecture: Aligning Principles, Practices, and Culture.* O'Reilly Media, Inc.

16 Mnih, V., Badia, A.P., Mirza, M. et al. (2016). Asynchronous methods for deep reinforcement learning. *International Conference on Machine Learning*, pp. 1928–1937.

17 Van Hasselt, H., Guez, A., and Silver, D. (2016). Deep reinforcement learning with double Q-learning. *13th AAAI Conference on Artificial Intelligence*.

18 Hessel, M., Modayil, J., Van Hasselt, H. et al. (2018). Rainbow: combining improvements in deep reinforcement learning. *32nd AAAI Conference on Artificial Intelligence*.

19 Arkian, H.R., Diyanat, A., and Pourkhalili, A. (2017). MIST: Fog-based data analytics scheme with cost-efficient resource provisioning for IoT crowdsensing applications. *Journal of Network and Computer Applications* 82: 152–165.

20 Brogi, A. and Forti, S. (2017). QoS-aware deployment of IoT applications through the fog. *IEEE Internet of Things Journal* 4 (5): 1185–1192.

21 Skarlat, O., Nardelli, M., Schulte, S., and Dustdar, S. (2017). Towards QoS-aware fog service placement. *2017 IEEE 1st International Conference on Fog and Edge Computing (ICFEC)*, IEEE, pp. 89–96.

22 Yasmeen, A., Javaid, N., Rehman, O.U. et al. (2018). Efficient resource provisioning for smart buildings utilizing fog and cloud based environment. *2018 14th International Wireless Communications & Mobile Computing Conference (IWCMC)*, IEEE, pp. 811–816.

23 Yao, J. and Ansari, N. (2019). Fog resource provisioning in reliability-aware IoT networks. *IEEE Internet of Things Journal* 6(5), pp. 8262–8269.

24 Chiti, F., Fantacci, R., Paganelli, F., and Picano, B. (2019). Virtual functions placement with time constraints in fog computing: a matching theory perspective. *IEEE Transactions on Network and Service Management* 16 (3): 980–989.

25 Goethals, T., De Turck, F., and Volckaert, B. (2019). FLEDGE: Kubernetes compatible container orchestration on low-resource edge devices. *SC2 2019, the 9th International Symposium on Cloud and Service Computing*, pp. 1–16.

26 Goethals, T., Volckaert, B., and De Turck, F. (2020). Adaptive fog service placement for real-time topology changes in kubernetes clusters. *CLOSER2020, The 10th International Conference on Cloud Computing and Services Science*, pp. 161–170.

27 Scalingi, A., Esposito, F., Muhammad, W., and Pescapé, A. (2019). Scalable provisioning of virtual network functions via supervised learning. *2019 IEEE Conference on Network Softwarization (NetSoft)*, IEEE, pp. 423–431.

28 Subramanya, T., Harutyunyan, D., and Riggio, R. (2020). Machine learning-driven service function chain placement and scaling in MEC-enabled 5G networks. *Computer Networks* 166: 106980.

29 Kim, H., Lee, D., Jeong, S. et al. (2019). Machine learning-based method for prediction of virtual network function resource demands. *2019 IEEE Conference on Network Softwarization (NetSoft)*, IEEE, pp. 405–413.

30 Kimovski, D., Ijaz, H., Saurabh, N., and Prodan, R. (2018). Adaptive nature-inspired fog architecture. *2018 IEEE 2nd International Conference on Fog and Edge Computing (ICFEC)*, IEEE, pp. 1–8.

31 Mao, H., Alizadeh, M., Menache, I., and Kandula, S. (2016). Resource management with deep reinforcement learning. *Proceedings of the 15th ACM Workshop on Hot Topics in Networks*, pp. 50–56.

32 Gai, K. and Qiu, M. (2018). Optimal resource allocation using reinforcement learning for IoT content-centric services. *Applied Soft Computing* 70: 12–21.

33 Li, J., Gao, H., Lv, T., and Lu, Y. (2018). Deep reinforcement learning based computation offloading and resource allocation for MEC. *2018 IEEE Wireless Communications and Networking Conference (WCNC)*, IEEE, pp. 1–6.

34 Troia, S., Alvizu, R., and Maier, G. (2019). Reinforcement learning for service function chain reconfiguration in NFV-SDN metro-core optical networks. *IEEE Access* 7: 167944–167957.

35 Chen, X., Leng, S., Zhang, K., and Xiong, K. (2019). A machine-learning based time constrained resource allocation scheme for vehicular fog computing. *China Communications* 16 (11): 29–41.

36 Santos, J., Wauters, T., Volckaert, B., and De Turck, F. (2020). Towards end-to-end resource provisioning in fog computing over low power wide area networks. *Journal of Network and Computer Applications* 175: 102915.

37 Watkins, C.J.C.H. and Dayan, P. (1992). Q-learning. *Machine Learning* 8 (3–4): 279–292.

38 Brockman, G., Cheung, V., Pettersson, L. et al. (2016). OpenAI Gym. *arXiv preprint arXiv:1606.01540.*

Part III

Management Functions and Applications

8

Designing Algorithms for Data-Driven Network Management and Control: State-of-the-Art and Challenges[1]

Andreas Blenk[1,2], Patrick Kalmbach[1], Johannes Zerwas[1], and Stefan Schmid[2]

[1]Chair of Communication Networks, Technical University of Munich, Munich, Germany
[2]Faculty of Computer Science, University of Vienna, Vienna, Austria

8.1 Introduction

The context: network management algorithms: Many components and protocols of computer networking rely on algorithms that solve an underlying optimization problem. These optimization problems range from routing in network topologies, over the placement of (softwarized) network functions, to the efficient operation of the deployed resources in different networks, such as data center networks or campus networks. Even more, with the recent trend of network programmability and virtualization, networked systems gained an additional dimension of flexibility. Both network programmability and virtualization introduced, e.g. new management interfaces and control protocols [1, 2]. Network algorithms revolve around the challenge of exploiting such flexibility, while optimizing runtime and solution quality [3].

The problem: computational hardness: The problems underlying network optimizations are often *computationally expensive*. Examples include management problems related to network planning or control problems related to traffic engineering. In more detail, problems related to the placement of caches, which feature connections to classic k-center clustering and facility location problems, or decision problems arising when testing the policy compliance of network configurations, are often hard to solve, i.e. computationally expensive. While some problems can be solved offline, other problems have to be

---

1 For brevity, when we talk about algorithms in this chapter, we always refer to algorithms solving network management problems. If we refer to another class of algorithms, e.g. for training neural networks, we will make this explicitly clear.

Communication Networks and Service Management in the Era of Artificial Intelligence and Machine Learning,
First Edition. Edited by Nur Zincir-Heywood, Marco Mellia, and Yixin Diao.
© 2021 The Institute of Electrical and Electronics Engineers, Inc. Published 2021 by John Wiley & Sons, Inc.

solved in an online fashion, without precise knowledge of future inputs or demands. In particular when solving problems online, the new configuration knobs as introduced by network programmability and virtualization demand faster algorithms. As both concepts introduce new configuration and operation opportunities, i.e. more optimization possibilities, algorithms simply have more options to choose from. Hence, the runtime of algorithms thus potentially increases.

The vision: data-driven algorithm design for network management and control: We believe that both artificial intelligence (AI) and machine learning (ML) provide new means to help tackle these challenges. Outside the networking community, the vision of data-driven algorithm design has already started to come to the fore [4], e.g. in different branches of optimization problems like learning admissible heuristics [5], or learning to branch [6] in solvers for mixed integer problems, etc. The underlying idea is to find a data-driven algorithm to solve a problem by looking at the available data produced by existing algorithms, which can be rendered as a statistical learning problem. As such algorithms are adapted to problem-specific instances, they potentially operate faster and more efficient. There is also a trend toward provably good algorithms learned by machines [4]. Hence, with the wealth of data available and the many problems still to be solved in networking, we believe that this paradigm has to be promoted more also in the networking community. We understand this chapter as an opportunity to get a quick overview of recent trends toward studying the data-driven algorithm design concept in the network management and control area.

Idea: utilizing the data produced by network algorithms: We believe that the wealth of data generated by algorithms solving network management and control problems can be a goldmine for improving algorithms themselves. When solving problems, algorithms produce what we call the (problem, solution)-pairs:, e.g. when solving the task to find a route, the pair consists of the two nodes (problem), and the resulting network route between both nodes (solution). In this chapter, we give an overview of methods used to learn from data and how algorithm executions could be accelerated by using learned patterns with ML. Furthermore, as an important research topic, we discuss how the methods should help researchers to even detect more challenging instances of problems (e.g. which are the pairs of nodes for which it is computationally most expensive in a network to find a route) to their algorithms during design time.

Challenges: finding problem coherences: Besides the production of valuable data, we see a main and novel challenge in finding efficient ways to represent and learn from data produced by network management and control algorithms. For instance, routing problems are represented as graphs; when solving such problems, algorithms can either decide on a per node basis or based on a global network view. Accordingly, as a next step toward establishing the idea, we need a better understanding of the structural properties of network optimization problems, in particular when represented as graphs. In general, we believe that it is important to go beyond considering network nodes in an isolated fashion. Recent research in ML and AI provides already valuable mechanisms, such as graph neural networks (GNRs) that help nodes to capture also their relations to their neighbors. Such approaches are data-driven and can in particular capture networks as a whole, i.e. nodes are not isolated anymore. Hence, in this book chapter, we overview different ways to represent network problems, i.e. to make them usable by ML and AI algorithms.

8.1.1 Contributions

This chapter makes the following contributions:

- We provide a short case study on exemplary networking management and control problems where data-driven design concepts can help (Section 8.1.2).
- We give an overview of used approaches, i.e. representations and concepts, to represent and handle (problem, solution)-pairs (Section 8.2).
- We survey work on data-driven learning of algorithm design in networking areas and beyond (Section 8.3).
- We take a look into the future and what can be done next, e.g. regarding adversarial input generation, ML- and AI-based approaches providing guarantees, etc. (Section 8.4).

Although many ML surveys and position papers have recently been proposed [7–12], these studies do not analyze the state-of-the-art with the focus as used in this book chapter: i.e. the data-driven algorithm learning paradigm and how ML-based methods can help to better predict *network solutions*, or even how they can even help to *solve networking problems*. We believe that such approaches are stepping stones toward realizing the vision of deploying the *data-driven algorithm design* [4] concept when facing network management and control problems.

8.1.2 Exemplary Network Use Case Study

Many network management and control problems rely on graph-based modeling, e.g. when placing network functions (management) or routing network traffic (control). Problems that can be represented as graphs provide a good opportunity to be used with ML, as valuable methods [13, 14] already exist to comprehensively represent graphs. Accordingly, we go from very abstract graph-based optimization problems (facility location problem), to virtualization-related graph-based problems (virtual network embedding [VNE]), to configuration problems on graphs (configuration verification in multiprotocol label switching [MPLS] networks). But note that the data-driven design is not limited to graph-based problems; hence, we briefly outline such trends also in Section 8.3.3. In the following, we put the potential application of ML in an italic formatting in order to emphasize the statements.

Facility location problems: Clustering and facility location problems are underlying many optimization problems in computer networks. For instance, the placement of caches in content distribution networks (CDNs), the simple determination of the best place of a web server in a Wide Area Network (WAN), or even the determination of places of base stations in cellular networks, have been modeled as variants of k-placement problems. In all cases, some functionality should be placed, e.g. "close" to the nodes using this functionality, like users need short communication timings with the caches providing their requested content. *In its simplest form, ML could be used to predict best locations.*

Virtual network embedding: With the slicing of communication networks such as 5G/6G networks, the VNE problem experiences a revival: the ability to determine slices with guaranteed resources becomes a must-have for latency critical applications such health care or remote control. Embedding whole slices, i.e. embedding virtual nodes and determining physical routes for virtual edges, is related to many general resource allocation and traffic engineering problems in computer networks. For guaranteed performance, a common theme is that virtual networks (or slices) need to be created (embedded) while adhering to hard resource constraints. Virtual nodes should not overutilize physical node resources; virtual edges should not overutilize physical edge resources to keep end-to-end guarantees. *A simple ML task can be to answer the question whether a physical network can actually accept a current virtual network request.* Such tasks might happen in mobile environments where multiple communication partners, like robots, are requesting slices for communication before starting their physical tasks. Without receiving the network resources as requested for their slices, robots might not reliably operate when working on their tasks.

Testing policy compliance and what-if analysis: With the increasingly stringent dependability requirements on communication networks, it is important to ensure that network configurations are policy compliant at any time. This, however, is challenging: already in medium-sized networks, existing network devices (e.g. switches, routers, middleboxes) may contain ten thousands of rules (e.g. forwarding rules, firewall rules, etc.), far too many for manual consideration. But even automated approaches to verify seemingly simple properties such as "Can A reach B?" or "Does the route from A to B traverse a location C?" can be computationally intensive, especially if the policy compliance properties must hold under failures [15–17]. In fact, even in scenarios where conducting such "what-if analysis" is possible in polynomial time (e.g. in MPLS networks whose stack-based labels can be described as push-down automata [17]), the runtimes can be relatively high. *ML-based applications could help to identify misconfigurations more quickly, e.g. after training with the exact verification tool, or by speeding up the tool itself* [18].

8.2 Technology Overview

This section first describes how ML/AI can help in the design and execution of algorithms. We then provide an overview of a subset of possibilities to represent algorithm data, i.e. (problem, solution)-pairs, in addition to highlighting the use of neural networks.

8.2.1 Data-Driven Network Optimization

Figure 8.1 gives a general view of the different optimization possibilities: the traditional optimization approach (Figure 8.1a), the ML/AI-enhanced way (Figure 8.1b), and our vision, a fully data-driven learned algorithm (Figure 8.1c). Traditionally, an optimization algorithm, like a routing algorithm, continuously faces problem instances as input. Here, the algorithm is always executed from scratch. However, routing requests between similar areas, e.g. sub-networks,

(a) ML/AI-enhanced problem optimization. Existing optimization algorithm simply take next problem instances and calculate the solutions; they neglect the data generated from previous runs.

(b) ML/AI sits between the problem instances and, e.g. existing optimization algorithms. ML/AI learn from data in order to provide information that helps to optimize the performance of traditional optimization algorithms.

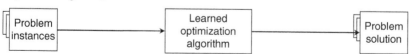

(c) The vision: an optimization algorithms is completely learned from data and replaces the original optimization algorithms, that produced the first starting data.

Figure 8.1 Comparison between (a) traditional and (b) ML/AI-based approaches. (c) The vision is to have a completely data-driven algorithm design. A trained algorithm replaces the original optimization algorithm.

might lead to similar solutions. Accordingly, an ML/AI-based algorithm could learn such information: the same sub-network might have the same representation over time, and the same requests between those networks might lead to the same solutions over time.

This idea is presented in Figure 8.1b. An ML-based model is trained with (problem, solution)-pairs created by an existing algorithm in the past. Note that the training samples are generated with any yet available algorithm. Plugging in the model between problem instances and the traditional optimization algorithm, ML could provide various opportunities to feed information that helps to increase the algorithm efficiency. In case of routing requests, a sub-graph could be extracted that reduces the search-space, or even a possible path could be offered in advance that is afterward improved by a local search procedure.

In contrast, as a final visionary step, an optimization pipeline could only use a completely learned algorithm (Figure 8.1c). Here, the ML algorithm is providing solutions to a given problem instance. No traditional optimization algorithm might be involved. Note, however, that many ML-based algorithms, such as Reinforcement Learning (RL), might not provide any guarantees for their solutions. Hence, recent research efforts, in particular, in the optimization community, focus on the design of ML/AI-based algorithms that can give guarantees [4]. We highlight this as a future research aspect also in Section 8.4 in more detail.

In all cases, we identify the efficient representation of (problem/solution)-pairs as an important challenge to tackle in order to integrate ML/AI in the algorithm design. Some representation concepts are outlined in Section 8.2.2.

8.2.2 Optimization Problems over Graphs

Many optimization problems in network management and control are formulated as mathematical programming problems. In their simplest form, a function taking input values has to be minimized or maximized: e.g. minimizing the path length when looking for the shortest path in a computer network. Such mathematical problems have strong relations to graphical representations, hence many general optimization problems and network optimization problems rely on the representation of problem instances as graphs (Figure 8.2): the Traveling Salesman Problem (TSP), the facility location problem, routing problems, congestion control problems, etc. To later assess the problem of representing graph data for ML or AI, we first take a step back and define the relation between optimization problems and their graphical presentation.

Graph definition: A graph is an abstract, discrete structure. This structure has a set of points (nodes, vertices) that are interconnected by a set of lines (links, edges). In this chapter, a network or graph $G = (\mathcal{V}, \mathcal{E})$ is a tuple consisting of the

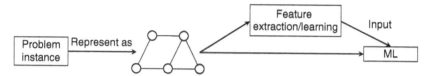

Figure 8.2 Problem instances are first represented as graph. The graph is then either directly fed into a ML algorithm, or first preprocessed to extract a feature representation suitable for ML techniques that cannot directly handle graph data as input.

set of nodes $\mathcal{V}$ and the set of links $\mathcal{E}$. Depending on the problem formulation, this graph can be directed, un-directed, have multi-edges, self-edges, can be weighted or have additional node and edge attributes. For example, in the VNE (see Section 8.1.2) each edge is assigned a bandwidth limitation and/or coordinates. Bidirectional communication in computer networks can be represented via undirected edges; directed edges are used, e.g. to describe hyperlinks of a graph representing a network of web pages.

The nodes can represent any object (even whole sub-networks); the links put two objects in relation with each other. Graphs themselves can be used to represent objects like the Internet. Nodes in the Internet graph could correspond to many Autonomous Systems (ASs). Links correspond to the possibility to send data from one AS to another AS. Interestingly, the ASs, i.e. the nodes in the Internet graph, are themselves networks.

Adjacency matrix: The adjacency matrix is a binary or real matrix that represents the edges in a graph. Rows and columns correspond to nodes in the graph. An element of the matrix indicates a relation between the two nodes. For example, in a binary adjacency matrix a value of one indicates that the edge exists. In a weighted adjacency matrix, a nonzero value can indicate e.g. the distance between two nodes, or the amount of traffic that flows between two nodes.

Combinatorial optimization over graphs: Optimization problems are generally based on specific tasks or questions: for example, what is the shortest path between two ASs?

As already stated, those problems can be represented as a graph. A graph then serves as the basis for the combinatorial optimization problem. From the graph, decision variables are derived and constraints formulated. The decision variables encode the answer to a question or task. The constraints encode the requirements that this answer must fulfill. For example, the routing problem has binary decision variables associated with each node that indicate whether this node is used in a path of a solution or not.

Learned algorithms vs. mathematical problem formulation: At a high level, data-driven (aka. learning) algorithms can be understood as a mapping from the graph, encoding the problem, to the answer, sidestepping any mathematical formulation. The input to learning algorithms is derived from the graph that is formed from the original question or task. The learning representation depends on the problem family and the learning algorithm used. The output data of learning algorithms are often related to or even are the assignments of decision variables. For example, the output in [19] corresponds to a solution (sequence of nodes) or it provides information for where to search for the best solution [20].

8.2.3 From Graphs to ML/AI Input

While graphs are a great tool for modeling and reasoning about complex systems, they might demand preprocessing in order to achieve generalization. This means transformation of graphs should be scale-free. That is, every graph of any size (e.g. in terms of number of nodes) is transformed into a representation of the same size (e.g. a vector having always the same length). For instance, different graphs can have a different amount of edges and nodes but should be used with the same ML model. A central question is thus: How to transform graphs to suitable input for ML and AI? The remainder of this section introduces different graph representations and which learning algorithms use them.

Adjacency matrix: Learning algorithms that use the adjacency matrix as input are, e.g. spectral methods and graph kernels [21].

Spectral methods: Spectral methods study the properties of the adjacency matrix, e.g. eigenvalues and eigenvectors. Spectral methods are popular since the calculated characteristics do not depend strongly on the labeling of the nodes. That is, the same graph can be represented with different adjacency matrices, e.g. by different sorting of the nodes. The adjacency matrices are thus different, their spectral properties, however, will be very similar [22]. The outcome of the spectral methods can be used as input for the learning algorithm.

Graph kernels: A common method to classify graphs is to employ *graph kernels*. For graph kernels, a function $k : \mathcal{G} \times \mathcal{G} \to \mathbb{R}$ is defined, where $\mathcal{G}$ is the set of all possible graphs for which a mapping $\varrho : \mathcal{G} \to \mathcal{H}$ into a Hilbert Space $\mathcal{H}$ exist, s.t. $k(G, G') = \langle \varrho(G), \varrho(G') \rangle$, $\forall G, G' \in \mathcal{G}$ [21]. Intuitively, graph kernels simply measure the similarity between two graphs. Accordingly, they help ML algorithms to work directly on the graph. Hence, graph kernels could be used to differentiate problem instances.

Graph structure: Other techniques that can be used to learn representations of a graph can operate directly on the graph structure itself. Examples here are graph features and latent space models.

Graph features: Graph and node measures such as degree of a node or the average node degree of a graph, were originally used to quantitatively argue about graphs and their nodes. Table 8.1 provides an overview about some

Table 8.1 Node and graph features used for problem representation and learning.

Node/graph feature	Computational complexity
Node degree	$O(n + m)$
Betweenness centrality	$O(n^3)$
Clustering coefficient	$O\left(\dfrac{m^2}{n}\right)$
Path length	$O(2n^2 + nm)$
Number of nodes	$O(n + m)$

Note that many more graph features exist; we just took one feature from every complexity class. The variable n constitutes the number of nodes; variable m the number of edges.

well-known graph and node measures and their computational complexities [23]. Each measure captures one or more features of a graph or a node, respectively. For instance, the betweenness centrality of a node can indicate important nodes in computer networks: many shortest paths are using a node with a high betweenness centrality. With respect to graphs, an average node degree can be used to quantify the general connectivity inside a graph: a high average node connectivity might yield a graph with many alternative routes.

Interestingly, some of the graph features themselves are target of learning approaches. For example, obtaining all possible path lengths is expensive. Thus, approaches exist that sample some paths and learn a model to estimate the actual value [5].

Beside their ability to help quantifying graphs, measures enable the representation of a graph or nodes as a fixed-length real-valued vector. For instance, graphs of similar sizes can initially and quite easily be compared by looking at their number of nodes or edges. Hence, it has been shown that they can be used by ML algorithms to classify graphs: they can even outperform graph kernels [23].

Latent space models: Latent space models take a graph as input and learn representations of the graph or its nodes in a latent features space. The latent space is usually a multi-dimensional Euclidean space but also other approaches exist [24]. Graph features can be interpreted similarly, calculating graph features maps nodes and graphs into an Euclidean space by calculating characteristic values. A latent space model is different in that usually a specific optimization goal exists based on which feature representations are learned. This optimization goal is not necessarily related to an actual use-case, though. For example, node representations are obtained and then used for a node classification task. That is, learning the representation and learning a classification model are two separate steps.

One such technique is node2vec [25]. node2vec learns node representations that maximize the probability of observing a specific neighborhood of a node. A neighborhood is not restricted to direct neighbors and can be sampled arbitrarily. The choice of the neighborhood influences the encoded information, for example, emphasizing structural or social patterns in the graph [25].

Usually, the purpose of latent space models is primarily to learn expressive feature representations that can be used by downstream ML tasks. Latent space models are thus considered as unsupervised or semi-supervised learning. Many approaches exist already [26–30].

Attributed graphs: In the case of attributed graphs, additional data is associated with nodes and edges that should be included for learning representations, or otherwise be made accessible to the learner. Additional data can be link bandwidth or the routing table capacity of a node. Thus, techniques are required that are able to consider both: The relational data encoded in the graph structure, and the node and edge attributes. Examples of techniques that leverage structural as well as node and edge attributes are Graph Convolution Networks (GCNs) [13], GNNs [14], sequence-based models [31], and engineering solutions [23] that transform attributed graphs into forms that can be consumed by traditional ML techniques.

Graph convolutional networks: GCNs take as input a $\mathcal{V} \times d$ matrix of d node features for each node and the adjacency matrix. GCNs then produce a node-level output, i.e. GCNs also embed nodes into a latent space. The main difference is the fact that GCNs can directly be integrated into an end-to-end differentiable model that solves a specific task. In contrast to traditional latent space models, GCNs learn representations that are tailored toward a specific task [32]. The main disadvantage of GCNs is that they have problems with graphs of varying size due to the integration of the adjacency matrix.

Graph neural networks: GNNs also learn node features. However, the underlying principle is different from GCNs. In GNNs, nodes continuously create latent representations through aggregating the current latent representations of their neighbors. This iterative process converges to a fixed point. The resulting features encode then neighborhood and structural information, similarly to GCNs. The principle underlying GNNs is the same as distance vector algorithms such as boarder gateway protocol (BGP) or routing information protocol (RIP) [14]. Similar to GCNs, GNNs can be used to learn task-specific representations.

Since their proposals, GNNs have evolved continuously. The main limitation of GNNs is that due to their iterative process, they have problems extracting hierarchical features. The aggregation function that is used inside the model is represented as a single neural network that is shared across all nodes and used in all iterations. Other approaches thus relax on the fixed point assumption and use a fixed sequence of iterations, which can then be represented with a multi-layer neural network [33, 34]. Those approaches are also able to generalize to previously unseen nodes. Node features are learned through shared functions that aggregate neighborhood information. Thus, nodes are represented through their neighbors' features, which can be calculated for new nodes as well.

Graph structuring: An example of a graph-structuring approach is PATCHY-SAN [13]. PATCHY-SAN is a framework to create fixed-size representations of attributed graphs that are suited for the use with Convolutional Neural Networks (CNNs) [35]. PATCHY-SAN samples local regions from arbitrary graphs. In contrast to images where the neighborhoods are implicitly given by the locations of the pixels, neighborhoods have to be defined for arbitrary graphs. PATCHY-SAN fills this gap by first, selecting a set of nodes from the graph for which neighborhoods should be constructed and second, normalizes the neighborhoods to obtain a fixed linear order. The result is a fixed-size structure that can be consumed by a CNN. It has been shown that PATCHY-SAN performs competitively to state-of-the-art graph kernels while being more efficient in terms of training duration [13]. Besides, extending the concepts of CNNs to arbitrary graphs enables feature visualization to gain structural insights.

8.2.4 End-to-End Learning

The following approaches fall under the category of the vision of fully learned optimization algorithms. Figure 8.1 shows the idea. Fully learned optimization algorithms target at learning the overall optimization problem in a data-driven manner. For instance, an ML model (e.g. a deep neural network [DNN]) takes as input the optimization problem, either formulated as a mathematical program or represented as a graph, and then outputs a solution directly. No additional optimization algorithm is needed here.

Mathematical formulation-based representations: The representation can be based directly on the mathematical formulation of the problem. Hopfield Neural Networks (HNNs) are one such example [36]. HNNs interconnect artificial neurons with weights and inputs representing optimization problems, i.e. the neurons, the input of the neurons, the weights of the interconnections of neurons, capture the constraints, and the optimization function of the problem. HNNs have a so-called energy function: the energy function describes in an abstract manner the current state of the neural network in relation to the optimization problem. A high energy might resemble a state where some constraints are not met or the optimal solution is not found yet. When minimizing the energy, the final states of neurons provide solutions to the optimization problems. Note that such approaches are seen as predecessors of end-to-end learning approaches.

Sequence-based models: Some optimization problems operate on a fully connected graph, e.g. routing problems. In such situations, GNNs and GCNs are not that useful since the graph does not provide meaningful structure, after all, each node is connected with every other node. In this specific niche, sequence-based models are used [31]. Those models read in a sequence of inputs and output an equally sized sequence. In the case of the TSP, the approach of [31] takes as input a sequence of nodes and returns as output a permuted sequence of nodes corresponding to the tour. Those networks work in principle on not-fully connected graphs as well and are then similar to [33].

8.3 Data-Driven Algorithm Design: State-of-the Art

In this section, we give the reader a brief overview of work that used ML and AI to advance optimization research. Note that for the first part, we only selected a subset of work as pointers. Then, we look at approaches using ML and AI to directly solve *network management and control* problems, and approaches that advance existing algorithms by improving their efficiency and runtimes.

8.3.1 Data-Driven Optimization in General

In general, there is much hope in tackling algorithmic problems by a data-driven design approach [4]. Table 8.2 summarizes papers reviewed within this section w.r.t. to the used graph representation.

One first example of data-driven optimization is on learning heuristics for search algorithms, such as A^* [39]. Considering the path finding or routing problem again, algorithms such as A^* use admissible heuristics to guide the search for an optimal solution. When A^* terminates, it is guaranteed that the found path is optimal. However, A^* can become computational expensive. In such cases, suboptimal search algorithms using inadmissible heuristics provide an alternative, although they provide only sub-optimal solutions. First approaches learned inadmissible heuristics for automated planning and problem solving [5]. Such heuristics are trained at runtime with (problem, solution) data. Similarly, Khalil et al. [37] automate the design of good heuristics or approximation algorithms for hard optimization problems. The authors also argue that real world

Table 8.2 Table provides overview of research work, general optimization problems and chosen representations.

Approach	Problem	Representation
[5]	Search heuristic	Graph-features
[20]	Branch-and-bound	
[6]	Branch-and-bound	
[36]	Traveling salesman problem	Attributed graph
[31]	Traveling salesman problem, Convex hull problem, Delaunay triangulation	
[19]	Traveling salesman problem, Knapsack problem	
[37]	Minimum vertex cover, maximum cut, traveling salesman problems	Latent-space
[38]	Maximum independent set, maximum vertex cover, maximum clique, satisfiability	GCN

demands of similar optimization problems are repeatedly solved, producing data, e.g. RL can exploit. Already a wide range of algorithms can actually be learned to solve, e.g. minimum vertex cover or maximum cut problems.

Another application field are branch-and-bound techniques: they are used, e.g. for global optimization of combinatorial problems [20]. Branch-and-bound guides solvers through a search tree of possible solutions to find optimal solutions. The way nodes are explored can be learned with ML [6]. Here, already solved problems can be used to train the approach.

The availability of large-scale cloud resources and the evolved computing technology and hardware drove recent breakthroughs in ML such as deep learning. Reference [31] applied supervised learning to train a pointer network, a special kind of neural network, to solve combinatorial optimization problems. Instead of applying a supervised learning approach, Bello et al. [19] use RL to train the pointer network architecture of [31] to solve combinatorial optimization problems such as knapsack problem or TSP. Here, a Recurrent Neural Network (RNN) finally learns to find the most likely route through cities. Li et al. [38] used GCNs to solve combinatorial optimization problems.

To summarize, there exist a wide range of works that apply ML and AI to either solve optimization problems or to help speeding up general optimization algorithms. Interestingly, many approaches perform best when they are applied to specific problem instances, i.e. real-world examples reflecting only a subset of all

Table 8.3 Table overviewing networking research work, general optimization problems and chosen representations.

Approach	Problem	Representation
[40]	VNE	Graph features
[41]	VNE, facility location	Graph features
[42]	SDN controller placement	Attributed graph
[43]	Virtual cluster embedding	Graph structuring
[18]	MPLS verification	Graph neural networks
[44]	Topology reconfiguration	Adjacency matrix (weighted)

possible problem instances. Accordingly, expert knowledge is still needed to cast, e.g. network management and control problems for ML- and AI-based approaches.

8.3.2 Data-Driven Network Optimization

In this section, we survey recent state-of-the-art that focus on improving the performance of algorithms solving network management and control problems by learning from available solution data (Table 8.3).

Learning from the data of algorithms solving network management and control problems: Efficient admission control for VNE problems (see Section 8.1.2) could filter out infeasible or low performing virtual network requests. Such admission control is implemented based on an RNN for VNE by Blenk et al. [40]. The idea is to check beforehand if the embedding request could be actually embedded before running a computational expensive embedding algorithm. Note that some problem instances cannot be easily decided beforehand. For instance, all nodes and edges could be embedded when treated isolated, however, altogether they cannot be embedded. Before running the embedding algorithm, hence, the RNN predicts whether a request can be accepted or not. To train the RNN efficiently, this system relies on graph and node features for the representation of virtual network requests and the substrate network. Simulation results show that it is indeed possible to predict the feasibility of requests: accordingly, the runtime has been improved by up to 91%.

The approach as presented by Blenk et al. [41] extends [40] to further problems. The presented system speeds up the execution time, e.g. for facility location problems (see Section 8.1.2). In addition, it extends the analysis of [40] to other ML algorithms. For instance, Blenk et al. [41] uses random forest algorithms, which allow to reason about the importance of features. Interestingly, quite simple features can help to improve the algorithm performance: number of nodes or edges. For the facility location problem, it can be shown that ML can reduce the search space by more than 75% while keeping the original performance of the algorithm without search space reduction.

Learning facility (controller) placements in dynamic traffic scenarios: Software-defined networks offer great opportunities with respect to flexibility; however, to fully exploit this flexibility, control plane mappings procedures need to adapt quickly to dynamic traffic scenarios. Reference [42] extend recent investigations (i.e. taking algorithm data into account) to dynamic traffic scenarios: here, the communication patterns of traffic flows within an software-defined networking (SDN) network change over time. For such changes, it is important to continuously adapt the locations of the network logic (i.e. the controllers) to always guarantee the best possible reaction time of control decisions. Based on the current traffic distribution of network flows among wide area topologies, an ML algorithm can solve a multi-label problem (multiple locations are chosen jointly among a vector of locations). Whereas the predicted solutions might not always be optimal, they can serve as good initial solutions, e.g. for approaches like local search. Here, ML can reduce the amount of executions to 50% on average.

Learning admission control in data centers: Zerwas et al. [45] initiated the application of data-driven algorithms to admission control in data centers. The authors propose AHAB, an admission control algorithm that strategically rejects individual requests, even if there are sufficient resources. AHAB "hunts" useful requests over time by using information about previous requests and embeddings. While AHAB already improves the overall cluster utilization by 13%, it is a time-consuming and data-intensive algorithm. To cope with this problem, the authors proposed ISMAEL, which extends AHAB by adding a ML component [43]. ISMAEL predicts the acceptance of virtual clusters by learning form network states and outcomes of acceptance decisions of the past. To achieve this, ISMAEL uses a fixed-size feature representation for graphs in combination with a CNN [13]. The algorithm achieves a high accuracy while significantly reducing the runtime.

Learning how to improve network verification: Geyer and Schmid [18] study how to speed up verification and synthesis of policy-compliant network configurations (see Section 8.1.2), in the context of MPLS. At the heart of their tool called DEEPMPLS lies a novel extension of graph-based neural networks: based on deep learning, the tool allows to predict counter examples (i.e. witness traces) to specific network properties (or queries), which can be verified fast. In the synthesis application, the idea is to predict which MPLS rules should be added, in order to re-establish certain properties. The tool is preliminary but shows the potential for overcoming the need to perform rigorous and time-consuming analyses.

Learning network reconfigurations in data centers: Wang et al. [44] train CNNs to optimize reconfigurable optical data-center networks on demand with the goal of setting connections between racks to minimize flow completion time. Their approach consists of multiple stages where they first predict solution quality of (demand, topology)-pairs based on data that artificially generated. This first CNN is then used to guide an algorithm searching for good solutions (not optimal ones). This search procedure generates the data used for training a second ML model. As runtime of this search is bounded but still large, they eventually learn the obtained (demand, topology)-pairs in a supervised fashion to online predict the topology configuration. To boost performance, their neural network is enhanced with a conditional random field which embeds prior human knowledge, e.g. about conflicting links (infeasible configurations).

8.3.3 Non-graph Related Problems

Besides the wide range of work in general and network optimization problems in particular, recent work has also applied the data-driven concept to non-graph-related problems. For comprehension, we briefly report on some of them here. For instance, mobile edge computing is an active research field, as the demands are high to support latency-sensitive and compute-intensive services. To achieve this, computing and storage capacities are deployed at the edge of a network. The assignment of such resources to service tasks needs to be solved fast and efficiently. Song et al. [46] train a DNN with data from optimal solutions. The proposed approach achieves always solutions that are near a 1.6-factor approximation of the optimal solutions for at least 99.5% of the evaluated problem instances.

Further, Song et al. [47] learns bitrate adaptation from solutions for a dynamic adaptive streaming over HTTP (DASH) video scenario. The authors use the solutions of an optimal algorithm to train a DNN to predict the best bitrate distribution among many clients. Using the DNN, the proposed scheme achieves 85% of the optimal solutions on all of the test problem instances.

Reference [48] look at the efficient operation of virtual network functions (VNFs) in a dynamic traffic scenario. In such scenarios, it is important to always determine the best number of currently operated VNFs to accommodate current and upcoming demands. The proposed ML-based approach uses monitoring data to adapt the current number of VNFs. In order to generate labeled data for their approach, the authors use labeled training data from an optimal algorithm. The approach achieved an accuracy of 73.8% when predicting the next decision.

8.4 Future Direction

The application of ML and AI to speed up algorithms for network management and control problems is a young field and opens many interesting avenues for future research. We will now identify and discuss some major research areas.

8.4.1 Data Production and Collection

The application of ML to speed up algorithms relies on the generation of data. This data generation, however, proposes challenges.

Fast and efficiently generated data: Monitoring and storing data demands efficient management solutions, such as in many other research fields applying ML; hence, it is important to implement systems and interfaces that allow an easy collection of data generated by algorithms. Interestingly, however, such data might not yet be available: for instance, control algorithms producing routing table entries in routers are still closed by vendors. Accordingly, new interfaces need to be created and standardized that (i) allow to collect data and (ii) to collect it efficiently.

Challenging data: In order to generate realistic data, new interfaces need to be created to export data from networks, e.g. from networking devices or from systems as a whole like data centers. Moreover, most data, e.g. from data centers, is still not available publicly for the research community, which renders it difficult to create learning systems for further study and comparison. Accordingly, researchers need to implement simulation and emulation frameworks. Such simulations or emulations frameworks are then used to benchmark management and control algorithms, e.g. based on traffic generated from simplified models or generated with human best guesses – the simulations do not create challenging problem settings. We envision ML and AI to provide help at another front: generating adversarial network data. Instead of only applying ML and AI on the solution front, they could also be applied when generating data. We envision a data generation system, that itself uses ML and AI to create adversarial data. The data generation component takes the solution quality of an algorithm as feedback, and tries to guide its search for more challenging input data. The algorithms consequently learn from data that is more challenging. Such an approach might help identify weak spots and make data-driven algorithms bullet proof for future networking demands.

Malicious data: Data used for learning might be compromised [49]; the data could be taken from systems that have been compromised by an attacker. Accordingly, taking such data for learning might lead to solutions that leave security holes and performance weak spots. We see interesting possibilities for future research: traditional systems that can help to detect such data as anomalies, however, with focus on algorithms for network management and control.

8.4.2 ML and AI Advanced Algorithms for Network Management with Performance Guarantees

While many interesting approaches are emerging to speed up and improve algorithms using techniques from ML and AI, much existing literature revolves around heuristics: While the proposed algorithms are shown to perform well in practice, these algorithms often do not provide hard and formal guarantees on their performance. That is, while ML/AI-based algorithms can perform very well under the most common and learned inputs, they may compute suboptimal solutions under "unusual inputs."

This lack of guarantees can be problematic especially in mission-critical environments: given stringent dependability requirements, the underlying algorithms may have to provide real-time guarantees and ensure a certain approximation ratio compared to the optimal solution, under arbitrary inputs.

Besides performance, many ML/AI approaches may also err in their decision with a certain probability. Going back to our example related to the verification of networks based on GNNs [18], the algorithm may provide fast and accurate answers in most cases, but may overlook specific scenarios. Depending on the nature of the error, e.g. whether it is one-sided or two-sided and whether the scenario may be identified quickly, one may simply employ postprocessing; other scenarios may be harder to fix, rendering the ML/AI solution inherently heuristic.

A promising approach to the design of algorithms that benefit from ML and AI while additionally providing hard correctness and performance guarantees in the worst case, could be to apply the heuristic algorithms *within certain bounds*. For example, exact and provably approximate algorithms often come with "flexibilities," e.g. feature knobs and parameters which can be tuned without losing their formal runtime, approximation, or correctness guarantees. One avenue for the design of ML/AI-based algorithms would be to optimize or even dynamically adapt these parameters and exploit the available flexibilities.

8.5 Summary

This chapter was motivated by the observation that algorithms generate much data, which could be exploited to improve network performance, using data-driven algorithm designs. We presented an overview of this vision and of the state-of-the-art technology, we identified application scenarios, and discussed research challenges. We hope that this chapter inspires more research in this area, for which we see much potential.

Acknowledgments

This work was funded by the Deutsche Forschungsgemeinschaft (DFG, German Research Foundation) – 438892507, and the Austrian Science Fund (FWF) joint D-A-CH project with Germany, I 4800-N (ADVISE), 2020-2023, and in part by the European Union's Horizon 2020 research and innovation program (grant agreement No. 647158-FlexNets).

Bibliography

1 Enns, R., Bjorklund, M., Schoenwaelder, J., and Bierman, A. (eds.) (2011). *Network Configuration Protocol (NETCONF)*. RFC 6241 (Proposed Standard). ISSN 2070-1721. Updated by RFCs 7803, 8526.

2 The Open Networking Foundation (2015). OpenFlow Switch Specification v1.5.1.

3 Kellerer, W., Kalmbach, P., Blenk, A. et al. (2019). Adaptable and data-driven softwarized networks: review, opportunities, and challenges. *Proceedings of the IEEE* 107 (4): 711–731.

4 Gupta, R. and Roughgarden, T. (2020). Data-driven algorithm design. *Communications of the ACM* 63 (6): 87–94.

5 Thayer, J.T., Dionne, A., and Ruml, W. (2011). Learning inadmissible heuristics during search. *Proceedings of International Conference on Automated Planning and Scheduling*, ICAPS'11, AAAI Press, pp. 250–257.

6 Khalil, E.B., Le Bodic, P., Song, L. et al. (2016). Learning to branch in mixed integer programming. In: *Proceedings of AAAI*. event-place, Phoenix, Arizona: AAAI Press, pp. 724–731.

7 Xie, J., Yu, F.R., Huang, T. et al. (2019). A survey of machine learning techniques applied to software defined networking (SDN): research issues and challenges. *IEEE Communication Surveys and Tutorials* 21 (1): 393–430.

8 Boutaba, R., Salahuddin, M.A., Limam, N. et al. (2018). A comprehensive survey on machine learning for networking: evolution, applications and research opportunities. *Journal of Internet Services and Applications* 9 (1): 16.

9 Latah, M. and Toker, L. (2019). Artificial intelligence enabled software defined networking: a comprehensive overview. *IET Networks* 8 (2): 79–99.

10 Feamster, N. and Rexford, J. (2018). Why (and how) networks should run themselves. *Proceedings of the Applied Networking Research Workshop*, Montreal, QC, Canada, ACM, p. 20.

11 Fadlullah, Z.Md., Tang, F., Mao, B. et al. (2017). State-of-the-art deep learning: evolving machine intelligence toward tomorrow's intelligent network traffic control systems. *IEEE Communication Surveys and Tutorials* 19 (4): 2432–2455.

12 Bengio, Y., Lodi, A., and Prouvost, A. (2021). Machine learning for combinatorial optimization: a methodological tour d'horizon. *European Journal of Operational Research* 290 (2): 405–421.

13 Niepert, M., Ahmed, M., and Kutzkov, K. (2016). Learning convolutional neural networks for graphs. In: *Proceedings of ICML* (ed. M.F. Balcan and K.Q. Weinberger), pp. 2014–2023. New York (20–22 June 2016).

14 Scarselli, F., Gori, M., Tsoi, A.C. et al. (2009). The graph neural network model. *IEEE Transactions on Neural Networks* 20 (1): 61–80.

15 Anderson, C.J., Foster, N., Guha, A. et al. (2014). NetKAT: semantic foundations for networks. *SIGPLAN Notices* 49 (1): 113–126.

16 Kazemian, P., Varghese, G., and McKeown, N. (2012). Header space analysis: static checking for networks. *Proceedings of USENIX NSDI*. San Jose, CA: USENIX Association, pp. 113–126.

17 Jensen, J.S., Krøgh, T.B., Madsen, J.S. et al. (2018). P-Rex: fast verification of MPLS networks with multiple link failures. *Proceedings of ACM CoNEXT*. Heraklion, Greece: ACM, pp. 217–227.

18 Geyer, F. and Schmid, S. (2019). DeepMPLS: fast analysis of MPLS configurations using deep learning. In *Proceedings of IFIP Networking*, Warsaw, Poland, pp. 1–9.

19 Bello, I., Pham, H., Le, Q.V. et al. (2017). Neural combinatorial optimization with reinforcement learning. *Proceedings of ICLR 2017*, Toulon, France, OpenReview.net.

20 He, H., Daume, H. III, and Eisner, J.M. (2014). Learning to search in branch and bound algorithms. In: *Advances in Neural Information Processing Systems 27* (ed. Z. Ghahramani, M. Welling, C. Cortes et al.), 3293–3301. Curran Associates, Inc.

21 Gärtner, T., Flach, P., and Wrobel, S. (2003). On graph kernels: hardness results and efficient alternatives. In: *Learning Theory and Kernel Machines* (ed. B. Schölkopf and M.K. Warmuth), 129–143. Springer.

22 von Luxburg, U. (2007). A tutorial on spectral clustering. *Statistics and Computing* 17 (4): 395–416.

23 Li, G., Semerci, M., Yener, B., and Zaki, M.J. (2012). Effective graph classification based on topological and label attributes. *Statistical Analysis and Data Mining: The ASA Data Science Journal* 5 (4): 265–283.

24 Matias, C. and Robin, S. (2014). Modeling heterogeneity in random graphs through latent space models: a selective review. *ESAIM: Proceedings and Surveys* 47: 55–74.

25 Grover, A. and Leskovec, J. (2016). node2vec: Scalable feature learning for networks. In: *SIGKDD 2016* (ed. B. Krishnapuram, M. Shah, A.J. Smola), 855–864. ACM: San Francisco, CA.

26 Cai, H.Y., Zheng, V.W., and Chang, K.C.-C. (2018). A comprehensive survey of graph embedding: problems, techniques, and applications. *IEEE Transactions on Knowledge and Data Engineering* 30 (9): 1616–1637.

27 Chen, H., Perozzi, B., Al-Rfou, R., and Skiena, S. (2018). A Tutorial on Network Embeddings. *arXiv:1808.02590 [cs]*, August 2018. http://arxiv.org/abs/1808.02590. arXiv: 1808.02590.

28 Goyal, P. and Ferrara, E. (2018). Graph embedding techniques, applications, and performance: a survey. *Knowledge-BASED SYSTEMS* 151: 78–94.

29 Zhang, D., Yin, J., Zhu, X., and Zhang, C. (2020). Network representation learning: a survey. *IEEE Transactions on Big Data* 6 (1): 3–28.

30 Hamilton, W.L., Ying, R., and Leskovec, J. (2017). Representation learning on graphs: methods and applications. *IEEE Data Engineering Bulletin* 40 (3): 52–74.

31 Vinyals, O., Fortunato, M., and Jaitly, N. (2015). Pointer networks. *Proceedings of Advances in Neural Information Processing Systems*, pp. 2692–2700.

32 Kipf, T.N. and Welling, M. (2017). Semi-supervised classification with graph convolutional networks. *Proceedings of ICLR (Poster)*. Toulon, France: OpenReview.net.

33 Veli?kovi?, P., Cucurull, G., Casanova, A. et al. (2018). Graph attention networks. *ICLR 2018*, Vancouver, BC, Canada, pp. 1–9.

34 Hamilton, W.L., Ying, Z., and Leskovec, J. (2017). Inductive representation learning on large graphs. In: *Proceedings NIPS 2017*. Curran Associates Inc., Red Hook, NY, USA, 1025–1035.

35 LeCun, Y., Bengio, Y., and Hinton, G. (2015). Deep learning. *Nature* 521 (7553): 436–444.

36 Hopfield, J.J. (1984). Neurons with graded response have collective computational properties like those of two-state neurons. *Proceedings of the National Academy of Sciences of the United States of America* 81 (10): 3088–3092.

37 Khalil, E.B., Dai, H., Zhang, Y. et al. (2017). Learning combinatorial optimization algorithms over graphs. In *Proceedings of NIPS*. Long Beach, CA, USA, 6348–6358.

38 Li, Z., Chen, Q., and Koltun, V. (2018). Combinatorial optimization with graph convolutional networks and guided tree search. *Advances in NIPS 30*, pp. 1–10.

39 Hart, P.E., Nilsson, N.J., and Raphael, B. (1968). A formal basis for the heuristic determination of minimum cost paths. *IEEE Transactions on Systems Science and Cybernetics* 4 (2): 100–107.

40 Blenk, A., Kalmbach, P., van der Smagt, P., and Kellerer, W. (2016). Boost online virtual network embedding: using neural networks for admission control. *Proceedings of CNSM*, pp. 10–18.

41 Blenk, A., Kalmbach, P., Kellerer, W., and Schmid, S. (2017). O'zapft is: tap your network algorithm's big data! *Proceedings of the Workshop on Big Data Analytics and Machine Learning for Data Communication Networks*, ACM, pp. 19–24.

42 He, M., Kalmbach, P., Blenk, A. et al. (2017). Algorithm-data driven optimization of adaptive communication networks. *Proceedings of IEEE 25th ICNP*, pp. 1–6.

43 Zerwas, J., Kalmbach, P., Schmid, S., and Blenk, A. (2019). Ismael: using machine learning to predict acceptance of virtual clusters in data centers. *IEEE Transactions on Network and Service Management* 16 (3): 950–964.

44 Wang, M., Cui, Y., Xiao, S. et al. (2018). Neural network meets DCN: traffic-driven topology adaptation with deep learning. *Proceedings of ACM on Measurement and Analysis of Computing Systems* 2 (2): 1–25.

45 Zerwas, J., Kalmbach, P., Fuerst, C. et al. (2018). AHAB: data-driven virtual cluster hunting. *Proceedings of IFIP Networking*, Zurich, Switzerland, pp. 1–9.

46 Song, T., Xu, W., Hu, W. et al. (2019). ARM: an accelerator for resource allocation in mobile edge computing. *Proceedings of IEEE GLOBECOM*, pp. 1–6.

47 Song, T., Hu, W., Xu, W. et al. (2019). Fair-area: a fast Ai-based joint optimization of rate adaptation and resource allocation for dash. *Proceedings of IEEE GLOBECOM*, pp. 1–6.

48 Lange, S., Kim, H., Jeong, S. et al. (2019). Machine learning-based prediction of VNF deployment decisions in dynamic networks. *Proceedings of the 2019 20th Asia-Pacific Network Operations and Management Symposium (APNOMS)*, pp. 1–6.

49 Meier, R., Holterbach, T., Keck, S. et al. (2019). (Self) driving under the influence: intoxicating adversarial network inputs. *Proceedings of ACM HotNets*. Princeton, NJ, USA: ACM, pp. 34–42.

9

AI-Driven Performance Management in Data-Intensive Applications

Ahmad Alnafessah[1], Gabriele Russo Russo[2], Valeria Cardellini[2], Giuliano Casale[1], and Francesco Lo Presti[2]

[1]Department of Computing, Imperial College London, London, UK
[2]Department of Civil Engineering and Computer Science Engineering, University of Rome Tor Vergata, Rome, Italy

9.1 Introduction

In recent years, the prominence of Big data has led to a growth in interest for developing intelligent data-intensive software systems in several application domains. Data-driven systems that can extract knowledge, plan, and adapt to events through processing, transformation, and analysis of datasets are thus increasingly widespread in both industry and society.

From a technical standpoint, data-driven software systems are often built by leveraging features such as batch analytics or streaming, now easily programmable through in-memory platforms such as Apache Spark, Hadoop/MapReduce, Storm, Flink, among others. We outline popular data processing platforms in Section 9.2. Although the combination of batch and streaming workloads enables richer functionalities, workload heterogeneity also means that achieving service levels objectives presents additional complexity in pinpointing causes of performance degradation and identifying tuning to address them. For example, performance metrics in data-driven software are difficult to predict as they often depend on data properties, such as volume or velocity, and frequently even on data type and content, making it difficult to reason about system performance at design time. Furthermore, the combination of batch and streaming features in a software means that different system components will strive to achieve different performance goals, i.e. high-throughput and high-utilization for analytic features and low-latency for stream processing operators, making the process of run-time performance tuning a fairly heterogeneous and complex exercise.

Communication Networks and Service Management in the Era of Artificial Intelligence and Machine Learning,
First Edition. Edited by Nur Zincir-Heywood, Marco Mellia, and Yixin Diao.
© 2021 The Institute of Electrical and Electronics Engineers, Inc. Published 2021 by John Wiley & Sons, Inc.

To support these challenges, the goal of this chapter is to overview artificial intelligence (AI) management techniques that are available in the literature to manage and tune the performance of data-intensive applications. AI methods offer considerable simplicity and flexibility in choosing the features that drive the management process, in spite of some opaqueness in presenting the way the models reach decisions.

Compared to traditional management methods, which either leverage low-level system characteristics or use mathematical modeling abstractions, AI management methods leverage learning on experimental datasets that reduce dependence on assumptions and shift the attention from conceptual modeling to data-collection and model training. This offers considerable potential to increase the effectiveness of management methods in situations where the system behaves according to complex and unpredictable logic, as it is often the case for systems driven by external data.

Summarizing, in this chapter we examine the applicability of AI methods in the context of data-intensive applications. We survey in particular studies that illustrate the versatility of AI models when applied to popular data streaming and batch analytic platforms. Our aim in particular is to cover a broad spectrum of AI methods, in order to inform the reader on the range of learning techniques that may be applicable to recurring management problems involved in data-driven systems. We look at common management tasks such as platform configuration, workload forecasting, resource scaling, monitoring, and detection of performance anomalies. We also give selected examples to build an intuition on their behavior, benefits, and limitations.

9.2 Data-Processing Frameworks

In this section, we overview the essential features of common execution platforms in use to define data-intensive applications. A summary of the key characteristics of each platform is shown in Table 9.1. The section also highlights key performance management challenges associated to each of these platforms.

9.2.1 Apache Storm

Apache Storm[1] [1] is a popular open source platform used for distributed real-time stream processing. The platform offers very low latency for dataflow processing, making it an ideal option for real-time processing [1]. Storm dataflow topologies involve two main node types: spouts and bolts. A spout is the source of the data

---

1 http://storm.apache.org/

Table 9.1 Summary of the key characteristics of data-processing platform.

Platform	Workloads		Processing style	
	Batch	Streaming	In-memory	Disk-heavy
Storm		✓	✓	
Hadoop/MR	✓			✓
Spark	✓	✓	✓	
Flink	✓	✓	✓	

stream at the input queue and may generate data by itself [2]. A bolt instead consumes the stream, operates transformations or computations, and ultimately produces an output stream as a result. Every task corresponds to one operating system thread.

Performance management of Storm applications frequently involves difficult decisions concerning optimal configuration options for spouts and bolts, ranging from decisions concerning buffer, message, and batch sizes, number of bolts, and selection of optimal waiting strategies. There is a limited understanding of the interplay between these parameters, posing intrinsic challenges for optimal system configuration. Moreover, the Storm system does not automatically manage load balancing and resource scaling, thus requiring ad hoc performance management techniques.

9.2.2 Hadoop MapReduce

Hadoop implements the MapReduce paradigm, and it is a well-known example of batch processing platform. It is used for intensive Big Data applications starting from a single server and can scale up to thousands of machines[2]. Usually, MapReduce uses an existing dataset that is stored in Hadoop Distributed File System (HDFS) before beginning to process batch data. Processing with native Hadoop can be paused or interrupted, but the dataset cannot be modified. This means that if current data is changed for any reason, the job needs to be run again.

Despite distinctive challenges arise in the area of optimal configuration of Hadoop platforms, over the years performance management has insisted in particular on the problem of detecting and handling straggler tasks, which falls into the general problem area of performance anomaly detection and mitigation. This is a result of the synchronizations between dataflow tasks that can block progress until straggler tasks complete their activities.

---

2 https://hadoop.apache.org/

9.2.3 Apache Spark

Apache Spark[3] is a large-scale in-memory processing framework that can support both batch and stream data processing, which can make it easy with a low cost to support different types of workloads on the same engine in a production environment, such as those arising from graph analysis and machine learning applications. Compared to older solutions such as Hadoop, the main goal of Apache Spark is to speed up batch processing by utilizing in-memory computation. Thanks to a reduced use of intermediate storage of processing results, Spark is orders of magnitude faster than Hadoop for in-memory analytics.

Spark can be deployed over Hadoop as an alternative to MapReduce, as well as on Amazon EC2, Apache Mesos, or as a standalone cluster. In addition, it can access many data sources, including HDFS, Cassandra, HBase, Hive, Tachyon, and any Hadoop data source. Spark provides a general purpose engine for different kinds of computation, including iterative algorithms, job batches, streaming, and interactive queries. These different types of computation were previously difficult to find in the same distributed system [3]. Beyond the ability to perform batch and stream processing, Spark also provides a rich library that is built on top of its core engine [4].

In terms of performance management, Apache Spark has around 200 configuration parameters (e.g. executors, CPU cores, memory, shuffle behavior, compression), which may significantly impact the overall Spark system performance [5]. The microarchitectural behaviors of Spark are different from those of other Big Data technologies. The Spark core data abstraction is the Resilient Distributed Dataset (RDD), which cannot be modified and RDD can be executed in parallel on different nodes. In addition, Spark needs more advanced auto tuning solutions to boost its performance within production environment. Therefore, precisely performance management to optimally manage and auto-tune Spark is needed to increase the performance efficiency of such a complex system and immediately gain advantages of cost and time saving.

9.2.4 Apache Flink

Apache Flink[4] [6] is another open source distributed processing engine designed for low-latency streaming computation. Analogously to Spark, Flink relies on in-memory computation and provides a unified application programming interface (API) for processing both bounded and unbounded datasets. However, differently from Spark, where batching has a primary role, Flink has been designed with streaming in mind. Indeed, Flink applications are built upon the

---

3 https://spark.apache.org/
4 http://flink.apache.org/

concepts of *streams* and *transformations*. Streams represent (possibly unbounded) data flows, while transformations are operations that given one or more streams as input, output one or more streams as the result (e.g. filtering). A few high-level libraries are built on top of these abstraction, easing the definition of common processing use cases (e.g. complex event processing, graph analytics).

At run-time, Flink applications are mapped to *streaming dataflows*, Directed Acyclic Graphs (DAGs) composed of processing nodes (often called *operators*), which implement transformations, connected by streams. For execution, Flink leverages a distributed architecture, designed according to the *master-worker* pattern. The master component is the *JobManager*, which coordinates distributed execution and is responsible for application scheduling, checkpointing, and recovery in case of failure. The TaskManagers (i.e. the workers) execute the application *tasks* (i.e. instances of operators) and manage the data transfers between them.

Performance management of Flink applications, which are often long-running, mainly involves run-time deployment and resource adaptation. First of all, varying infrastructure conditions may require migrating operator tasks between computing nodes during execution. Flink supports migrating both stateless and stateful tasks through the *savepoint* mechanism, which ensures no loss of information. Moreover, workload variability requires dynamically scaling the parallelism of Flink applications and balancing the load across the cluster to keep consistent performance levels over time. To this end, load prediction techniques can be helpful to proactively adapt application configuration.

9.3 State-of-the-Art

We review in this section the existing techniques for dealing with the most relevant performance management issues in the context of data-intensive applications, with particular emphasis on AI-based approaches.

9.3.1 Optimal Configuration

As a consequence of the availability of numerous configuration parameters, their optimization is a critical task in the domain of data-intensive systems. The goal of configuration optimization is to find the ideal configuration with respect to the system performance. Various automated parameter tuning methods have been proposed in the literature, which are discussed in Sections 9.3.1.1 and 9.3.1.2.

9.3.1.1 Traditional Approaches

The authors of [10] propose a parameter tuning framework based on design of experiments (DOE) approach. Their goal is to find an initial range of parameter values for automated tuning using a factorial experiment design to screen and rank all the parameters, so as to focus the search on the most influencing parameters. In addition, the authors examine response surface methodology, which is a model-based approach within DOE that can be used to quantify the effect of each parameter to find the most promising initial range for the vital parameter values. Their approach can be integrated with existing automated parameter tuning configuration, called ParamILS and randomized convex search (RCS). Their method seems promising for both discrete and continuous parameter configuration settings.

The authors in [11] introduce MRTuner from IBM, which is a tool to enable holistic optimization for MapReduce jobs. Their design uses an efficient search algorithm (grid-based Search) to find the optimal execution plan. Around 20 configuration parameters are investigated to understand the relationships that have a noticeable impact on MapReduce performance. The tool is evaluated using HiBench on two Hadoop clusters. Their results show MRTuner has low latency and can find accurate execution plans.

Bilal and Canini [12] examine an automatic parameter tuning framework for stream processing platforms. Gray-Box, Black-box analysis, and a rule-based optimization method are combined, and configuration parameters are initialized using Latin hypercube sampling. Hill climbing algorithm is used to explore the configuration space. The authors evaluate using three benchmark applications within the Apache Storm streaming system. They find that rule-based can converge up to five times faster than other approaches, making it suitable for parameter tuning within stream processing platforms.

9.3.1.2 AI Approaches

A machine learning approach is used by Chen et al. [13] to appropriately tune configurate the parameters of Hadoop. Their approach has two stages, which are the prediction stage to estimate the performance of a MapReduce job and the optimization stage to repeatedly search for the optimal configuration parameters. The authors claim that their method can improve Hadoop performance up to eight times compared with traditional methods.

Wang et al. [14] introduce a parameter tuning method based on binary classification and multi-classification for Apache Spark systems. Decision trees are used for auto-tuning of configurations with four different types of workloads, which are Sort, Wordcount, Grep, and NavieBayes workloads from *BigDataBench* benchmark. Their experimental results show that the proposed method can improve Spark performance on average by 36% compared to default Spark configuration.

Hernández et al. [15] optimize parallelism for data-intensive platforms using machine learning. Boosted regression trees are used as the authors claim that they have the lowest variance compared with other algorithms. In addition, they argue that decision trees are interpretable, which means that it is possible to quantify the impact of collected features on the overall performance. They evaluate proposed solution using a benchmark of 15 different Spark applications running on YARN. The results show that their task parallelization method is capable of improving the performance of Spark by 51%.

Bayesian optimization (BO) is an effective and efficient method for auto-tuning systems and machine learning algorithms. Joy et al. [16] propose a framework that uses BO to tune hyperparameters of data-intensive applications. Their idea is dividing the data into small chunks with the same size to boost the search by applying BO tuning in parallel. To validate the performance of their framework, they use the proposed method to tune two machine learning algorithms, deep neural networks (DNNs) and support vector machines (SVMs). BO offers effective hyperparameters tuning with less computational overhead.

Jamshidi and Casale [17] tackle the challenging issue of finding optimal configurations for a data-intensive streaming system by proposing auto-tuning methods that can help systems administrators to determine the near-optimal configurations with a limited budget of experiments. Their solution revolves around BO for configuration optimization, which utilizes Gaussian processes (GPs) to continuously capture posterior distributions of configuration space for the application. Their method works in a way that the optimal configurations will eventually be discovered. The authors validate the proposed method using a Storm cluster in the cloud.

Yigitbasi et al. [18] examine and explore a machine learning model to tune the configuration parameters of Hadoop and MapReduce. They use support vector regression (SVR) to a smart search algorithm in terms of the effectiveness of parameter space exploration. Their results show that SVR obtains higher accuracy than Starfish auto-tuner, which uses a cost-based search model.

Liao et al. [19] illustrate that the Hadoop platform has hundreds of configuration parameters that have very complicated interactions. This wide configuration space makes it time-consuming for system administrators to optimally tune the Hadoop parameters. They provide an evaluation to automate the tuning process of Hadoop based on a cost-based and machine learning approach (neural network, SVR, multiple linear regression, and decision trees).

Di Sanzo et al. [20] provide a study about auto-tuning cloud-based in-memory transactional data grids configuration by using a machine learning black-box approach. They use artificial neural networks (ANNs) to optimize the dynamic selection of the amount of cache servers and the replication level of data objects to reduce the cost of cloud system operations. They conduct preliminary experiments based on a synthetic benchmark and a real data grid system that is run on Amazon EC2 virtual servers. The authors conclude that the ANN-based approach is effective for tuning transactional data grids. There are some additional works related to optimal configuration using classification and regression trees

(CARTs) [21], long short-term memory (LSTM) [22], regression trees, nearest neighbor [23], and reinforcement learning (RL) [24].

9.3.1.3 Example: AI-Based Optimal Configuration

In this section, we illustrate the effectiveness of BO for finding optimal configurations by searching through the configuration space. The goal of BO is to utilize the prior knowledge and evidence to optimize the posterior at each evaluation step, to reduce the gap between the actual global optimization and expected optimization for the model [25]. Compared with traditional search algorithms (grid search, random search, and manual tuning), in [26] we show how BO can facilitate parameter tuning with more parameters and fewer number of experiments to find optimal configurations.

BO is an ideal choice to find the optimal training dataset size and configuration parameters to efficiently and effectively train the anomaly detection model to achieve high F-score in a short period of time. Before applying BO, there are two main choices that need to be carefully made, which is the prior over functions and type of acquisition function [27]. We use GPs, which are stochastic processes defined by the property that any finite set of N points induces a multivariate Gaussian distribution [28, 29]. They are efficient for uncertainty estimation. We use the *expected improvement* acquisition function that [28] provides for configuration space with high uncertainty and high estimated value to evaluate a point x to sample based on the posterior distribution function to guide exploration. It can trade-off between exploration of the configuration search space and exploitation of current promising subspace.

Figure 9.1 shows a comparison between BO and random search in achieving the high performance training of the machine learning algorithm (ANNs in our case) to efficiently detect anomalies (CPU, cache thrashing, and context switching) within Apache Spark Streaming datasets that [26] provides with predefined

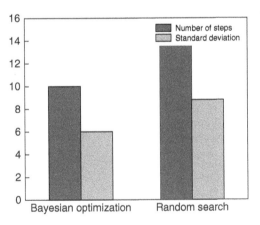

Figure 9.1 Comparison between Bayesian optimization and random search to reach the highest F-score.

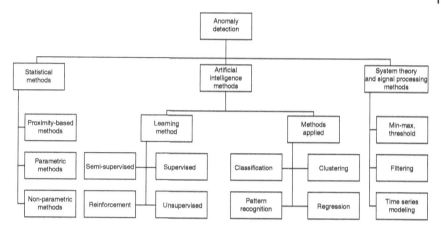

Figure 9.2 A taxonomy of anomaly detection techniques generalizing the ones in [34–36]. Source: Adapted from Sebestyen et al. [36], Hodge and Austin [34], Chandola et al. [35].

F-score. For CPU anomaly detection, the BO can optimally train the anomaly detection model using 10 combinations of configurations, whereas the random search needs 14 combinations of configurations on average. In addition, BO outperforms the random search in training the AI model for detecting CPU, cache, and context switching anomalies.

9.3.2 Performance Anomaly Detection

System performance is often described in terms of the time taken to process a set of tasks with a given amount of computing resources that are consumed within a given observation period [30]. The growing complexity and dynamicity of cloud systems and data-intensive technologies requires significantly higher levels of automation and significant attention [31]. Performance anomalies have become a major concern for developers and academic researchers particularly for Big Data and AI technologies over cloud computing systems. Anomalous performance can occur as a result of service operator faults [32], system failures, user errors [33], environmental issues, and security violations [30], among others.

Many anomaly detection studies are generic, while others are specifically conducted for certain application domains (e.g. data-intensive applications, networking, web-based application, etc.). There are studies that provide an overview about techniques that have been developed in traditional statistical approaches and machine learning techniques for anomaly identification [30, 34, 35]. Figure 9.2 shows a taxonomy of existing anomaly detection techniques that are built on common taxonomy based on [34–36].

9.3.2.1 Traditional Approaches

Statistical techniques are the earliest approaches used for performance anomaly detection. Lu et al. [37] use a statistical offline approach to detect abnormal Apache Spark tasks and analyze the root causes based on statistical spatial-temporal analysis. They use some features related to execution time, memory usage, garbage collection, and data locality of each Spark task to determine the degree of abnormal tasks. They use mean and standard deviation of all tasks in each stage to decide the threshold and to get information about macro-awareness on the task execution time. They analyze performance issues using factor combination criteria for every performance anomaly based on weighted factors. They validate their method on a private Spark cluster and SparkBench [7].

Kelly [38] examines how to get insights about performance issues of globally distributed systems using simple queueing theoretic observations together with standard optimization methods. The author obtains extensive empirical results from three distributed commercial production systems that serve real customers.

Yang et al. [39] propose an anomaly detection and diagnosis solution within grid environments using statistical and signal processing approaches. Their work extends the traditional window-based strategy by using signal processing to filter out recurring background variations and determine which resource is the probable cause of an anomalous performance in a system. They use window averaging, which is a widely used statistical anomaly identification technique because it is simple and efficient. The anomalies are injected into three grid systems (Cactus, GridFTP, and Sweep3d) at random time intervals to evaluate the proposed technique. The results show that their method can classify 75% of anomalies causes within the grid environment.

9.3.2.2 AI Approaches

Research on automated anomaly detection is essential in practice because any late detection or slow manual resolution of performance anomalies in a real production environment may cause prolonged service-level agreement violations and significant financial penalties [8, 9]. This leads to a demand for performance anomaly detection solutions in cloud computing and data intensive systems that are both dynamic and proactive in nature [30]. The need to develop these methods for production environment with very different characteristics means that AI is ideally positioned for system diagnosis to automatically identify performance anomalies. These techniques provide the capability to quickly learn baseline performance characteristics through a large monitoring metrics space in order to distinguish normal and anomalous patterns [40].

Classification techniques aim to determine whether the instances in a given feature space belong to a particular class or multiple classes [30]. There are popular classification techniques for anomaly identification, such as ANNs, SVM, and nearest neighbor. The classification technique is significantly affected by the accuracy of the labeled data and algorithms that have been used. For example, the training and testing processes for decision trees algorithms are usually faster than SVM, which involve quadratic optimization.

Alnafessah and Casale [41] propose an ANN-driven methodology for anomaly identification, particularly for Apache Spark. The authors use a machine learning approach to quickly sift through Spark logs and system monitoring metrics to precisely detect and classify anomalous behaviors. The authors evaluate the proposed method against three popular machine learning algorithms, decision trees, nearest neighbor, and SVM, as well as against four different monitoring datasets. Their results show that the recommended method has ability to classify overlapped anomalies and outperforms other methods by obtaining 98–99% F-scores, and offering much higher performance than alternative techniques to detect both the period in which anomalies occurred and their type.

Lu et al. [42] utilize convolutional neural networks (CNNs) for performance anomaly diagnosis for Big Data system logs, and specifically for the HDFS logs. They implement the proposed model with different filters to automatically train model on the relationships among events. The CNN is configured to have *logkey2vec* embeddings, three 1D convolutional layers, dropout layer, fully connected Softmax layer, and max-pooling. The authors provide a comparison between CNNs and other well-known networks such as LSTM networks and multilayer perceptron (MLP). The experimental results show that the CNN model is more accurate and faster in detecting anomalies than LSTM and MLP for HDFS logs.

Fulp et al. [43] predict system failures using SVM algorithm for binary classification based on system log files. The proposed approach utilizes advantages of the sequential nature of logs and uses a sliding window of messages to predict the likelihood of system failure within that has 1024 computing nodes. The SVM associates the messages to a class of normal or abnormal event. Their results show that the proposed solution can predict hard disk failure with 73% accuracy. Fu et al. [44] propose a hybrid anomaly identification framework using one-class and two-class SVM algorithms. They claim that their approach does require prior knowledge about system failure history and offers self-adapt learning from observing system failure within cloud environment.

There are several other studies in the literature that deal with stragglers. Yadwadkar and Choi [45] introduce a proactive straggler avoidance regression decision tree model that periodically learns correlations between node level status and task execution time for MapReduce logs. The authors justify the choice of regression trees by showing the fast prediction of stragglers. They apply their method on a trace from Facebook Hadoop system and Berkeley EECS department's local Hadoop cluster (icluster). Qi et al. [46] use a white-box model that utilizes classification and regression trees for root causes analyses for Spark logs and hardware sampling tools to train their model. A special type of tree called a CART tree (*classification and regression tree*) is used to mitigate overfitting issues. They use a customized prune method for several iterations to improve analysis accuracy and the classification performance metrics are checked for each node and its leaves. The authors apply their method on Spark with HiBench benchmark.

Based on the local neighborhoods of event, the neighbor-based technique uses unsupervised learning to analyze data instances. This technique can distinguish an anomalous instance among normal instances because normal instances usually occur in dense neighborhoods, whereas anomalous instances occur far from their closest neighbors [35]. Huang et al. [47] propose a special type of neighbor-based technique, called local outlier factor (LOF), for an anomaly identification that can learn system behaviors during training and detecting time within cloud computing environment. They argue that their method is adaptive to changes, detects contextual anomalies, and requires less effort for collecting performance metrics for training process.

9.3.2.3 Example: ANNs-Based Anomaly Detection

Classification techniques aim to determine whether the instances in a given feature space belong to a particular class or multiple classes [30]. ANNs algorithms are the most popular classification technique for anomaly identification. This is because ANNs represent a data-driven, nonlinear, and self-adaptive method that can adjust itself to the given datasets without requiring prior knowledge about the distribution or function of the used model, and generalize the models even to input data that has never been seen before [28]. These advantages have caused ANNs to be considered a universal functional approximation.

One of the well-known critical issues for ANNs and other classifiers is feature selection. The objective of feature selection is to discover the smallest set of appropriate input features and at the same time achieve the desirable predictive performance. It is crucial to reduce the number of input features for the classifier to achieve satisfactory accuracy with reduced computation in the model [28].

We use backpropagation and conjugate gradients to train ANNs, that is to update values of weights and biases in the network. We use scaled conjugate because it is often fast [48], especially for time-dependent applications. *Sigmoid* transfer function is often used as an activation function in the hidden layer because it exists between (0–1), where 0 means absence of the feature and 1 means its presence. In addition, we use *Softmax* transfer function in the output layer to handle classification problems with multiple classes (e.g. normal, CPU anomaly, cache thrashing anomaly, context switching anomaly). For a cost function, we use *cross-entropy* to evaluate the performance and compare the actual output error results with the

Figure 9.3 The performance of ANNs models that is trained with training dataset that have been collected for 1, 5, 30, and 60 minutes.

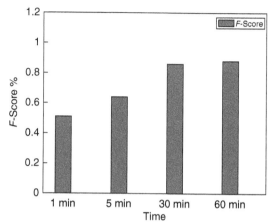

desired output values (labeled data). We use *cross-entropy* because it has significant practical advantages over squared-error cost functions [49].

The input layer contains a number of neurons equal to the number of input features. The size of the hidden layer is determined by using a "trial and error" method, by trying all the possible numbers between the sizes of input neurons and output neurons [50]. The output layer contains a number of neurons equal to the number of target classes (normal + types of anomalies).

Figure 9.3 shows a sensitivity analysis for the size of collected datasets to train the neural network algorithms to learn complex nonlinear relationships among performance metrics and detect the anomalous performance within Apache Spark systems. It is clear that the size of collected training data significantly impacts the performance of ANN model. It is challenging to find the optimal size of the training data. The small size of training dataset causes unacceptable F-score, whereas a large dataset may lead to a waste of computing resources. Figure 9.3 shows that collecting the Apache Spark performance dataset for 30 minutes is ideal for training the neural networks.

9.3.3 Load Prediction

Data streaming applications usually deal with unbounded data flows, meaning that they are kept in execution indefinitely. As a consequence, these applications likely face different working conditions over time (e.g. varying workloads), hence requiring dynamic resource management solutions to keep acceptable performance. Indeed, researchers have spent a lot of effort aiming to enhance streaming systems with online adaptation capabilities and, in particular, *elasticity*, that is the ability to dynamically acquire and release computing resources [51] as needed. To this end, there are two main research directions so far, revolving around

(i) application load prediction and (ii) auto-scaling policies. In this section, we focus initially on load prediction.

9.3.3.1 Traditional Approaches

Given the variability that often characterizes streaming workloads, predicting the application processing load in the future (e.g. application input data rate) is a difficult yet important task for driving resource management with foresight. To this end, traditional time series forecasting methods can be useful in the context of streaming applications. For instance, Imai et al. [52] use the well-known ARIMA model for workload forecasting. They additionally consider an online regression approach for predicting the maximum sustainable throughput of streaming applications, and scale the number of virtual machines (VMs) allocated to the system accordingly. Kombi et al. [53] instead exploit regression techniques to predict the input rate of Storm operators, and drive the application auto-scaling. They consider three prediction models, respectively based on linear, logarithmic, and exponential regression, and select the best model to use at run-time based on fitting accuracy observed in the previous iteration.

9.3.3.2 AI Approaches

AI techniques often allow to outperform traditional forecasting approaches for workload and resource utilization prediction. For instance, Zacheilas et al. [54] use GPs for predicting the future input rate and processing latency of operators, and hence drive horizontal elasticity of *complex event processing* applications running on top of Storm. Their elasticity algorithm exploits the uncertainty estimation provided by GPs to avoid making auto-scaling decisions whenever the uncertainty level is considered too high. Hu et al. [55] instead use SVR for predicting resource usage of Spark Streaming applications, and allocate virtual machines for the cluster so as to meet SLA requirements. Runsewe and Samaan [56] also target Spark Streaming and leverage layered hidden Markov models to predict the resource usage of multiple applications running on a Spark cluster. Based on the obtained predictions, they scale the Spark cluster as needed.

A few works have investigated the use of NNs to predict the future load of data streaming applications. Lombardi et al. [57] propose ELYSIUM, a multi-level elasticity solution for Storm, which controls both the operator parallelism and the number of worker nodes in the Storm cluster. ELYSIUM relies on NNs for predicting both (i) the application input rate in the near future and (ii) the CPU utilization of each application operator based on the input rate.

Mu et al. [58] use DNNs for multi-step operator performance prediction. They define two prediction strategies based on DNNs, to be used, respectively, on offline and online collected metrics. They use ensemble learning techniques to merge the offline and online predictions, and obtain the final prediction, which can be used to drive auto-scaling. Xu et al. [59] rely on DNNs as well, to predict operator performance online. Specifically, they exploit recurrent DNNs to make accurate performance predictions, which also account for interference due to co-located operators (i.e. operators deployed in the same worker node). They integrate this solution in Storm, and their experiments show that it outperforms ARIMA- and SVR-based approaches for prediction. Mixture density networks are used instead in [60] to estimate resource usage of streaming applications as probability density functions. They show how the resulting distribution-based workload prediction can be applied to drive both auto-scaling and application admission control in presence of SLAs.

9.3.4 Scaling Techniques

Data-intensive systems largely exploit parallelism to efficiently process high-volume datasets. For streaming applications, whose datasets are collected in real time and hence are not known at deployment time, scaling the amount of computing resources at run-time is fundamental to avoid the risk of under- or over-provisioning resources.

9.3.4.1 Traditional Approaches

As extensively surveyed in [61], researchers so far have investigated a large number of approaches to devise auto-scaling policies for streaming systems, including, e.g. queueing theory, control theory, state-space based methods, and, recently, AI. Existing solutions can be classified as either reactive and proactive. *Reactive* solutions make auto-scaling decisions in response to observed changes (e.g. increase in the application input data rate), whilst *proactive* approaches try to adapt the application deployment before observing changes, based on predictions.

Among the reactive approaches, threshold-based policies are widely adopted. According to these policies, auto-scaling actions are triggered whenever one or more observed metrics (e.g. resource utilization, throughput) violate predefined threshold values [62, 63]. Other works instead rely on models to periodically evaluate the expected application performance or resource utilization, and trigger scaling actions accordingly. For instance, Lohrmann et al. [64] rely on queueing theory to model application performance, and make horizontal scaling decisions so as to meet response time requirements.

Among the proactive auto-scaling solutions, several works present policies that exploit load prediction to make scaling decisions. Indeed, most the prediction solutions mentioned in Section 9.3.3, are complemented with auto-scaling mechanisms. A different approach is considered in [65], where *model predictive control* is used to proactively scale streaming operators, also combining horizontal and vertical elasticity.

9.3.4.2 AI Approaches

The behavior of traditional auto-scaling solutions often depends on manually configured parameters, and AI-based approaches aim at overcoming this limitation. For instance, RL is a class of methods allowing agents (e.g. resource managers) to learn policies by direct interaction with their environment (e.g. managed applications). RL has been adopted by several works to derive auto-scaling policies for streaming applications at run-time. For instance, Heinze et al. [66] use the SARSA algorithm, relying on a reward function that captures the difference between current operator CPU utilization and target utilization values. Similarly, Cheng et al. [67] rely on the well-known Q-learning algorithm to adapt the amount of resources allocated to jobs running in Spark Streaming. They consider a performance-oriented reward function that accounts for throughput and latency. Lombardi et al. [57] also use Q-learning to automatically tune the parameters for a threshold-based auto-scaling algorithm.

Russo Russo et al. [68] consider the auto-scaling problem in the presence of heterogeneous computing resources to host parallel operator instances. They use linear function approximation to deal with the large model state space, and investigate model-based initialization to speed up the learning process. Their reward function accounts for the amount of allocated resources, the adaptation cost, and a service level objective (SLO) violation penalty.

9.3.5 Example: RL-Based Auto-scaling Policies

RL agents learn by experience the *actions* to perform in order to maximize a cumulative *reward* over time [69]. We define the task faced by RL agents as an infinite-horizon, discrete-time Markov decision process (MDP), where agents perform an action at every time step, selected according to their *policy* and the observed current *state*. Following action execution, agents get a reward and possibly enter a new state. Their goal is maximizing the (discounted) cumulative reward over the infinite time horizon.

The auto-scaling problem for a streaming operator could be modeled as follows. Considering a slotted time model, we define the state at time step i, s_i, as the pair (k_i, λ_i), where $1 \leq k_i \leq K_{max}$ denotes the operator parallelism, and λ_i the monitored input rate (discretized using a suitable quantum). Actions in this model represent scaling operations that alter the parallelism level, hence they are selected from the set $\mathcal{A} = \{-1, 0, +1\}$, except for the states where no further scale-out (or, scale-in) is allowed.

In the context of resource management, it is often convenient to reason in terms of *cost* instead of reward. Therefore, we define the cost $c(s, a, s')$ paid for operating the system in state s' after taking action a in s, and let the reward $r(s, a, s') = -c(s, a, s')$. Depending on the specific scenario, the cost function may capture different aspects. We define it as a weighted sum of three cost components: resources cost (proportional to the operator parallelism level), adaptation cost (capturing the overhead due to scaling), and performance violation cost (paid whenever the chosen configuration does not satisfy performance requirements).

Most RL algorithms rely on the so-called *Q-function* $Q(s, a)$, which estimates the cumulative discounted reward obtained in the long-term when choosing action a in s. The most popular RL algorithm is Q-learning, which performs a single Q update at every time step:

$$Q^{new}(s_i, a_i) \longleftarrow (1 - \alpha)\, Q^{old}(s_i, a_i) + \alpha \left(r_i + \gamma \max_{a'} Q^{old}(s_{i+1}, a') \right) \quad (9.1)$$

where r_i is the reward obtained at time i, and $\alpha \in (0, 1)$ is the learning rate. Q-Learning is easy to implement, and guarantees convergence to the optimal policy as infinite exploration is provided. However, to achieve faster convergence, in practice it is worth including any available knowledge about the model in the learning algorithm (*model-based* RL). For instance, the concept of *post-decision state* (PDS) [70] can be used to separate the known system dynamics (e.g. impact of a scaling action on the parallelism) from the unknown ones (e.g. input rate variations), and let the agent only learn the latter. To demonstrate the benefits provided by PDS, we simulated the execution of a data streaming operator under varying input data rate, using the RL-based auto-scaling policy. Figure 9.4 shows a sample of the workload we used, and the average reward accumulated over time by the RL agent, in the case of plain Q-learning and Q-learning with PDS. We can note that the PDS-based agent clearly outperforms plain Q-learning, as it exploits the available knowledge about the system and needs to learn fewer parameters.

This model can be extended to account for other adaptation mechanisms (e.g. operator migration), infrastructure characteristics, or to include more general

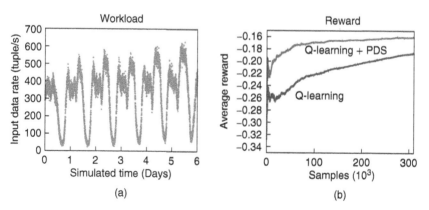

Figure 9.4 Streaming workload in a simulated experiment (a) and average reward obtained by RL-based auto-scaling agents (b).

workload characterizations. As more complex models are used, however, the state space often grows significantly, requiring, e.g. function approximation techniques for manipulating the Q-function. To this end, Deep RL is receiving growing interest, where DNNs are used to approximate Q.

9.4 Conclusion and Future Direction

Table 9.2 summarizes the mapping of AI models to management and resource allocation activities we have referenced through this chapter. Our analysis reveals that AI-driven optimal configuration and anomaly detection in data-intensive applications have received considerable attention already, covered a broad range of methods. In spite of this, certain techniques, such as RL, still present open room for further investigation.

As mentioned, performance management for data streaming systems benefits from run-time load prediction, which is often coupled with auto-scaling or admission control mechanisms. Neural networks, and in particular DNNs, have received the largest share of attention so far, as they have been shown to outperform other techniques, like SVMs and GPs. The adoption of other approaches, already used for forecasting in other domains, including, e.g. LSTM models and regression trees, could be the subject of future investigation.

As regards the definition of auto-scaling control policies for data-intensive applications, only RL techniques have been exploited so far. Nevertheless, as the class of RL methods is quite large and complex, this research direction is far from being completely explored. Among the main issues to be tackled when adopting RL algorithms, state space explosion is critical in the context of performance management,

Table 9.2 Summary of the AI techniques considered for performance management of data-intensive systems

AI method	Optimal configuration	Anomaly detection	Load prediction	Scaling
Neural networks	✓	✓	✓	
Boosted regr. trees	✓			
CART	✓	✓		
Decision trees	✓	✓		
Gaussian processes	✓		✓	
LSTM	✓	✓		
Nearest neighbor	✓	✓		
RL	✓			✓
SVMs	✓	✓	✓	

as it limits the granularity and completeness of the application and performance models used by the agents. To overcome this issue, *Deep Reinforcement Learning* (DRL) algorithms exploit DNNs for approximate learning in otherwise intractable tasks. For instance, Li et al. [71] and Ni et al. [72] have recently applied DRL to control the placement of streaming operators over a cluster of machines. Further investigations are needed in the literature to understand the potential of DRL in the management of data-intensive systems.

In conclusion, the chapter has shown that AI methods are already subject to intense research work across various management areas, with the areas of optimal configuration and anomaly detection being the most mature. The richness and the rapid evolution of the AI research landscape offer considerable opportunities to further raise the maturity of AI methods. Our analysis reveals that, although significant work already exists in this area, load prediction and scaling techniques would particularly benefit from a broader research investigation.

Bibliography

1 Toshniwal, A., Taneja, S., Shukla, A. et al. (2014). Storm@Twitter. *Proceedings of ACM SIGMOD '14*, ACM, pp. 147–156.

2 Shahrivari, S. (2014). Beyond batch processing: towards real-time and streaming big data. *Computers* 3 (4): 117–129.

3 Karau, H., Konwinski, A., Wendell, P., and Zaharia, M. (2015). *Learning Spark: Lightning-fast Big Data Analysis*. O'Reilly Media, Inc. ISBN 144935906X.

4 Meng, X., Bradley, J., Yavuz, B. et al. (2016). MLlib: machine learning in Apache Spark. *Journal of Machine Learning Research* 17 (1): 1235–1241.

5 Herodotou, H., Chen, Y., and Lu, J. (2020). A survey on automatic parameter tuning for big data processing systems. *ACM Computing Surveys* 53 (2): 1–37.

6 Carbone, P., Katsifodimos, A., Ewen, S. et al. (2015). Apache Flink: stream and batch processing in a single engine. *IEEE Data Engineering Bulletin* 38 (4): 28–38.

7 Li, M., Tan, J., Wang, Y. et al. (2015). SparkBench: a comprehensive benchmarking suite for in memory data analytic platform Spark. *2015 12th ACM CF*, ACM, pp. 1–8.

8 Dean, D.J., Nguyen, H., and Gu, X. (2012). UBL: unsupervised behavior learning for predicting performance anomalies in virtualized cloud systems. *Proceedings of ACM ICAC '12*, ACM, pp. 191–200.

9 Tan, Y., Nguyen, H., Shen, Z. et al. (2012). PREPARE: predictive performance anomaly prevention for virtualized cloud systems. *Proceedings of IEEE ICDCS '12*, pp. 285–294.

10 Gunawan, A. and Lau, H.C.L. (2011). Fine-tuning algorithm parameters using the design of experiments approach. *LION 2011*, volume 6683 of *LNCS*, Springer, pp. 278–292.

11 Shi, J., Zou, J., Lu, J. et al. (2014). MRTuner: a toolkit to enable holistic optimization for MapReduce jobs. *Proceedings of the VLDB Endowment* 7 (13): 1319–1330.

12 Bilal, M. and Canini, M. (2017). Towards automatic parameter tuning of stream processing systems. *Proceedings of SoCC '17*, ACM, pp. 189–200.

13 Chen, C.-O., Zhuo, Y.-Q., Yeh, C.-C. et al. (2015). Machine learning-based configuration parameter tuning on Hadoop system. *Proceedings of IEEE Big Data Congress '15*, IEEE, pp. 386–392.

14 Wang, G., Xu, J., and He, B. (2016). A novel method for tuning configuration parameters of Spark based on machine learning. *Proceedings of IEEE HPCC/SmartCity/DSS '16*, IEEE, pp. 586–593.

15 Hernández, A.B., Perez, M.S., Gupta, S., and Muntés-Mulero, V. (2018). Using machine learning to optimize parallelism in big data applications. *Future Generation Computer Systems* 86: 1076–1092.

16 Joy, T.T., Rana, S., Gupta, S., and Venkatesh, S. (2016). Hyperparameter tuning for big data using Bayesian optimisation. *Proceedings of ICPR '16*, IEEE, pp. 2574–2579.

17 Jamshidi, P. and Casale, G. (2016). An uncertainty-aware approach to optimal configuration of stream processing systems. *Proceedings of IEEE MASCOTS '16*, IEEE, pp. 39–48.

18 Yigitbasi, N., Willke, T.L., Liao, G., and Epema, D. (2013). Towards machine learning-based auto-tuning of MapReduce. *Proceedings of IEEE MASCOTS '13*, IEEE, pp. 11–20.

19 Liao, G., Datta, K., and Willke, T.L. (2013). Gunther: Search-based auto-tuning of mapreduce. *Proceedings of Euro-Par '13*, volume 8097 of *LNCS*, Springer, pp. 406–419.

20 Di Sanzo, P., Rughetti, D., Ciciani, B., and Quaglia, F. (2012). Auto-tuning of cloud-based in-memory transactional data grids via machine learning. *Proceedings of NCCA '12*, IEEE, pp. 9–16.

21 Nguyen, N., Khan, M.M.H., and Wang, K. (2018). Towards automatic tuning of Apache Spark configuration. *Proceedings of IEEE CLOUD '18*, pp. 417–425.

22 Fang, H., Lu, W., Li, Q. et al. (2019). Predictive analytics based knowledge-defined orchestration in a hybrid optical/electrical datacenter network testbed. *Journal of Lightwave Technology* 37 (19): 4921–4934.

23 Berral, J.L., Poggi, N., Carrera, D. et al. (2015). ALOJA-ML: a framework for automating characterization and knowledge discovery in Hadoop deployments. *Proceedings of ACM SIGKDD '15*, pp. 1701–1710.

24 Peng, C., Zhang, C., Peng, C., and Man, J. (2017). A reinforcement learning approach to map reduce auto-configuration under networked environment. *International Journal of Security and Networks* 12 (3): 135–140.

25 Brochu, E., Cora, V.M., and De Freitas, N. (2010). A tutorial on Bayesian optimization of expensive cost functions, with application to active user modeling and hierarchical reinforcement learning. *arXiv preprint arXiv:1012.2599*.

26 Alnafessah, A.S. and Casale, G. (2020). TRACK: optimizing artificial neural networks for anomaly detection in Spark Streaming systems. *Proceedings of ACM VALUETOOLS '20*, pp. 188–191.

27 Snoek, J., Larochelle, H., and Adams, R.P. (2012). Practical Bayesian optimization of machine learning algorithms. *26th Annual Conference on Neural Information Processing Systems 2012, NIPS 2012*, pp. 2951–2959.

28 Zhang, G.P. (2000). Neural networks for classification: a survey. *IEEE Transactions on Systems, Man, and Cybernetics* 30 (4): 451–462.

29 Shahriari, B., Swersky, K., Wang, Z. et al. (2015). Taking the human out of the loop: a review of Bayesian optimization. *Proceedings of the IEEE* 104 (1): 148–175.

30 Ibidunmoye, O., Hernández-Rodriguez, F., and Elmroth, E. (2015). Performance anomaly detection and bottleneck identification. *ACM Computing Surveys* 48 (1): 1–35.

31 Fu, S. (2011). Performance metric selection for autonomic anomaly detection on cloud computing systems. *Proceedings of IEEE GLOBECOM '11*, IEEE, pp. 1–5.

32 Oppenheimer, D., Ganapathi, A., and Patterson, D.A. (2003). Why do internet services fail, and what can be done about it? *Proceedings of USITS '03*, USENIX.

33 Pertet, S. and Narasimhan, P. (2005). Causes of Failure in Web Applications. *Technical Report CMU-PDL-05-109*. Carnegie Mellon University.

34 Hodge, V.J. and Austin, J. (2004). A survey of outlier detection methodologies. *Artificial Intelligence Review* 22 (2): 85–126.

35 Chandola, V., Banerjee, A., and Kumar, V. (2009). Anomaly detection: a survey. *ACM Computing Surveys* 41 (3): 15.

36 Sebestyen, G., Hangan, A., Czako, Z., and Kovacs, G. (2018). A taxonomy and platform for anomaly detection. *Proceedings of IEEE AQTR '18*, IEEE, pp. 1–6.

37 Lu, S., Rao, B., Wei, X. et al. (2017). Log-based abnormal task detection and root cause analysis for Spark. *Proceedings of IEEE ICWS '17*, IEEE, pp. 389–396.

38 Kelly, T. (2005). Detecting performance anomalies in global applications. *Proceedings of WORLDS '05*, volume 5, USENIX, pp. 42–47.

39 Yang, L., Liu, C., Schopf, J.M., and Foster, I. (2007). Anomaly detection and diagnosis in grid environments. *Proceedings of ACM/IEEE SC '07*, ACM, p. 33.

40 Rogers, S. and Girolami, M. (2016). *A First Course in Machine Learning*, 2e. Chapman and Hall/CRC.

41 Alnafessah, A.S. and Casale, G. (2018). A neural-network driven methodology for anomaly detection in Apache Spark. *Proceedings of QUATIC '18*, IEEE, pp. 201–209.

42 Lu, S., Wei, X., Li, Y., and Wang, L. (2018). Detecting anomaly in big data system logs using convolutional neural network. *2018 IEEE 16th DASC/PiCom/DataCom/CyberSciTech*, IEEE.

43 Fulp, E.W., Fink, G.A., and Haack, J.N. (2008). Predicting computer system failures using support vector machines. *WASL '08*. USENIX Association.

44 Fu, S., Liu, J., and Pannu, H. (2012). *A hybrid anomaly detection framework in cloud computing using one-class and two-class support vector machines*. *Proceedings of ADMA '12*, Springer, pp. 726–738.

45 Yadwadkar, N.J. and Choi, W. (2012). Proactive straggler avoidance using machine learning. *White paper, University of Berkeley*.

46 Qi, W., Li, Y., Zhou, H. et al. (2017). Data mining based root-cause analysis of performance bottleneck for big data workload. *Proceedings of IEEE HPCC/SmartCity/DSS '17*, IEEE, pp. 254–261.

47 Huang, T., Zhu, Y., Zhang, Q. et al. (2013). An LOF-based adaptive anomaly detection scheme for cloud computing. *2013 IEEE COMPSACW*, IEEE, pp. 206–211.

48 Møller, M.F. (1993). A scaled conjugate gradient algorithm for fast supervised learning. *Neural Networks* 6 (4): 525–533.

49 Kline, D.M. and Berardi, V.L. (2005). Revisiting squared-error and cross-entropy functions for training NN classifiers. *Neural Computing Applications* 14 (4): 310–318.

50 Sheela, K.G. and Deepa, S.N. (2013). Review on methods to fix number of hidden neurons in neural networks. *Mathematical Problems in Engineering* 2013: 425740. 10.1155/2013/425740.

51 Hummer, W., Satzger, B., and Dustdar, S. (2013). Elastic stream processing in the cloud. *WIREs Data Mining and Knowledge Discovery* 3 (5): 333–345.

52 Imai, S., Patterson, S., and Varela, C.A. (2018). Uncertainty-aware elastic virtual machine scheduling for stream processing systems. *2018 18th IEEE/ACM CCGRID*, pp. 62–71.

53 Kombi, R.K., Lumineau, N., Lamarre, P. et al. (2019). DABS-Storm: a data-aware approach for elastic stream processing. In: *Transactions on Large-Scale Data and Knowledge-Centered Systems XL*, 58–93. Springer: Berlin, Heidelberg.

54 Zacheilas, N., Kalogeraki, V., Zygouras, N. et al. (2015). Elastic complex event processing exploiting prediction. *Proceedings of IEEE Big Data '15*, pp. 213–222.

55 Hu, Z., Kang, H., and Zheng, M. (2019). Stream data load prediction for resource scaling using online support vector regression. *Algorithms* 12 (2): 37.

56 Runsewe, O. and Samaan, N. (2017). Cloud resource scaling for big data streaming applications using a Layered Multi-dimensional Hidden Markov Model. *Proceedings of IEEE/ACM CCGRID '17*, pp. 848–857.

57 Lombardi, F., Aniello, L., Bonomi, S., and Querzoni, L. (2018). Elastic symbiotic scaling of operators and resources in stream processing systems. *IEEE TPDS* 29 (3): 572–585.

58 Mu, W., Jin, Z., Liu, F. et al. (2019). OMOPredictor: an online multi-step operator performance prediction framework in distributed streaming processing. *2019 IEEE ISPA/BDCloud/SocialCom/SustainCom*.

59 Xu, J., Tang, J., Xu, Z. et al. (2019). A deep recurrent neural network based predictive control framework for reliable distributed stream data processing. *Proceedings of IEEE IPDPS '19*, pp. 262–272.

60 Khoshkbarforoushha, A., Ranjan, R., Gaire, R. et al. (2017). Distribution based workload modelling of continuous queries in clouds. *IEEE Transactions on Emerging Topics in Computing* 5 (1): 120–133.

61 Röger, H. and Mayer, R. (2019). A comprehensive survey on parallelization and elasticity in stream processing. *ACM Computing Surveys* 52 (2): 36:1–36:37.

62 Fernandez, R.C., Migliavacca, M., Kalyvianaki, E., and Pietzuch, P. (2013). Integrating scale out and fault tolerance in stream processing using operator state management. *Proceedings of ACM SIGMOD '13*, ACM, pp. 725–736.

63 Gedik, B., Schneider, S., Hirzel, M., and Wu, K. (2014). Elastic scaling for data stream processing. *IEEE TPDS* 25 (6): 1447–1463.

64 Lohrmann, B., Janacik, P., and Kao, O. (2015). Elastic stream processing with latency guarantees. *Proceedings of IEEE ICDCS '15*, pp. 399–410.

65 De Matteis, T. and Mencagli, G. (2016). Keep calm and react with foresight: strategies for low-latency and energy-efficient elastic data stream processing. *SIGPLAN Not.* 51 (8): 1–12.

66 Heinze, T., Pappalardo, V., Jerzak, Z., and Fetzer, C. (2014). Auto-scaling techniques for elastic data stream processing. *Proceedings of ICDEW '14*, pp. 296–302.

67 Cheng, D., Zhou, X., Wang, Y., and Jiang, C. (2018). Adaptive scheduling parallel jobs with dynamic batching in Spark Streaming. *IEEE Transactions on Parallel and Distributed Systems* 29 (12): 2672–2685.

68 Russo Russo, G.R., Cardellini, V., and Lo Presti, F. (2019). Reinforcement learning based policies for elastic stream processing on heterogeneous resources. *Proceedings of ACM DEBS '19*, pp. 31–42.

69 Sutton, R.S. and Barto, A.G. (2018). *Reinforcement Learning - An Introduction*, 2e. MIT Press.

70 Mastronarde, N. and van der Schaar, M. (2011). Fast reinforcement learning for energy-efficient wireless communication. *IEEE Transactions on Signal Processing* 59 (12): 6262–6266.

71 Li, T., Xu, Z., Tang, J., and Wang, Y. (2018). Model-free control for distributed stream data processing using deep reinforcement learning. *Proceedings of the VLDB Endowment* 11 (6): 705–718.

72 Ni, X., Li, J., Yu, M. et al. (2020). Generalizable resource allocation in stream processing via deep reinforcement learning. *AAAI* 34 (01): 857–864.

10

Datacenter Traffic Optimization with Deep Reinforcement Learning

Li Chen, Justinas Lingys, Kai Chen, and Xudong Liao

Department of Computer Science and Engineering, ISING Lab, Hong Kong University of Science and Technology, Hong Kong SAR, China

10.1 Introduction

Datacenter traffic optimizations (TOs, e.g. flow/coflow scheduling [1–11], congestion control [12–14], load balancing and routing [15, 16], see Section 10.2.3 for more details.) have significant impact on application performance. Currently, TO is dependent on hand-crafted heuristics for varying traffic load, flow size distribution, traffic concentration, etc. When parameter setting mismatches traffic, TO heuristics may suffer performance penalty. For example, in PIAS [1], thresholds are calculated based on a long-term flow size distribution, and is prone to mismatch the current/true size distribution in run-time. Under mismatch scenarios, performance degradation can be as much as 38.46% [1]. Pfabric [5] shares the same problem when implemented with limited switch queues: for certain cases, the average flow completion time (FCT can be reduced by over 30% even if the thresholds are carefully optimized. Furthermore, in coflow scheduling, fixed thresholds in Aalo [10] depend on the operator's ability to choose good values upfront, since there is no run-time adaptation.

Apart from parameter-environment mismatches, the turn-around time of designing TO heuristics is long – at least weeks. Because they require operator insight, application knowledge, and traffic statistics collected over a long period of time. A typical process includes: first, deploying a monitoring system to collect end-host and/or switch statistics; second, after collecting enough data, operators analyze the data, design heuristics, and test it using simulation tools and optimization tools to find suitable parameter settings; finally, tested heuristics

Communication Networks and Service Management in the Era of Artificial Intelligence and Machine Learning,
First Edition. Edited by Nur Zincir-Heywood, Marco Mellia, and Yixin Diao.

are enforced[1] (with application modifications [8, 9], OS kernel module [1, 2], switch configurations [14], or any combinations of the above).

Automating the TO process is thus appealing, and we desire an automated TO agent that can adapt to voluminous, uncertain, and volatile datacenter traffic, while achieving operator-defined goals. In this work, we investigate reinforcement learning (RL) techniques [17], as RL is the subfield of machine learning (ML) concerned with decision-making and action control. It studies how an agent can learn to achieve goals in a complex, uncertain environment. An RL agent observes previous environment states and rewards, then decides an action in order to maximize the reward. RL has achieved good results in many difficult environments in recent years with advances in deep neural networks (DNNs): DeepMind's Atari results [18] and AlphaGo [19] used deep reinforcement learning (DRL) algorithms which make few assumptions about their environments, and thus can be generalized in other settings. Inspired by these results, we are motivated to enable DRL for automatic datacenter TO.

We started by verifying DRL's effectiveness in TO. We implemented a flow-level centralized TO system with a basic DRL algorithm, policy gradient [17]. However, in our experiments (Section 10.2.4), even this simple algorithm running on current machine learning software frameworks,[2] and advanced hardware (GPU) cannot handle TO tasks at the scale of production datacenters (>10^5 servers). The crux is the computation time (~100 ms): short flows (which constitute the majority of the flows) are gone before the DRL decisions come back, rendering most decisions useless.

Therefore, in this work, we try to answer the key question: *How to enable DRL-based automatic TO at datacenter-scale?* To make DRL scalable, we first need to understand the long-tail distribution of datacenter traffic [12, 23, 24]: most of the flows are short flows,[3] but most of the bytes are from long flows. Thus, TO decisions for short flows must be generated quickly; whereas decisions for long flows are more influential as they take longer time to finish.

We present AuTO, an end-to-end DRL system for datacenter-scale TO that works with commodity hardware. AuTO is a two-level DRL system, mimicking the peripheral and central nervous systems in animals. Peripheral systems (PSs) run on all end-hosts, collect flow information, and make instant TO decisions locally for short flows. PS's decisions are informed by the central system (CS), where global traffic information is aggregated and processed. CS further makes

---

1 After the heuristic is designed, its parameters can usually be computed in a short time for average scenarios: minutes [1, 2, 9] or hours [8]. AuTO seeks to automate the entire TO design process, rather than just parameter selection.
2 E.g. TensorFlow [20] PyTorch [21], Ray [22].
3 The threshold between short and long flows is dynamically determined in AuTO based on current traffic distribution (Section 10.3.5).

individual TO decisions for long flows which can tolerate longer processing delays.

The key to AuTO's scalability is to detach time-consuming DRL processing from quick action-taking for short flows. To achieve this, we adopt multiple level feedback queue (MLFQ) [1] for PS to schedule flows guided by a set of thresholds. Every new flow starts at the first queue with highest priority, and is gradually demoted to lower queues after its sent bytes pass certain thresholds. Using MLFQ, AuTO's PS makes per-flow decisions instantly upon local information (bytes-sent and thresholds)[4], while the thresholds are still optimized by a DRL algorithm in the CS over a relatively longer period of time. In this way, global TO decisions are delivered to PS in the form of MLFQ thresholds (which is more delay-tolerant), enabling AuTO to make globally informed TO decisions for the majority of flows with only local information. Furthermore, MLFQ naturally separates short and long flows: short flows complete in the first few queues, and long flows descend down to the last queue. For long flows, CS centrally processes them individually using a different DRL algorithm to determine routing, rate limiting, and priority.

We have implemented an AuTO prototype using Python. AuTO is thus compatible with popular learning frameworks, such as Keras/TensorFlow. This allows both networking and machine learning community to easily develop and test new algorithms, because software components in AuTO are reusable in other RL projects in datacenter.

We further build a testbed with 32 servers connected by two switches to evaluate AuTO. Our experiments show that, for traffic with stable load and flow size distribution, AuTO's performance improvement is up to 48.14% compared to standard heuristics (shortest-job-first [SJF] and least-attained-service-first [LAS]) after eight hours of training. AuTO is also shown to learn steadily and adapt across temporally and spatially heterogeneous traffic: after only eight hours of training, AuTO achieves 8.71% (9.18%) reduction in average (tail) FCT compared to heuristics.

10.2 Technology Overview

In this section, we first overview the RL background. Later, we show the emerging trend in applying machine learning techniques to networking problems. Next, we review the documented TO algorithms. Then, we describe and apply a basic RL algorithm, policy gradient, to enable flow scheduling in TO.

---

4 For short flows, AuTO relies on equal-cost multi-path routing (ECMP) [25] (which is also not centrally controlled) for routing/load-balancing and makes no rate-limiting decisions.

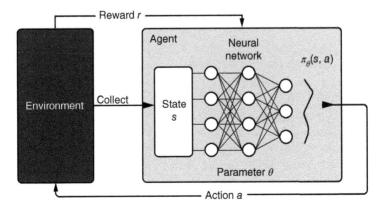

Figure 10.1 A general reinforcement learning setting using neural network as policy representation.

10.2.1 Deep Reinforcement Learning (DRL)

Recently, Machine Learning (ML) techniques have shown superhuman break-throughs in various application felids, such as computer vision, speech interpretation and machine translation. Amid machine learning, algorithms could be generated by learning from data without following predefined rules. Existing machine learning methods consist of three categories: supervised learning (SL), unsupervised learning (USL), and RL. More detailedly, SL algorithms target at addressing regression or classification tasks, while USL algorithms concentrate on clustering data that are not labeled. In RL area, agents learn to build a control policy that can generate an optimal action series achieving the highest cumulative reward by interacting with the environment. A more comprehensive introduction of RL is as follows.

As shown in Figure 10.1, *environment* is the surroundings of the *agent* with which the *agent* can interact through observations, actions, and feedback (rewards) on actions [17]. Specifically, in each time step t, the agent observes state s_t, and chooses action a_t. The state of the environment then transits to s_{t+1}, and the agent receives reward r_t. The state transitions and rewards are stochastic and Markovian [26]. The objective of learning is to maximize the expected cumulative discounted reward $E[\sum_{t=0}^{\infty} \gamma^t r_t]$ where $\gamma_t \in (0, 1]$ is the discounting factor.

The RL agent takes actions based on a *policy*, which is a probability distribution of taking action a in the state s: $\pi(s, a)$. For most practical problems, it is infeasible to learn all possible combinations of state-action pairs, thus function approximation [27] technique is commonly used to learn the policy. A function approximator $\pi_\theta(s, a)$ is parameterized by θ, whose size is much smaller (thus mathematically tractable) than the number of all possible state-action pairs. Function approximator can have many forms, and recently, DNNs have been shown to solve practical, large-scale dynamic control problems similar to flow scheduling. Therefore, we also use DNN as the representation of function approximator in AuTO.

With function approximation, the agent learns by updating the function parameters θ with the state s_t, action a_t, and the corresponding reward r_t in each time period/step t. We focus on one class of updating algorithms that learn by performing gradient-descent on the policy parameters. The learning involves updating the parameters (link weights) of a DNN so that the aforementioned objective could be maximized.

$$\theta \leftarrow \theta + \alpha \sum_t \nabla_\theta \log \pi_\theta \left(s_t, a_t\right) v_t \tag{10.1}$$

Training of the agent's DNN adopts a variant of the well-known REINFORCE algorithm [28]. This variant uses a modified version of Eq. (10.1), which alleviates the drawbacks of the algorithm: convergence speed and variance. To mitigate the drawbacks, Monte Carlo Method [29] is used to compute an empirical reward, v_t, and a baseline value (the cumulative average of experienced rewards per server) is used for reducing the variance [30]. The resultant update rule (Eq. (10.2)) is applied to the policy DNN, due to its variance management and guaranteed convergence to at least a local minimum [28]:

$$\theta \leftarrow \theta + \alpha \sum_t \nabla_\theta \log \pi_\theta \left(s_t, a_t\right) \left(v_t - baseline\right) \tag{10.2}$$

10.2.2 Applying ML to Networks

Dealing with complex problems, which are somehow difficult to model, is one of the most significant advantages of machine learning. Since the network field often sees complex problems that were not solved efficiently with the traditional methods, it is promising to rethink them coupled with machine learning for achieving higher performance. It is possible to build new applications and solutions by incorporating machine learning into network design. The newly designed ML libraries and frameworks (e.g. TensorFlow, PyTorch, Ray, and Spark) reinforce the possibility of probing the potential of ML in networked systems.

The following gives the reasons why ML could be fit and efficient for tackling networking problems. First of all, machine learning is the most well-known to solve classification and prediction tasks, which is aligned with intrusion detection and performance prediction [31]. Especially, RL can help decision making, which could make enrich network scheduling task, such as flow scheduling, congestion control, and job scheduling. Second, computer network problems build solution by interacting with complex environments, while they are always not easy to be efficiently modeled. When the real environment conditions deviate from input assumptions (built models), the model-based solution is prone to degraded performance [32, 33]. Luckily, we can learn to build algorithms by model-free RL without the amount of efforts to construct accurate models.

Next, we review the recent breakthroughs of applying machine learning in the network community. Existing efforts have brought considerable advance in different subfields of networking.

Congestion control: Aurora [34] placed the first hand of RL on congestion control. Aurora leverages vanilla DRL to determine the sending rates among the transmission process. Different from Aurora, Orca [35] proposed a more pragmatic RL-based congestion control for the Internet. The key insight of Orca is coupling RL method with underlying TCP congestion control algorithms, which can significantly reduce the computing overhead and contribute to quicker convergence and stabler results.

Traffic prediction: An accurate estimation of traffic volume is of great importance and beneficial to a lot of problems, such as congestion control, resource allocation, and high-level video streaming applications. For example, the work in [36] tries to predict traffic volume by learning the dependence between the flow counts and flow volume. Another work [37], which is motivated by end-to-end deep learning solutions, attempts to learn a pattern that takes some easily obtained information in the packet header as input and predicts the flow volume.

Packet classification: Packet classification is a fundamental problem in computer networking. The goal of packet classification is to match a given packet to a rule from a set of predefined rules. To achieve a better trade-off between computation overhead and state complexity, the work in [38] proposes a DRL approach to solve the problem. Their solution, NeuroCuts, adopts succinct representations to encode state and action space, and efficiently explore candidate decision trees to optimize for a global objective (classification time and/or memory footprint).

Job scheduling: Efficient job scheduling on distributed computing clusters is important and thus requires efficient algorithms. Traditional solutions use simple, generalized heuristics but ignoring the workload-specific information because it is untractable to tune the scheduling policy for each workload. The work [39] uses RL and neural networks to learn workload-specific scheduling algorithms without any hardwired instruction to fully optimize for a high-level objective (such as job completion time), which shows that modern machine learning techniques can generate highly efficient policies automatically.

10.2.3 Traffic Optimization Approaches in Datacenter

There have been continuous efforts on TO in datacenters. In general, three categories of mechanisms are explored: load balancing, congestion control, and flow scheduling. We focus on flow scheduling approaches.

Most previous flow scheduling proposals [4, 5, 40, 41] assume prior and full knowledge of accurate flow information, such as flow sizes or deadlines, to obtain outstanding performance. For example, PDQ [4], pFabric [5], and PASE [40], all make the assumption that flow size is known as a priori, and thus uses it to approximate SJF policy, which is the theoretically optimal scheduling policy for minimizing the average FCT over a single link. However, it is commonly known that gathering flow information is inherently difficult, since it requires modifications to application, environments, and sometimes OS kernel network stack. Therefore, these limitations make it hard for them to implement in practice.

Motivated by some approaches to reduce FCT without depending on the flow size information, such as DCTCP [12], HULL [42], etc., the community found that FCT could also be improved by maintaining low queue occupancy through queuing policies such as explicit congestion notification (ECN), and pacing. PIAS [1] is a practical information-agnostic flow scheduling algorithm for minimizing FCTs in datacenter networks. PIAS can emulate SJF with no priori information. At its core, PIAS leverages multiple priority queues available in existing commodity switches to implement a MLFQ, in which each PIAS flow is gradually demoted from higher priority queue to lower priority queue based on the bytes it has sent. In this way, PIAS can ensure short flows, which dominates in the famous datacenter workload, are prioritized over long flows.

The above approaches are of general purposes to optimize FCT. However, datacenter flows are naturally diverse and have other demands. For example, user-facing datacenter applications such as web search (WS), often have stringent latency boundaries, and thus generate flows with strict deadlines [12]. Flows do not meet their deadlines will be excluded from the results and eventually hurt the user's experience. Therefore, to better serve different kinds of flows, datacenter flow scheduler needs to take these requirements into consideration. Some solutions [1, 4, 40] do not consider the mix-flow scenario, while pFabric [5] simply prioritizes deadline flows over non-deadline flows, which will hurt FCT of other kinds of flows. At this end, Karuna [2] is developed to schedule mix-flows, where the deadline meet rate is maximized for deadline flows and FCT is minimized for non-deadline flows simultaneously. The key insight of Karuna is that the FCT for non-deadline flows should be minimally impacted when scheduling to meet deadlines.

CODA [8] uses unsupervised clustering algorithm to identify flow information without application modifications. However, its scheduling decisions are still made by a heuristic algorithm with fixed parameters.

Routing and load balancing on the Internet have employed RL-based techniques [43] since 1990s. However, they are switch-based mechanisms, which are difficult to implement at line rate in modern datacenters with >10 GbE links. Machine learning techniques [32] have been used to optimize parameter setting for congestion control. The parameters are fixed given a set of traffic distributions, and there is no adaptation of parameters at run-time.

10.2.4 Example: DRL for Flow Scheduling

As an example, we formulate the problem of flow scheduling in datacenters as a DRL problem, and describe a solution using the policy gradient (PG) algorithm based on Eq. (10.2).

10.2.4.1 Flow Scheduling Problem
We consider a datacenter network connecting multiple servers. For simplicity, we adopt the *big-switch* assumption by previous works in flow scheduling [2, 5], where the network is non-blocking with full-bisection bandwidth and proper load-balancing. Instead of focusing on the individual switches, we model the whole datacenter fabric to be a big switch. The ingress queues into the *big-switch* are at the network interface cards (NICs) and the egress queues out of it are at the last-hop top of rack (TOR) switches. In this context, the scheduling problem over datacenter fabric can be translated to find the best schedule to minimize the average FCT over the backplane of the *big-switch*. Following this assumption, the flow scheduling problem is simplified to the problem of deciding the sending order of flows. We consider an implementation that enables preemptive scheduling of flows using strict priority queueing. We create K priority queues for flows in each server [44] and enforce strict priority queuing among them. K priority queues are also configured in the switches, similar to [1]. The priority of each flow can be changed dynamically to enable preemption. The packet of each flow is tagged with its current priority number, and will be placed in the same queue throughout the entire datacenter fabric.

10.2.4.2 DRL Formulation
Action space: The action provided by the agent is a mapping from active flows to priorities: for each active flow f, at time step t, its priority is $p_t(f) \in [1, K]$.

State space: The big-switch assumption allows for a much simplified state space. As routing and load balancing are out of our concern, the state space only includes the flow states. In our model, states are represented as the set of all active flows, F_a^t, and the set of all finished flows, F_d^t, in the entire network at current time step t. Each flow is identified by its 5-tuple [1, 45]: source/destination IP, source/destination port numbers, and transport protocol. Active flows have an additional attribute, which is its priority; while finished flows have two additional attributes: FCT and flow size[5].

Rewards: Rewards are feedback to the *agent* on how good its actions are. The reward can be obtained after the completion of a flow, thus is computed only on the set of finished flows F_d^t for time step t. The average throughput of each finished flow f is $Tput_f = \frac{Size_f}{FCT_f}$. We model the reward as the ratio between the average throughput of two consecutive time steps.

$$r_t = \frac{\sum_{f^t \in F_d^t} Tput_f^t}{\sum_{f^{t-1} \in F_d^{t-1}} Tput_f^{t-1}} \tag{10.3}$$

It signals if the previous actions have resulted in a higher per-flow throughput experienced by the agent, or it has degraded the overall performance. The objective is to maximize the average throughput of the network as a whole.

10.2.4.3 DRL Algorithm

We use the update rule specified by Eq. (10.2). The DNN residing on the agent computes probability vectors for each new state and updates its parameters by evaluating the action that resulted in the current state. The evaluation step compares the previous average throughput with the corresponding value of the current step. Based on the comparison and Eq. (10.3), an appropriate reward (either negative or positive) is produced which is added to the baseline value. Thus, we can ensure that the function approximator improves with time and can converge to a local minimum by updating DNN weights in the direction of the gradient. The update which follows (10.2) ensures that poor flow scheduling decisions are discouraged for similar states in the future, and the good ones become more probable for similar states in the future. When the system converges, the policy achieves a sufficient flow scheduling mechanism for a cluster of servers.

10.3 State-of-the-Art: AuTO Design

10.3.1 Problem Identified

First, we show the problem of an RL system running PG using testbed experiments, motivating AuTO, and elaborate the naive method of applying RL to traffic control is problematic.

Using the DRL problem of flow scheduling described in Section 10.2.4 as an example, we implement PG using popular machine learning frameworks:

---

5 Flow size and FCT can be measured when the flow ends using either OS utility [46] or application layer mechanisms [8, 47, 48].

Keras/TensorFlow, PyTorch, and Ray. We simplify the DRL agent to have only one hidden layer. We use two servers: DRL agent resides in one, and the other sends mock traffic information (states) to the agent using an remote procedure call (RPC) interface. We set the sending rate of the mock server to 1000 flows per second (fps). We measure the processing latency of different implementations at the mock server: the time between finish sending the flow information and receiving the action. The servers are Huawei Tecal RH1288 V2 servers running 64-bit Debian 8.7, with 4-core Intel E5-1410 2.8 GHz CPU, NVIDIA K40 GPU, and Broadcom 1 Gbps NICs.

As shown in Figure 10.2, even for small flow arrival rate of 1000 fps and only one hidden layer, the processing delays of all implementations are more than 60 ms, during which time any flow within 7.5 MB would have finished on a 1 Gbps link. For reference, using the well-known traffic traces of a WS application and a data mining (DM) application collected in Microsoft datacenters[1, 12, 49], a 7.5 MB flow is larger than 99.99% and 95.13% of all flows, respectively. This means, most of the DRL actions are useless, as the corresponding flows are already gone when the actions arrive.

Summary Current DRL systems' performance is not enough to make online decisions for datacenter-scale traffic. They suffer from long processing delays even for simple algorithms and low traffic load.

10.3.2 Overview

The key problem of current DRL systems is the long latency between collection of flow information and generation of actions. In modern datacenters with ≥10 Gbps link speed, to achieve flow-level TO operations, the round-trip latency of actions should be at least sub-millisecond. Without introducing specialized hardware, this is unachievable (Section 10.2.4). Using commodity hardware, the processing latency of DRL algorithm is a hard limit. Given this constraint, how to scale DRL for datacenter TO?

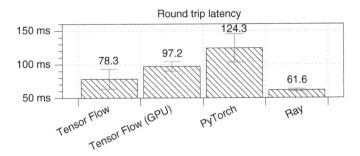

Figure 10.2 Current DRL systems are insufficient.

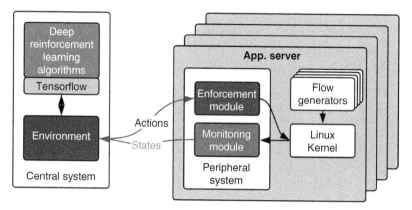

Figure 10.3 AuTO overview.

Recent studies [12, 23, 24] have shown that most datacenter flows are short flows, yet most traffic bytes are from long flows. Informed by such long-tail distribution, our insight is to delegate most short flow operations to the end-host, and formulate DRL algorithms to generate long-term (sub-second) TO decisions for long flows.

We design AuTO as a two-level system, mimicking the peripheral and central nervous systems in animals. As shown in Figure 10.3, PSs run on all end-hosts, collect flow information, and make TO decisions locally with minimal delay for short flows. CS makes individual TO decisions for long flows that can tolerate longer processing delays. Furthermore, PS's decisions are informed by the CS where global traffic information are aggregated and processed.

10.3.3 Peripheral System

The key to AuTO's scalability is to enable PS to make globally informed TO decisions on short flows with only local information. PS has two modules: an enforce module and a monitoring module (MM).

10.3.3.1 Enforcement Module

To achieve the above goal, we adopt (MLFQ, introduced in PIAS [1]) to schedule flows without centralized per-flow control. Specifically, PS performs packet tagging in the differentiated services code point (DSCP) field of IP packets at each end-host as shown in Figure 10.4. There are K priorities, $P_i, 1 \leq i \leq K$, and $(K - 1)$ demotion thresholds, $\alpha_j, 1 \leq j \leq K - 1$. We configure all the switches to perform strict priority queueing based on the DSCP field. At the end-host, when a new flow is initialized, its packets are tagged with P_1, giving them the highest priority in the network. As more bytes are sent, the packets of this flow will be tagged

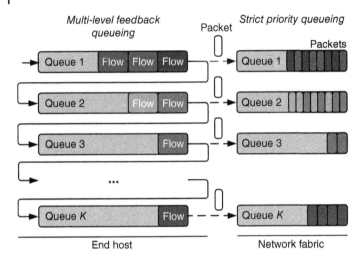

Figure 10.4 Multi-level feedback queuing.

with decreasing priorities P_j $(2 \leq j \leq K)$, thus they are scheduled with decreasing priorities in the network. The threshold to demote priority from P_{j-1} to P_j is α_{j-1}. With MLFQ, PS has the following properties:

- It can make instant per-flow decisions based only on local information: bytes-sent and thresholds.
- It can adapt to global traffic variations. To be scalable, CS must not directly control small flows. Instead, CS optimizes and sets MLFQ thresholds with global information over a longer period of time. Thus, thresholds in PS can be updated to adapt to traffic variations. In contrast, PIAS [1] requires weeks of traffic traces to update the thresholds.
- It naturally separates short and long flows. As shown in Figure 10.5, short flows finished in the first few queues, and long flows drop to the last queue. Thus, CS can centrally process long flows individually to make decisions on routing, rate limit, and priority.

10.3.3.2 Monitoring Module
For CS to generate thresholds, the MM collects flow sizes and completion times of all finished flows, so that CS can update flow size distribution. The MM also reports on-going long flows that have descended into the lowest priority on its end-host, so that CS can make individual decisions.

10.3.4 Central System

The CS is composed of two DRL agents (RLA): short flow Reinforcement Learning Agent (RLA) (sRLA) is for optimizing thresholds for MLFQ and long flow

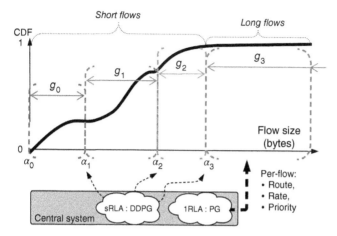

Figure 10.5 AuTO: A four-queue example.

RLA (lRLA) is for determining rates, routes, and priorities for long flows. sRLA attempts to solve an FCT minimization problem, and we develop a Deep Deterministic Policy Gradient (DDPG) algorithm for this purpose. For lRLA, we use a PG algorithm (Section 10.2.4) to generate actions for the long flows. In the next section, we describe the two DRL problems and solutions.

10.3.5 DRL Formulations and Solutions

Next, we describe the two DRL algorithms in CS.

10.3.5.1 Optimizing MLFQ Thresholds

We consider a datacenter network connecting multiple servers. Scheduling of flows is imposed by using K strict priority queues at hosts and network switches (Figure 10.4) by setting the DCSP field in each of the IP headers. The longer the flow is, the lower priority is assigned to the flow as it is demoted through host priority queues in order to approximate SJF. The packet's priority is preserved throughout the entire datacenter fabric till it reaches the destination.

One of the challenges of MLFQ is the calculation of the optimal demotion thresholds for the K priority queues at the host. Prior works [1, 2, 50] provide mathematical analysis and models for optimizing the demotion thresholds: $\{\alpha_1, \alpha_2, \ldots, \alpha_{K-1}\}$. Bai et al. [50] also suggest weekly/monthly recomputation of the thresholds with collected flow-level traces. AuTO takes a step further and proposes a DRL approach to optimizing the values of the α's. Unlike prior works that used machine learning in datacenter problems [26, 51, 52], AuTO is unique due to its target – optimization of real values in continuous action space. We formulate the threshold optimization problem as a DRL problem and try to explore the capabilities of DNN for modeling the complex datacenter network for computing the MLFQ thresholds.

As shown in Section 10.2.4, PG is a basic DRL algorithm. The agent follows a policy $\pi_\theta (a|s)$ parameterized by a vector θ and improves it with experience. However, REINFORCE and other regular PG algorithms only consider stochastic policies, $\pi_\theta (a|s) = P[a|s; \theta]$, that select action a in state s according to the probability distribution over the action set $\mathcal{A}$ parameterized by θ. PG cannot be used for value optimization problem, as a value optimization problem computes real values. Therefore, we apply a variant of *Deterministic Policy Gradient* (DPG) [53] for approximating optimal values $\{a_0, a_1, \dots, a_n\}$ for the given state s such that $a_i = \mu_\theta(s)$ *for* $i = 0, \dots, n$. Figure 10.6 summarizes the major differences between stochastic and deterministic policies. DPG is an actor-critic [54] algorithm for deterministic policies, which maintains a parameterized actor function μ_θ for representing current policy and a critic neural network $Q(s, a)$ that is updated using the Bellman equation (as in Q-learning [18]). We describe the algorithm with Eqs. (10.4)–(10.6) as follows: The actor samples the environment and has its parameters θ updated according to Eq. (10.4). The result of Eq. (10.4) follows from the fact that the objective of the policy is to maximize the expected cumulative discounted reward Eq. (10.5) and its gradient can be expressed in the following form Eq. (10.5). For more details, please refer to [53].

$$\theta^{k+1} \leftarrow \theta^k + \alpha E_{s \sim \rho^{\mu^k}} \left[\nabla_\theta \mu_\theta(s) \nabla_a Q^{\mu^k}(s, a) \Big|_{a = \mu_\theta(s)} \right] \tag{10.4}$$

where ρ^{μ^k} is the state distribution at time k.

$$J(\mu_\theta) = \int_S \rho^\mu(s) r(s, \mu_\theta(s)) ds$$
$$= E_{s \sim \rho^\mu} [r(s, \mu_\theta(s))] \tag{10.5}$$

$$\nabla_\theta J(\mu_\theta) = \int_S \rho^\mu(s) \nabla_\theta \mu_\theta(s) \nabla_a Q^{\mu^k}(s, a) \Big|_{a = \mu_\theta(s)} ds$$
$$= E_{s \sim \rho^\mu} \left[\nabla_\theta \mu_\theta(s) \nabla_a Q^{\mu^k}(s, a) \Big|_{a = \mu_\theta(s)} \right] \tag{10.6}$$

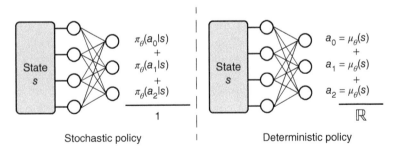

Figure 10.6 Comparison of deep stochastic and deep deterministic policies.

Algorithm 10.1 DDPG Actor-Critic Update Step

Sample a random mini-batch of N transitions $\left(s_i, a_i, r_i, s_{i+1}\right)$ from buffer

Set $y_i = r_i + \gamma Q'_{\theta Q'}(s_{i+1}, \mu'_{\theta \mu'}(s_{i+1}))$

Update critic by minimizing the loss: $L = \frac{1}{N} \sum_{i=1}^{N} (y_i - Q_{\theta Q}(s_i, a_i))^2$

Update the actor policy using the sampled policy gradient:

$$\nabla_{\theta \mu} J \approx \frac{1}{N} \sum_{i=1}^{N} \nabla_{\theta \mu}(s_i) \mu_{\theta Q}(s_i) \nabla_{a_i} Q_{\theta Q}(s_i, a_i) \Big|_{a_i = \mu_{\theta Q}(s_i)}$$

Update the target networks:

$$\theta^{Q'} \leftarrow \tau \theta^Q + (1 - \tau)\theta^{Q'}$$

$$\theta^{\mu'} \leftarrow \tau \theta^\mu + (1 - \tau)\theta^{\mu'}$$

where γ and τ are small values for stable learning

DDPG [55] is an extension of DPG algorithm, which exploits deep learning techniques [18]. We use DDPG as our model for the optimization problem and explain how it works in the following. Same as DPG, DDPG is also an actor-critic [54] algorithm, and it maintains four DNNs. Two DNNs, critic $Q_{\theta Q}(s, a)$ and actor $\mu_{\theta \mu}(s)$ with weights θ^Q and θ^μ, are trained on sampled mini-batches of size N, where an item represents an experienced transition tuple $\left(s_i, a_i, r_i, s_{i+1}\right)$, while the agent interacts with the environment. The DNNs are trained on random samples, which are stored in a buffer, in order to avoid correlated states which cause the DNNs to diverge [18]. The other two DNNs, target actor μ'_θ and target critic $Q'_{\theta Q'}(s, a)$, are used for smooth updates of the actor and critic networks, respectively (Algorithm 10.1 [55]). The update steps stabilize training the actor-critic networks and achieve state-of-the-art results on continuous space actions [55]. AuTO applies DDPG for optimizing threshold values to achieve better flow scheduling decisions.

DRL formulation Next, we show that the optimization of thresholds can be formulated as an actor-critic DRL problem solvable by DDPG. We first develop an optimization problem of choosing an optimal set of thresholds $\{\alpha_i\}$ to minimize the average FCT of flows. Then, we translate this problem into DRL problem to be solved using DDPG algorithm.

Denote the cumulative density function of flow size distribution as $F(x)$, thus $F(x)$ is the probability that a flow size is no larger than x. Let L_i denote the number of packets a given flow brings in queue Q_i for $i = 1, \ldots, K$. Thus, $E[L_i] \leq (\alpha_i - \alpha_{i-1})(1 - F(\alpha_{i-1}))$. Denote flow arrival rate as λ, then the packet arrival rate to queue Q_i is $\lambda_i = \lambda E[L_i]$. The service rate for a queue depends on whether the queues with higher priorities are all empty. Thus, P_1 (highest priority) has capacity $\mu_1 = \mu$ where μ is the service rate of the link. The idle rate of Q_1 is $(1 - \rho_1)$ where $\rho_i = \lambda_i/\mu_i$ is the utilization rate of Q_i. Thus, the service rate of Q_2 is $\mu_2 = (1 - \rho_1)\mu$ since its service rate is μ (the full link capacity) given that P_1 is empty. We have $\mu_i = \Pi_{j=0}^{i-1}(1 - \rho_j)\mu$, with $\rho_0 = 0$. Thus, $T_i = 1/(\mu_i - \lambda_i)$ which is the average delay of queue i assuming M/M/1 queues. For a flow with size in $[\alpha_{i-1}, \alpha_i)$, it experiences the delays in different priority queues up to the i-th queue. Denote T_i as the average time spent in the i-th queue. Let $i_{max}(x)$ be the index of the smallest demotion threshold larger than x. Therefore, the average FCT for a flow with size x, $T(x)$, is upper-bounded by: $\sum_{i=1}^{i_{max}(x)} T_i$.

Let $g_i = F(\alpha_i) - F(\alpha_{i-1})$ denote the percentage of flows with sizes in $[\alpha_{i-1}, \alpha_i)$. Thus, g_i is the gap between two consecutive thresholds. Using g_i to equivalently express α_i, we can formulate the FCT minimization problem as[6]:

$$\min_{\{g\}} \quad \mathcal{T}(\{g\}) = \sum_{l=1}^{K}\left(g_l \sum_{m=1}^{l} T_m\right) = \sum_{l=1}^{K}\left(T_l \sum_{m=l}^{K} g_m\right) \tag{10.7}$$

$$\text{subject to} \quad g_i \geq 0, \quad i = 1, \dots, K - 1$$

We proceed to translate Problem (10.7) into a DRL problem.

State space: In our model, states are represented as the set of the set of all finished flows, F_d, in the entire network in the current time step. Each flow is identified by its 5-tuple[1, 45]: source/destination IP, source/destination port numbers, and transport protocol. As we report only finished flows, we also record the FCT and flow size as flow attributes. In total, each flow has seven features.

Action space: The action space is computed by a centralized agent, sRLA. At time step t, the action provided by the agent is a set of MLFQ threshold values $\{\alpha_i^t\}$.

Rewards: Rewards are delayed feedback to the *agent* on how good its actions are for the previous time step. We model the reward as the ratio between objective functions of two consecutive time steps: $r_t = \frac{\mathcal{T}^{t-1}}{\mathcal{T}^t}$. It signals if the previous actions have resulted in a lower average FCT, or it has degraded the overall performance.

DRL algorithm We use the update rule specified by Eq. (10.4) (Algorithm 10.1). The DNN computes g_i's for each newly received state from a host, and stores a tuple: (s_t, a_t, r_t, s_{t+1}) in its buffer for later learning. Reward r_t and the next state s_{t+1} are only known when the next update comes from the same host, so the agent buffers s_t and a_t until all needed information is received. Updates of parameters are performed in random batches to stabilize learning and to reduce probability of divergence [18, 55]. The reward r_t is computed at a host at step t and is compared to

---

6 For a solution to this problem, e.g. $\{g_i'\}$, we can retrieve the thresholds $\{\alpha_i'\}$ with $\alpha_i' = F^{-1}(\sum_{j=1}^{l} g_j)$, where $F^{-1}(\cdot)$ is the inverse of $F(\cdot)$.

the previous average FCT. Based on the comparison, an appropriate reward (either negative or positive) is produced which is sent to the agent as a signal for evaluating action a_t. By following Algorithm 10.1, the system can improve the underlying actor–critic DNNs and converge to a solution for Problem (10.7).

10.3.5.2 Optimizing Long Flows

The last threshold, α_{K-1}, separates long flows from short flows by sRLA, thus α_{K-1} is updated dynamically according to current traffic characteristics, in contrast to prior works with fixed threshold for short and long flows [3, 56]. For long flows and lRLA, we use a PG algorithm similar to the flow scheduling problem in Section 10.2.4, and the only difference is in the action space.

Action space: For each active flow f, at time step t, its corresponding action is $\{Prio_t(f), Rate_t(f), Path_t(f)\}$, where $Prio_t(f)$ is the flow priority, $Rate_t(f)$ is the rate limit, and $Path_t(f)$ is the path to take for flow f. We assume the paths are enumerated in the same way as in XPath [57].

State space: Same as Section 10.2.4, states are represented as the set of all active flows, F_a^t, and the set of all finished flows, F_d^t, in the entire network at current time step t. Apart from its 5-tuple [1, 45], each active flow has an additional attribute: its priority; each finished flow has two additional attributes: FCT and flow size.

Rewards: The reward is obtained for the set of finished flows F_d^t. Choices for the reward function can be: difference or ratios of sending rate, link utilization, and throughput in consecutive time steps. For modern datacenters with at least 10 Gbps link speed, it is not easy to obtain timely flow-level information for active flows. Therefore, we choose to compute reward with finished flows only, and use the ratio between the average throughputs of two consecutive time steps as reward, as in Eq. (10.3). The reward is capped to achieve quick convergence [55].

10.4 Implementation

In this section, we describe the implementation. We develop AuTO in Python 2.7. The language choice facilitates integration with modern deep learning frameworks [20, 58, 59], which provide excellent Python interfaces [59]. The current prototype uses the *Keras* [58] deep learning library (with TensorFlow as backend).

10.4.1 Peripheral System

PS is a daemon process running on each server. It has a MM and an enforcement module (EM). The MM thread collects information about flows including recently finished flows and the presently active long flows (in the last queue of MLFQ). At the end of each period, the MM aggregates collected information, and sends to CS. The PS's EM thread performs tagging based on the MLFQ thresholds on currently active flows, as well as routing, rate limiting, and priority tagging for long flows. We implement a RPC interface for communications between PS and CS. CS uses RPC to set MLFQ thresholds and to perform actions on active long flows.

10.4.1.1 Monitoring Module (MM):

For maximum efficiency, the MM can be implemented as a Linux kernel module, as in PIAS[1]. However, for the current prototype, since we are using a flow generator (as seen in [1, 14, 60]) to produce workloads, we choose to implement the MM directly inside the flow generator. This choice allows us to obtain the ground truth and get rid of other network flows that may interfere with the results.

For long flows (flows in the last queue of MLFQ), every T seconds, MM merges n_l active long flows (each with six attributes), and m_l finished long flows (each with seven attributes) into a list. For short flows (in the first few queues of MLFQ) in the same period, MM collects m_s finished flows (each with seven attributes) into an list. Finally, MM concatenates two lists and sends them to CS as an observation of the environment.

AuTO's parameters, $\{n_l, m_l, m_s\}$, are determined by traffic load and T: For each server, n_l (m_l) should be the upper-bound of number of active (finished) long flows within T, and m_s should also be the upper-bound of finished short flows. In the case that the actual number of active (finished) flow is less than $\{n_l, m_l, m_s\}$, the observation vector is zero-padded to the same size of the corresponding agent's DNN(s). We make this design choice because the number of input neurons of the DNN in CS is fixed, therefore can take only fixed-sized inputs. We leave dynamic DNN and recurrent neural network structure as future work. For the current prototype and experiments on the prototype, since we control the flow generator, it is easy to comply with this constraint. We choose $\{n_l = 11, m_l = 10, m_s = 100\}$ in the experiments.

10.4.1.2 Enforcement Module (EM):

EM receives actions from CS periodically. The actions include new MLFQ thresholds, and TO decisions on local long flows. For MLFQ thresholds, EM builds upon the PIAS [1] kernel module, and adds dynamic configuration of demotion thresholds.

For short flows, we leverage ECMP [25] for routing and load-balancing, which does not require centralized per-flow control, and DCTCP [12] for congestion control.

For long flows, the TO actions include priority, rate limiting, and routing. EM leverages the same kernel module for priority tagging. Rate limiting is done using hierarchical token bucket (HTB) queueing discipline in Linux traffic control (tc). EM is configured with a parent class in HTB with outbound rate limit to represent the total outbound bandwidth managed by CS on this node. When a flow descends into the last queue in MLFQ, EM creates an HTB filter matching the exact 5-tuple for that flow. When EM receives rate allocation decisions from the CS, EM updates the child class of the particular flow by sending Netlink messages to Linux kernel: the rate of the traffic control (TC class is set as the rate that centralized scheduler decides, and its ceiling is set to the smaller of the original ceiling and twice of the rates from CS.

10.4.2 Central System

CS runs RL agents (sRLA and lRLA) to make optimized TO decisions. Our implemented CS follows a SEDA-like architecture [61] when handling incoming updates and sending actions to the flow generating servers. The architecture is subdivided into different stages: http request handling, deep network learning/processing, and response sending. Each stage has its own process(es) and communicate through queues to pass required information to the next stage. Such an approach ensures that multiple cores of the CS server are involved in handling the requests from the hosts and load is distributed. The multiprocessing architecture has been adopted due to the Global lock problem [62] in the CPython implementation of the Python programming language. The states and actions are encapsulated at the CS as an "environment" (similar to [63]), with which the RL agents can interact directly and programmatically.

10.4.2.1 sRLA

As discussed in Section 10.3.5.1, we use Keras to implement the sRLA running the DDPG algorithm with the aforementioned DNNs (actor, critic, target actor, and target critic).

Actors: The actors have two fully connected hidden layers with 600 and 600 neurons, respectively, and the output layer with $K - 1$ output units (one for each threshold). The input layer takes states (700 features per-server ($m_s = 100$)) and outputs MLFQ thresholds for a host server for time step t. Note that policy-based algorithms do not need to concern too much about the *neural network architecture* because regardless of the policy network's express ability, policy-based algorithms will always optimize for the reward objective over the space of policies that the neural network can express.

Critics: The critics are implemented with three hidden layers, thus the networks are a bit more complicated as compared to the actor network. Since the critic is supposed to "criticize" the actor for bad decisions and "compliment" for good ones, the critic neural network also takes as its input the outputs of the actor. However, as [53] suggests, the actor outputs are not direct inputs, but are only fed into the critic's network at a hidden layer. Therefore, the critic has two hidden layers same as the actor and one extra hidden layer which concatenates the actor's outputs with the outputs of its own second hidden layer, resulting in one additional hidden layer. This hidden layer eventually is fed into the output layer consisting of one output unit – approximated value for the observed/received state.

The neural networks are trained on a batch of observations periodically by sampling from a buffer of experience: $\{s_t, a_t, r_t, s_{t+1}\}$. The training process is described in Algorithm 10.1.

10.4.2.2 lRLA

For lRLA, we also use Keras to implement the PG algorithm with a fully connected NN with 10 hidden layers of 300 neurons. The RL agent takes a state (136 features per-server ($n_l = 11$, $m_l = 10$)) and outputs probabilities for the actions for all the active flows.

Summary The hyper-parameters (structure, number of layer, height, and width of DNN) are chosen based on a few empirical training sessions. Our observation is that more complicated DNNs with more hidden layers and more parameters took longer to train and did not perform much better than the chosen topologies. Overall, we find that such RLA configurations lead to good system performance and is rather reasonable considering the importance of computation delay, as we reveal next in the evaluations.

10.5 Experimental Results

In this section, we evaluate the performance of AuTO using real testbed experiments. We seek to understand: (i) With stable traffic (flow size distribution and traffic load are fixed), how does AuTO compare to standard heuristics? (ii) For varying traffic characteristics, can AuTO adapt? (iii) how fast can AuTO respond to traffic dynamics? and (iv) what are the performance overheads and overall scalability?

Summary of results (grouped by scenarios):

- *Homogeneous*: For traffic with fixed flow size distribution and load, AuTO-generated thresholds converge, and demonstrate similar or better performance compared to standard heuristics, with up to 48.14% average FCT reduction.
- *Spatially heterogeneous*: We divide the servers into four clusters; each is configured to generate traffic with different flow size distribution and load. In these

experiments, AuTO-generated thresholds also converge, with up to 37.20% average FCT reduction.

- *Spatially and Temporally Heterogeneous*: Building upon the above scenario, we then change the flow size distribution and load periodically for each cluster. For time-varying flow size distributions and traffic load, AuTO exhibits learning and adaptation behavior. Compared to fixed heuristics that excel only for certain combinations of traffic settings, AuTO demonstrates steady performance improvement across all combinations.

- *System overhead*: The current AuTO implementation can respond to state updates within 10 ms on average. AuTO also exhibits minimal end-host overhead in terms of CPU utilization and throughput degradation.

10.5.1 Setting

We deploy AuTO on a small-scale testbed (Figure 10.7) that consists of 32 servers. Our switch supports ECN and strict priority queuing with at most eight class of service queues[7] Each server is a Dell PowerEdge R320 with a 4-core Intel E5-1410 2.8 GHz CPU, 8G memory, and a Broadcom BCM5719 NetXtreme Gigabit Ethernet NIC with 4×1 Gpbs ports. Each server runs 64-bit Debian 8.7 (3.16.39-1 Kernel). By default, advanced NIC offload mechanisms are enabled to reduce the CPU overhead. The base round-trip time (RTT) of our testbed is 100 μs.

We adopt the traffic generator [60] used in prior works [15, 50, 64, 65] that produces network traffic flows based on given flow size distribution and traffic load. We use two realistic workloads (Figure 10.8): WS workload [1] and DM workload [14]. Fifteen servers hosting flow generators are called application servers, and the remaining one hosts the CS. Each application server is connected to a data

Figure 10.7 Testbed topology.

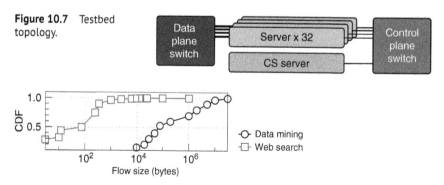

Figure 10.8 Traffic distributions in evaluation.

---

7 As in most production datacenters [1, 2, 14], some queues are reserved for other services, such as latency-sensitive traffic and management traffic[64].

plane switch using three of its ports, as well as to a control plane switch to communicate with the CS server using the remaining port. The three ports are configured to different subnets, forming three paths between any pair of application servers. Both switches are Pronto-3297 48-port Gigabit Ethernet switch. States and actions are sent on the control plane switch (Figure 10.7).

10.5.2 Comparison Targets

We compare with two popular heuristics in flow scheduling: SJF and LAS. The main difference between the two is that SJF schemes [3–5] require flow size at the start of a flow, while LAS schemes [1, 2, 66] do not. For these algorithms to work, sufficiently enough data should be collected before calculating their parameters (thresholds). The shortest period to collect enough flow information to form an accurate and reliable flow size distribution is an open research problem [2, 50, 67, 68], and we note that previously reported distributions are all collected over periods of at least weeks (Figure 10.8), which indicates the turn-around time are also at least weeks for these algorithms.

In the experiments, we mainly compare with quantized version of SJF and LAS with four priority levels. The priority levels are enforced both in the server using Linux qdisc [44] and in the data plane switch using strict priority queueing [1]:

- **Quantized SJF (QSJF):** QSJF has three thresholds: $\alpha_0, \alpha_1, \alpha_2$. We can obtain flow size from the flow generator at its start. For flow size s, if $x \leq \alpha_0$, it is given highest priority; if $x \in (\alpha_0, \alpha_1]$, it is given the second priority; and so on. In this way, shorter flows are given higher priority, similar to SJF.
- **Quantized LAS (QLAS):** QLAS also has thresholds: $\beta_0, \beta_1, and \beta_2$. All the flows are given high priority at the start. If a flow sends more than β_i bytes, it is then demoted to the $(i + 1)$-th priority. In this way, longer flows gradually drop to lower priorities.

The thresholds for both schemes can be calculated using methods described in [2] for "type-2/3 flows," and they are dependent on the flow size distribution and traffic load. In each experiment, unless specified, we use the thresholds calculated for DCTCP distribution at 80% load (i.e. the total sending rate is at 80% of the network capacity).

10.5.3 Experiments

10.5.3.1 Homogeneous Traffic

In these scenarios, the flow size distribution and the load generated from all 32 servers are fixed. We choose WS and DM distributions at 80% load. These two distributions represent different group of flows: a mixture of short and long flows

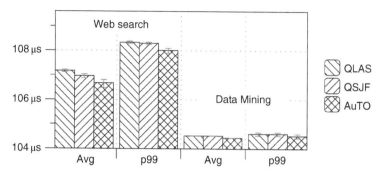

Figure 10.9 Homogeneous traffic: Average and p99 FCT.

(WS) and a set of short flows (DM). The average and 99th percentile (p99) FCT are shown in Figure 10.9. We train AuTO for eight hours and use the trained DNNs to schedule flows for another hour (shown in Figure 10.9 as AuTO). We make the following observations:

- For a mixture of short and long flows (WS), AuTO outperforms the standard heuristics, achieving up to 48.14% average FCT reduction. This is because it can dynamically change priority of long flows, avoiding the starvation problem in the heuristics.

- For distribution with mostly short flows (DM), AuTO performs similar to the heuristics. Since AuTO also gives any flow highest priority when it starts, AuTO performs almost the same as QLAS.

- Training the RL network results in average FCT reduction of 18.31% and 4.12% for WS¢DM distribution respectively, which demonstrates AuTO is capable to learn and adapt to traffic characteristics overtime.

- We further isolate the incast traffic [69] from the collected traces, and we find that they are almost the same with both QLAS and QSJF. This is because incast behavior is best handled by the congestion control and parameter setting. DCTCP [12], which is the transport we used in the experiments, already handles incast very well with appropriate parameter settings [12, 50].

10.5.3.2 Spatially Heterogeneous Traffic

We proceed to divide the servers into four clusters to create spatially heterogeneous traffic. We configure the flow generators in each cluster with different distribution and load pairs: <WS, 60%>, <WS, 80%>, <DM, 60%>, <DM, 60%>. We use AuTO to control all four clusters, and plot the average and p99 FCTs in Figure 10.10. For the heuristics, we compute the thresholds for each cluster individually according to its distribution and load. We observe similar results compared to the homogeneous scenarios. Compared to QLAS (QSJF), AuTO is shown to reduce the average

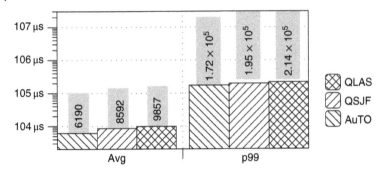

Figure 10.10 Spatially heterogeneous traffic: Average and p99 FCT.

FCT by 37.20% (27.95%) and p99 FCT by 19.78%(11.98%). This demonstrates that AuTO can adapt to spatial traffic variations.

10.5.3.3 Temporally and Spatially Heterogeneous Traffic

In these scenarios, we change the flow size distribution and network load every hour: The load value is chosen from $\{60\%, 70\%, 80\%\}$, and the distribution is randomly chosen from the ones in Figure 10.8. We ensure that the same distribution/load does not appear in consecutive hours. The experiment runs for eight hours.

The average and p99 FCTs are plotted against time in Figures 10.11 and 10.12. We can see:

- For heuristics with fixed parameters, when the traffic characteristics match the parameter setting, both average and p99 FCTs outperform the other schemes. But when mismatch occurs, the FCTs sharply drop. This shows that heuristics with fixed parameter setting cannot adapt to dynamic traffic well. Their parameters are usually chosen to perform well in the average case, but in practice, the traffic characteristics always change [1].

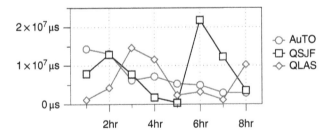

Figure 10.11 Dynamic scenarios: average FCT.

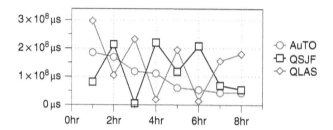

Figure 10.12 Dynamic scenarios: p99 FCT.

- AuTO is shown to steadily learn and adapt across time-varying traffic character-istics, in the last hour, AuTO achieves 8.71% (9.18%) reduction in average (p99) FCT compared to QSJF. This is because that AuTO, using two DRL agents, can dynamically change the priorities of flows in different environments to achieve better performance. Without any human involvement, this process can be done quickly and scalably.

Considering AuTO, a constant decline in FCTs indicates learning behavior, which eventually, as we have discussed in Section 10.3.5, lead to convergence to a local optimum for the dynamic traffic generation process. Figures 10.11 and 10.12 confirms our assumption that datacenter traffic scheduling can be converted into a RL problem, and DRL techniques (Section 10.3.5) can be applied to solve it.

10.5.4 Deep Dive

In the following, we inspect the design components of AuTO.

10.5.4.1 Optimizing MLFQ Thresholds using DRL

We first examine the MLFQ thresholds generated by sRLA. In Figure 10.13, we compare the MLFQ thresholds generated by sRLA and those by an optimizer [2, 50]. We obtain a set of three thresholds (for four queues) from sRLA in CS after eight hours of training for each flow size distribution at 60% load. We observe that both sets of thresholds are similar in the thresholds of the first three queues, and the main difference is in the last queue. For example, the last sRLA threshold (α_3) for WS distribution is 64 packets, while α_3 from optimizer is 87 packets. The same is true for DM distribution. However, the discrepancy does not reflect in significant difference in terms of performance. We plot the average and p99 FCT results for both sets of thresholds in Figures 10.14 and 10.15. The results are grouped by flow size. For sRLA generated thresholds and optimizer-generated thresholds, we observe that the difference in FCT is small in all groups of flow sizes. We conclude that, after eight hours of training, sRLA generated thresholds are similar to optimizer-generated ones in terms of performance.

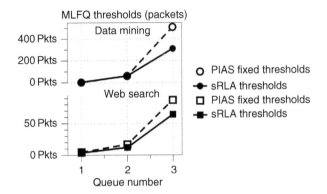

Figure 10.13 MLFQ thresholds from sRLA vs. optimal thresholds.

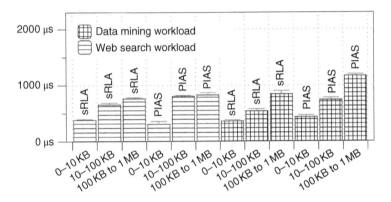

Figure 10.14 Average FCT using MLFQ thresholds from sRLA vs. optimal thresholds.

10.5.4.2 Optimizing Long Flows using DRL

Next, we look at how lRLA optimizes long flows. During the experiments in Section 10.5.3.3, we log the number of long flows on each link for five minutes in lRLA. Denote L as the set of all links, $N_l(t)$ as the number of long flows on link $l \in L$ at time t, and $N(t) = \{N_l(t), \forall l\}$. We plot $max(N(t)) - min(N(t)), \forall t$ in Figure 10.16, which is the difference in number of long flows on the link that have the most long flows and the link that have the least. This metric is an indicator of load imbalance. We observe that this metric is less than 10 most of the time. When temporary imbalance occurs, as shown in the magnified portion of Figure 10.16 (from 24 to 28 seconds), lRLA reacts to the imbalance by routing the excess flows onto the less congested links. This is because, as we discussed in Section 10.2.4, the reward of the PG algorithm is directly linked to the throughput: when long flows share a link, the total throughput is less than when they are using different

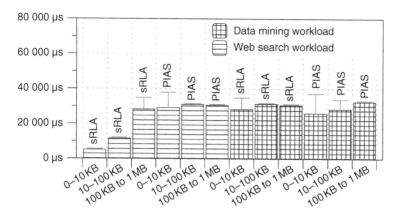

Figure 10.15 p99 FCT using MLFQ thresholds from sRLA vs. optimal thresholds.

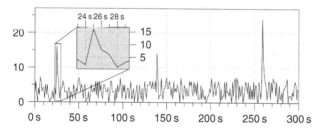

Figure 10.16 Load balancing using lRLA (PG algorithm): difference in number of long flows on links.

links. lRLA is rewarded when it places long flows on different links, thus it learns to load balance long flows.

10.5.4.3 System Overhead

We proceed to investigate the performance and overheads of AuTO modules. First, we look at the response latency of CS, as well as its scalability. Then, we examine the overheads of the end-host modules in PS.

CS Response Latency During experiments, response delay of the CS server (Figure 10.17) is measured as follows: t_u is the time instant of CS receiving an update from one server, and t_s is the time instant of CS sending the action to that server, so the response time is $t_s - t_u$. This metric directly shows how fast can the scheduler adapt to traffic dynamics reported by PS. We observe that CS can respond to an update within 10 ms on average for our 32-server testbed. This latency is mainly due to the computation overhead of DNN, as well as the queueing delay of servers' updates at CS. AuTO currently only uses CPU. To

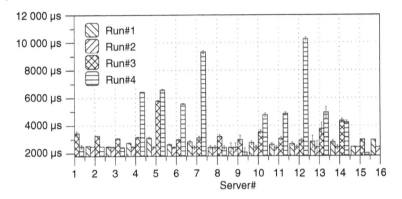

Figure 10.17 CS response latency: Traces from four runs.

reduce this latency, one promising direction is CPU–GPU hybrid training and serving [70], where CPUs handle the interaction with the environment, while GPUs train the models in the background.

Response latency also increases with computation complexity of DNN. In AuTO, the network size is defined by $\{n_l, m_l, m_s\}$. Since long flows are few, the increment of n_l, m_l are expected to be moderate even for datacenters with high load. We increase $\{n_l, m_l\}$ from $\{11, 10\}$ to $\{1000, 1000\}$, and find the average response time for lRLA becomes 81.82 ms (median 25.14 ms). However, m_s can increase significantly in high load scenarios, and we conduct an experiment to understand the impact on response latency of sRLA. In Figure 10.18, we vary m_s from 100 (used in the above experiments) to 1000, and measure the response latency. We find that the average response time only slightly increase for larger m_s. This is because, m_s determines the input layer size, which only affects the matrix size of link weights between the input layer and the first hidden layer. Moreover, if in the future AuTO employs more complex DNNs, we can reduce the response latency with parallelization techniques proposed for DRL [71–74].

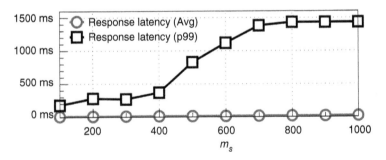

Figure 10.18 CS response latency: Scaling short flows (m_s)

CS Scalability Since our testbed is small, the NIC capacity of CS server is not fully saturated. Using the same parameter settings as in the experiments (Section 10.5.3.3), the bandwidth of monitoring flows is 12.40 Kbps per server. Assuming 1 Gbps network interface, the CS server should support 80.64K servers, which should be able to handle the servers in typical production datacenters [75, 76]. We also intend to achieve higher scalability in the following ways: (i) 1 Gbps link capacity is chosen to mimic the experiment environment, and in current production datacenters, the typical bandwidth of server is usually 10 Gbps or above [75, 76]; (ii) we expect CS to have GPUs or other hardware accelerators [70], so that the computation can complete faster; and (iii) we can reduce the bandwidth requirement of monitoring flows by implementing compression and/or sampling in PS.

PS Overhead End-host overhead refers to the additional work done for each flow to collect information and enforce actions. The overhead can be measured by CPU utilization and reduction in throughput when PS is running. We measured both metric during the experiments, and rerun the flows without enabling MM and EM. We find that the throughput degradation is negligible, and the CPU utilization is less than 1%. Since EM is similar to the tagging module in PIAS [1], our results confirm that both the throughput and CPU overhead are also minimal as PIAS.

10.6 Conclusion and Future Directions

Inspired by recent successes of DRL techniques in solving complex online control problems, in this work, we attempted to enable DRL for automatic TO. However, our experiments show that the latency of current DRL systems is the major obstacle to TO at the scale of current datacenters. We solved this problem by exploiting long-tail distribution of datacenter traffic. We developed a two-level DRL system, AuTO, mimicking the peripheral and central nervous systems in animals, to solve the scalability problem. We deployed and evaluated AuTO on a real testbed, and demonstrated its performance and adaptiveness to dynamic traffic in datacenters. AuTO is a first step toward automating datacenter TO, and we hope many software components in AuTO can be reused in other DRL projects in datacenters.

By summarizing our experience, we identify two key challenges of applying RL (or other machine learning-based methods) to networks.

Unseen environments One major concern about RL methods is its performance on unseen environments. The intuitive solution is to let RL learn on the changed settings. While effective, this approach might undergo the well-known sample-inefficiency problem of RL and thus make this solution inefficient. For example, Aurora [34] will produce serious under-utilization or overutilized when placed in a rapidly changing network [35]. But training Aurora until the new convergence point is time-consuming. Datacenter may suffer more from this problem. In the multi-service settings, the real-time traffic might mismatch the premeasured workload under training phase, causing performance degradation. One possible solution is to make RL be more robust and quick-adaptive through other learning techniques such as transfer learning. We leave it as future work.

Computing overhead Another concern about the fully learning-based solutions is the computing overhead. As we mentioned in Section 10.3.1, *naive* application of Machine learning will be not enough to produce timely online decisions for scheduling short flows in datacenter. In this work, we address this challenge by leveraging MLFQ in the end-hosts, which are informed by the central agent of global demotion thresholds, to cut down the waiting time. Similarly, Orca [35] also turn to couple learning based methods with heuristic algorithms. To this end, we think the promising approach to utilize the benefits of learn-based methods without causing unacceptable pressure on computing may lay on the combination of learning methods with traditional designs.

AuTO is still an evolving system, and has many limitations. In addition to the potential improvements we mentioned in Sections 10.4 and 10.5, we intend to: (i) implement the asynchronous communication in a compiled language to boost system responsiveness; (ii) implement CM and EM as kernel module for efficiency and application-transparency; (iii) investigate new RL models and algorithms for datacenter TO.

Our current prototype is constrained by an emulated system where the AuTO daemon can access active flows and send a rather precise state of the server flows. It is not the ultimate goal of the project and a few significant enhancements have already been identified as the next steps toward a functional system. Operating at a low level of abstraction of flows is an essential feature that our framework is missing. Taking this into account, we are exploring various ways of recording currently active flows and recently finished ones.

Another potential improvement is flow scheduling at higher speed and throughput. Currently, the Linux priorities are used from user space to schedule flows. However, a scalability concern arises since we observe delays for updates during experiments and sluggishness in processing updates. This implies a need for a fast and light communication interface with the kernel. A special kernel module seems as an adequate solution to solve the aforementioned problem as such an approach can ensure faster updates of scheduling at the kernel level and alleviation of system overload. Although a kernel module would provide what our framework needs, it would also raise a concern about modifying the kernel and would make the framework inflexible since a special module would be needed to be installed. Taking this into account, we are looking for a scalable way of controlling flows in the user space as benefits such as portability, safety, and avoidance of modifying the kernel, we believe, outweigh the scalability and speed concerns. Our testbed results (10.5) demonstrated that flows can be scheduled in the user space rather effectively.

Furthermore, as it has been mentioned before, we are not favoring moving our *RL server* away from the Python programming language due to the benefits the language provides. Nevertheless, the core of the server (incoming/outgoing communication handling) is being revised and might be rewritten in a compiled language to boost system responsiveness.

In the end, the AuTO daemon is a global process that has access to each of our generated flows and can directly interact with every flow through a globally shared structure. Due to this, the future direction of the project will focus on removing the abstraction of the flows by implementing a system that can trace real system flows, which coincidentally can also be done with kernel module implementation.

For future work, while this work focuses on employing RL to perform flow scheduling and load balancing, RL algorithms for congestion control and task scheduling can be developed. In addition to the potential improvements we mentioned in Sections 10.4 and 10.5, we also plan to investigate applications of RL beyond datacenters, such as WAN bandwidth management.

Bibliography

1 Bai, W., Chen, L., Chen, K. et al. (2015). Information-agnostic flow scheduling for commodity data centers. *12th USENIX Symposium on Networked Systems Design and Implementation (NSDI 15)*. Oakland, CA: USENIX Association, pp. 455–468. ISBN 978-1-931971-218.

2 Chen, L., Chen, K., Bai, W., and Alizadeh, M. (2016). Scheduling mix-flows in commodity datacenters with Karuna. *Proceedings of the 2016 ACM SIG-COMM Conference*, SIGCOMM '16. New York, NY, USA: Association for Computing Machinery, pp. 174–187. https://doi.org/10.1145/2934872.2934888, ISBN 9781450341936.

3 Al-Fares, M., Radhakrishnan, S., Raghavan, B. et al. (2010). Hedera: dynamic flow scheduling for data center networks. *7th USENIX Symposium on Networked Systems Design and Implementation, (NSDI 10)*. San Jose, CA, USA: USENIX Association, pp. 281–296.

4 Hong, C.-Y., Caesar, M., and Godfrey, P.B. (2012). Finishing flows quickly with preemptive scheduling. *Proceedings of the ACM SIGCOMM 2012 Conference on Applications, Technologies, Architectures, and Protocols for Computer*

Communication, SIGCOMM '12. New York, NY, USA: Association for Computing Machinery, pp. 127–138. https://doi.org/10.1145/2342356.2342389, ISBN 9781450314190.

5 Alizadeh, M., Yang, S., Sharif, M. et al. (2013). pFabric: Minimal near-optimal datacenter transport. *Proceedings of the ACM SIGCOMM 2013 Conference on SIGCOMM*, SIGCOMM '13. New York, NY, USA: Association for Computing Machinery, pp. 435–446. https://doi.org/10.1145/2486001.2486031, ISBN 9781450320566.

6 Li, Z., Bai, W., Chen, K. et al. (2017). Rate-aware flow scheduling for commodity data center networks. *IEEE INFOCOM 2017-IEEE Conference on Computer Communications*, IEEE, pp. 1–9.

7 Li, Z., Zhang, Y., Li, D. et al. (2016). OPTAS: decentralized flow monitoring and scheduling for tiny tasks. *IEEE INFOCOM 2016-The 35th Annual IEEE International Conference on Computer Communications*, IEEE, pp. 1–9.

8 Zhang, H., Chen, L., Yi, B. et al. (2016). CODA: toward automatically identifying and scheduling coflows in the dark. *Proceedings of the 2016 ACM SIGCOMM Conference*, SIGCOMM '16. Association for Computing Machinery, pp. 160–173. https://doi.org/10.1145/2934872.2934880, ISBN 9781450341936.

9 Chowdhury, M., Zhong, Y., and Stoica, I. (2014). Efficient coflow scheduling with Varys. *Proceedings of the 2014 ACM Conference on SIGCOMM*, SIGCOMM '14. New York, NY, USA: Association for Computing Machinery, pp. 443–454. https://doi.org/10.1145/2619239.2626315, ISBN 9781450328364.

10 Chowdhury, M. and Stoica, I. (2015). Efficient coflow scheduling without prior knowledge. *Proceedings of the 2015 ACM Conference on Special Interest Group on Data Communication*, SIGCOMM '15. New York, NY, USA: Association for Computing Machinery, pp. 393–406. https://doi.org/10.1145/2785956.2787480.

11 Susanto, H., Jin, H., and Chen, K. (2016). Stream: decentralized opportunistic inter-coflow scheduling for datacenter networks. *2016 IEEE 24th International Conference on Network Protocols (ICNP)*, IEEE, pp. 1–10.

12 Alizadeh, M., Greenberg, A., Maltz, D.A. et al. (2010). Data center TCP (DCTCP). *Proceedings of the ACM SIGCOMM 2010 Conference*, SIGCOMM '10. New York, NY, USA: Association for Computing Machinery, pp. 63–74. https://doi.org/10.1145/1851182.1851192, ISBN 9781450302012.

13 Vamanan, B., Hasan, J., and Vijaykumar, T.N. (2012). Deadline-aware datacenter TCP (D2TCP). *ACM SIGCOMM Computer Communication Review* 42 (4): 115–126.

14 Bai, W., Chen, L., Chen, K., and Wu, H. (2016). Enabling ECN in multi-service multi-queue data centers. *13th USENIX Symposium on Networked Systems Design and Implementation (NSDI 16)*. Santa Clara, CA, USA: USENIX Association, pp. 537–549. ISBN 978-1-931971-29-4.

15 Alizadeh, M., Edsall, T., Dharmapurikar, S. et al. (2014). CONGA: distributed congestion-aware load balancing for datacenters. *Proceedings of the 2014 ACM Conference on SIGCOMM*, SIGCOMM '14. New York, NY, USA: Association for Computing Machinery, pp. 503–514. https://doi.org/10.1145/2619239.2626316, ISBN 9781450328364.

16 Zhang, H., Zhang, J., Bai, W., Chen, K., and Chowdhury, M. (2017). Resilient datacenter load balancing in the wild. *Proceedings of the Conference of the ACM Special Interest Group on Data Communication*, pp. 253–266.

17 Sutton, R.S. and Barto, A.G. (1998). *Introduction to Reinforcement Learning*, vol. 135. Cambridge: MIT Press.

18 Mnih, V., Kavukcuoglu, K., Silver, D. et al. (2013). Playing Atari with deep reinforcement learning. *arXiv preprint arXiv:1312.5602*.

19 Silver, D., Huang, A., Maddison, C.J. et al. (2016). Mastering the game of go with deep neural networks and tree search. *Nature* 529 (7587): 484–489.

20 TensorFlow. API documentation: TensorFlow. https://www.tensorflow.org/api_docs/ (accessed 18 April 2017).

21 Paszke, A., Gross, S., Massa, F. et al. (2019). PyTorch: an imperative style, high-performance deep learning library. *Advances in Neural Information Processing Systems 32*. Curran Associates, Inc., pp. 8024–8035.

22 Moritz, P., Nishihara, R., Wang, S. et al. (2018). Ray: a distributed framework for emerging AI applications. *13th USENIX Symposium on Operating Systems Design and Implementation (OSDI 18)*. Carlsbad, CA, USA: USENIX Association, pp. 561–577. ISBN 978-1-939133-08-3.

23 Kandula, S., Sengupta, S., Greenberg, A. et al. (2009). The nature of data center traffic: measurements & analysis. *Proceedings of the 9th ACM SIGCOMM Conference on Internet Measurement*, IMC '09. New York, NY, USA: Association for Computing Machinery, pp. 202–208. https://doi.org/10.1145/1644893.1644918, ISBN 9781605587714.

24 Benson, T., Akella, A., and Maltz, D.A. (2010). Network traffic characteristics of data centers in the wild. *Proceedings of the 10th ACM SIGCOMM Conference on Internet Measurement*, IMC '10. New York, NY, USA: Association for Computing Machinery, pp. 267–280. https://doi.org/10.1145/1879141.1879175, ISBN 9781450304832.

25 Hopps, C. (2000). Analysis of an equal-cost multi-path algorithm. *RFC 2992*.

26 Mao, H., Alizadeh, M., Menache, I., and Kandula, S. (2016) Resource management with deep reinforcement learning. *Proceedings of the 15th ACM workshop on hot topics in networks*. pp. 50–56.

27 Hornik, K. (1991). Approximation capabilities of multilayer feedforward networks. *Neural Networks*. 4 (2): 251–257

28 Sutton, R.S., McAllester, D.A., Singh, S.P., and Mansour, Y. (2012). Policy gradient methods for reinforcement learning with function approximation.

29 Hastings, W.K. (1970). Biometrika. http://www.jstor.org/stable/2334940 (accessed 20 April 2021).

30 Schulman, J., Levine, S., Moritz, P. et al. (2015). Trust region policy optimization. *CoRR*.

31 Sun, Y., Yin, X., Jiang, J. et al. (2016). CS2P: improving video bitrate selection and adaptation with data-driven throughput prediction. *Proceedings of the 2016 ACM SIGCOMM Conference*, pp. 272–285.

32 Winstein, K. and Balakrishnan, H. (2013). TCP ex machina: Computer-generated congestion control. *ACM SIGCOMM Computer Communication Review* 43 (4): 123–134.

33 Dong, M., Meng, T., Zarchy, D. et al. (2018). {PCC} Vivace: online-learning congestion control. *15th USENIX Symposium on Networked Systems Design and Implementation (NSDI '18)*, pp. 343–356.

34 Jay, N., Rotman, N.H., Godfrey, P.B., Schapira, M., and Tamar, A. (2019) A deep reinforcement learning perspective on internet congestion control. *Proceedings of the 36th International Conference on Machine Learning*, PMLR. pp. 3050–3059.

35 Abbasloo, S., Yen, C.-Y., and Chao, H.J. (2020) Classic meets modern: a pragmatic learning-based congestion control for the internet. *Proceedings of the Annual conference of the ACM Special Interest Group on Data Communication on the applications, technologies, architectures, and protocols for computer communication.* pp. 632–647.

36 Chen, Z., Wen, J., and Geng, Y. (2016). Predicting future traffic using Hidden Markov models. *2016 IEEE 24th International Conference on Network Protocols (ICNP)*, IEEE, pp. 1–6.

37 Poupart, P., Chen, Z., Jaini, P. et al. (2016). Online flow size prediction for improved network routing. *2016 IEEE 24th International Conference on Network Protocols (ICNP)*, IEEE, pp. 1–6.

38 Liang, E., Zhu, H., Jin, X., and Stoica, I. (2019). Neural packet classification. *ACM SIGCOMM '19*, pp. 256–269.

39 Mao, H., Schwarzkopf, M., Venkatakrishnan, S.B. et al. (2019). Learning scheduling algorithms for data processing clusters. *Proceedings of the 2019 ACM SIGCOMM Conference*, pp. 270–288.

40 Munir, A., Baig, G., Irteza, S.M. et al. (2014). Friends, not foes: synthesizing existing transport strategies for data center networks. *Proceedings of the 2014 ACM Conference on SIGCOMM*, SIGCOMM '14. New York, NY, USA: Association for Computing Machinery, pp. 491–502. https://doi.org/10.1145/2619239.2626305, ISBN 9781450328364.

41 Wilson, C., Ballani, H., Karagiannis, T., and Rowtron, A. (2011). Better never than late: meeting deadlines in datacenter networks. *ACM SIGCOMM Computer Communication Review* 41 (4): 50–61.

42 Alizadeh, M., Kabbani, A., Edsall, T. et al. (2012) Less is more: trading a little bandwidth for ultra-low latency in the data center. *9th {USENIX} Symposium on Networked Systems Design and Implementation ({NSDI} 12)*. pp. 253–266.

43 Boyan, J.A. and Littman, M.L. (1994). Packet routing in dynamically changing networks: a reinforcement learning approach. *Advances in Neural Information Processing Systems*, CiteSeerX, pp. 671–678.

44 Linux Foundation. Priority qdisc - Linux man page. https://linux.die.net/man/8/tc-prio (accessed 17 April 2017).

45 McKeown, N., Anderson, T., Balakrishnan, H. et al. (2008). OpenFlow: enabling innovation in campus networks. *ACM SIGCOMM Computer Communication Review* 38 (2): 69–74. https://doi.org/10.1145/1355734.1355746.

46 Netfilter.Org. The netfilter.org project. https://www.netfilter.org/ (Accessed 17 April 2017).

47 Peng, Y., Chen, K., Wang, G. et al. (2014) HadoopWatch: a first step towards comprehensive traffic forecasting in cloud computing. *IEEE INFOCOM 2014-IEEE Conference on Computer Communications*, IEEE. pp. 19–27.

48 Peng, Y., Chen, K., Wang, G. et al. (2015). Towards comprehensive traffic forecasting in cloud computing: design and application. *IEEE/ACM Transactions on Networking* 24 (4): 2210–2222.

49 Greenberg, A., Hamilton, J.R., Jain, N. et al. (2009). Vl2: a scalable and flexible data center network. *Proceedings of the ACM SIGCOMM 2009 Conference on Data Communication*, SIGCOMM '09. New York, NY, USA: Association for Computing Machinery, pp. 51–62. https://doi.org/10.1145/1592568.1592576, ISBN 9781605585949.

50 Bai, W., Chen, L., Chen, K. et al. (2017). PIAS: practical information-agnostic flow scheduling for commodity data centers. *IEEE/ACM Transactions on Networking (TON)* 25 (4): 1954–1967.

51 Arzani, B., Ciraci, S., Loo, B.T. et al. (2016). Taking the blame game out of data centers operations with netpoirot. *ACM SIGCOMM'16*.

52 Yadwadkar, N.J., Ananthanarayanan, G., and Katz, R. (2014) Wrangler: predictable and faster jobs using fewer resources. *Proceedings of the ACM Symposium on Cloud Computing*. pp. 1–14.

53 Silver, D., Lever, G., Heess, N. et al. (2014). Deterministic policy gradient algorithms. *Proceedings of the 31st International Conference on International Conference on Machine Learning - Volume 32*, ICML'14. JMLR.org, pp. I-387–I-395. http://dl.acm.org/citation.cfm?id=3044805.3044850 (accessed 21 April 2021).

54 Bhatnagar, S., Ghavamzadeh, M., Lee, M., and Sutton, R.S. (2008). Incremental natural actor-critic algorithms.

55 Lillicrap, T.P., Hunt, J.J., Pritzel, A. et al. (2015). Continuous control with deep reinforcement learning. *CoRR*, abs/1509.02971. http://arxiv.org/abs/1509.02971 (accessed 20 April 2021).

56 Farrington, N., Porter, G., Radhakrishnan, S. et al. (2010). Helios: a hybrid electrical/optical switch architecture for modular data centers. *Proceedings of the ACM SIGCOMM 2010 Conference*, SIGCOMM '10. New York, NY, USA: Association for Computing Machinery, pp. 339–350. https://doi.org/10.1145/1851182.1851223, ISBN 9781450302012.

57 Hu, S., Chen, K., Wu, H. et al. (2015). Explicit path control in commodity data centers: design and applications. *12th USENIX Symposium on Networked Systems Design and Implementation (NSDI 15)*. Oakland, CA, USA: USENIX Association, pp. 15–28. ISBN 978-1-931971-218.

58 Chollet, F. Keras documentation. https://keras.io/ (accessed 18 April 2017).

59 NVIDIA. Deep learning frameworks. https://developer.nvidia.com/deep-learning-frameworks (accessed 18 April 2017).

60 Cisco. Simple client-server application for generating user-defined traffic patterns. https://github.com/datacenter/empirical-traffic-gen (accessed 24 April 2017).

61 Welsh, M., Culler, D., and Brewer, E. (2001). SEDA: an architecture for well-conditioned, scalable internet services. *Proceedings of the 18th ACM Symposium on Operating Systems Principles*, SOSP '01. New York, NY, USA: Association for Computing Machinery, pp. 230–243. https://doi.org/10.1145/502034.502057, ISBN 1581133898.

62 Python Software Foundation. Global interpreter lock. https://wiki.python.org/moin/GlobalInterpreterLock (accessed 18 April 2017).

63 OpenAI. Openai gym. https://gym.openai.com/ (accessed 24 April 2017).

64 Chen, L., Xia, J., Yi, B., and Chen, K. (2018). PowerMan: an out-of-band management network for datacenters using power line communication. *15th USENIX Symposium on Networked Systems Design and Implementation (NSDI 18)*. Renton, WA, USA: USENIX Association, pp. 561–578. ISBN 978-1-939133-01-4.

65 Bai, W., Chen, K., Chen, L. et al. (2016). Enabling ECN over generic packet scheduling. *Proceedings of the 12th International on Conference on Emerging Networking EXperiments and Technologies*, CoNEXT '16. New York, NY, USA: Association for Computing Machinery, pp. 191–204. https://doi.org/10.1145/2999572.2999575, ISBN 9781450342926.

66 Munir, A., Qazi, I.A., Uzmi, Z.A. et al. (2013). Minimizing flow completion times in data centers. *Proceedings IEEE INFOCOM 2013*, pp. 2157–2165.

67 Dell, R.B., Holleran, S., and Ramakrishnan, R. (2002). Sample size determination. *ILAR Journal*. 43 (4): 207–213.

68 Lenth, R.V. (2001). Some practical guidelines for effective sample size determination. *The American Statistician*. 55 (3): 187–193

69 Chen, Y., Griffith, R., Liu, J. et al. (2009). Understanding TCP Incast throughput collapse in datacenter networks. *Proceedings of the 1st ACM Workshop* on *Research on Enterprise Networking*, WREN '09. New York, NY, USA: Association for Computing Machinery, pp. 73–82. https://doi.org/10.1145/ 1592681.1592693, ISBN 9781605584430.

70 Nvlabs. Hybrid CPU/GPU implementation of the A3C algorithm for deep rein-forcement learning. https://github.com/Nvlabs/GA3C (accessed 13 June 2018).

71 Mnih, V., Badia, A.P., Mirza, M. et al. (2016). Asynchronous methods for deep reinforcement learning. *Proceedings of The 33rd International Conference on Machine Learning (ICML '16)*, pp. 1928–1937.

72 Babaeizadeh, M., Frosio, I., Tyree, S. et al. (2016). *4th International Conference on Learning Representations, ICLR*. Reinforcement learning through asyn-chronous advantage actor-critic on a GPU.

73 Gu, S., Holly, E., Lillicrap, T., and Levine, S. (2017). Deep reinforcement learn-ing for robotic manipulation with asynchronous off-policy updates. *Proceedings of 2017 IEEE International Conference on Robotics and Automation (ICRA)*.

74 Frans, K. and Hafner, D. (2016). Parallel trust region policy optimization with multiple actors.

75 Singh, A., Ong, J., Agarwal, A. et al. (2015). Jupiter rising: a decade of clos topologies and centralized control in Google's datacenter network. *Proceedings of the 2015 ACM Conference on Special Interest Group on Data Communication, SIGCOMM '15*. New York, NY, USA: Association for Computing Machinery, pp. 183–197. https://doi.org/10.1145/2785956.2787508, ISBN 9781450335423.

76 Roy, A., Zeng, H., Bagga, J. et al. (2015). Inside the social network's (datacenter) network. *Proceedings of the 2015 ACM Conference on Special Interest Group on Data Communication, SIGCOMM '15*. New York, NY, USA: Association for Computing Machinery, pp. 123–137. https://doi.org/10.1145/ 2785956.2787472, ISBN 9781450335423.

11

The New Abnormal: Network Anomalies in the AI Era

Francesca Soro[1], Thomas Favale[1], Danilo Giordano[1], Luca Vassio[1], Zied Ben Houidi[2], and Idilio Drago[3]

[1]*Politecnico di Torino, 10129, Torino, Corso Duca degli Abruzzi 24, Italy*
[2]*Huawei Technologies, 92100, Boulogne-Billancourt, 18 Quai du Point du Jour, France*
[3]*University of Turin, 10149, Torino, Corso Svizzera 185, Italy*

11.1 Introduction

Authors of [1] define *anomaly detection* as the "problem of finding patterns in data that do not conform to expected behavior." In computer networks anomaly detection techniques have been employed in several tasks, such as finding nodes compromised by malware, triggering alerts in network monitoring systems, and pinpointing faults reported in service logs.

Most anomaly detection algorithms used in networking problems are based on techniques proposed for other scenarios. Methods to perform anomaly detection are indeed researched and exploited since decades before the development of the Internet itself – from the study of outliers in probability distributions to the search for frauds in pre-Internet systems, e.g. banking systems. The surge on data coming from networked applications (e.g. social networks, IoT devices, cyber-physical systems) together with the measurements needed to operate these applications have pushed anomaly detection further. Anomaly detection more than ever requires techniques and algorithms able to uncover anomalous behaviors on datasets that are large, complex, and diverse.

Initial approaches ported to the network anomaly detection problem have been strongly rooted in rules-of-thumb, statistics, information theory, and machine learning. Threshold-based anomaly detection, for instance, has been widely adopted in cyber-security for the detection of port scans and distributed denial-of-service (DDoS) attacks. Similarly, diverse statistical solutions have been

Communication Networks and Service Management in the Era of Artificial Intelligence and Machine Learning,
First Edition. Edited by Nur Zincir-Heywood, Marco Mellia, and Yixin Diao.

employed to identify anomalies on time series exported by Internet telemetry systems. As more and more data became available, data-driven solutions gained momentum. Machine learning algorithms for prediction, clustering, and classification have been applied on anomaly detection thanks to their good capabilities to automatically learn patterns from data. The trend has been exacerbated in recent years: the continuous growth on data availability, the unprecedented increase on computing resources, and breakthroughs on artificial intelligence (AI) research have allowed data-driven solutions to solve new complex problems on various fields. Some of these breakthroughs have potential to revolutionize the research on anomaly detection too.

This chapter summarizes ongoing recent progress on anomaly detection research. We first introduce the anomaly detection problem. Starting from a comprehensive survey [1], we summarize the classic techniques and key applications of anomaly detection on network monitoring. Building upon the taxonomy proposed by the authors of [1], we evaluate how developments on AI algorithms bring new possibilities for anomaly detection. More concretely, we here answer the following questions:

- How have anomaly detection techniques evolved in the last 10 years?
- What are key tools implementing anomaly detection? Are they profiting from recent advances in AI and deep learning?

This chapter provides a picture of anomaly detection algorithms that are emerging from advances on AI research. We here do not aim at providing a comprehensive survey on the topic, but instead illustrate the progress on the field discussing significant and recent works. In the following, Section 11.2 defines anomaly detection, introduces the taxonomy used as basis for our discussion, and reviews classic anomaly detection methods. Section 11.3 describes how recent AI developments are influencing the anomaly detection landscape. Section 11.4 summarizes key tools and frameworks implementing anomaly detection, and discusses whether they profit from the identified AI-based approaches. Section 11.5 concludes the chapter with our view on a possible future development with a case study on anomaly detection on graphs.

11.2 Definitions and Classic Approaches

Anomaly detection in networked applications are studied since early days of the Internet. Many surveys have summarized the developments in the field [2–5]. In this section we introduce the definitions and the taxonomy used throughout the chapter, which is based on [1]. We then conclude the section with a brief summary of classical anomaly detection approaches.

11.2.1 Definitions

The term *Anomaly detection* aggregates many different, yet related, tasks. As said before, we adhere to the somehow loose definition by authors of [1] and consider an *anomaly* any unexpected behavior in a data variable. Novelty detection, outlier detection, and rare event detection are some of the related tasks that are commonly found in the literature, which we group together as anomaly detection.

Outliers are values detached from the remaining samples. For example, given a random variable, an outlier can be a value that should not happen because it is out of the acceptable variable range, or because it falls far from the expected value. *Rare events* are usually defined similarly – points falling far from expected values. However, they represent events that are known to happen rarely. As such, some anomaly detection algorithms may consider such events as *normal*, even if they deviate from common patterns.

Finally, *novelty* represents a behavioral (and possibly permanent) change of a variable. Unlike outlier detection, where deviating points have been seen before (in training), novelty detection aims at capturing whether a new sample is an outlier compared to the past or not. Here again the surge of a novelty can be considered an anomaly, as the novel points diverge from the usual patterns. Some novelty detection algorithms however try to identify whether points deviating from expected patterns represent indeed such a change in behavior, thus tagging the change as novelty, rather than an anomaly.

11.2.2 Anomaly Detection: A Taxonomy

The authors of [1] characterize anomaly detection techniques from two different angles. We reproduce and extend this taxonomy in Figure 11.1 and use it in the remainder of the chapter to position novel AI-based anomaly detection algorithms. According to the taxonomy, anomaly detection can be characterized by the application domain and the problem characteristics from one side; and, from the other side, by the research area leading to the algorithm.

In terms of **application domains**, in contrast to [1], we borrow the *Functional Areas* of the well-known taxonomy for network and service management proposed by IFIP.[1] This taxonomy is a convenient way to characterize applications in telecommunications, thus suiting perfectly our scope as we discuss anomalies in computer networks only. It groups network management problems in fault, configuration, accounting, performance, security, service level (e.g. QoS), and network events. When illustrating new anomaly detection techniques, we will provide examples using these domains.

---

1 http://wg66.ifip.org/taxonomy.html

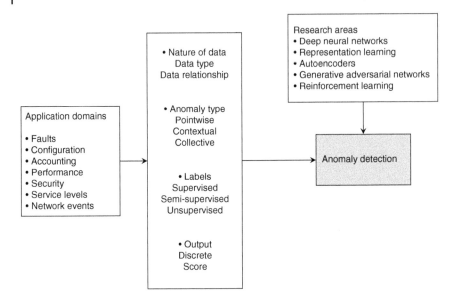

Figure 11.1 A taxonomy for anomaly detection problems and techniques. Source: Modified from Chandola et al. [1].

Each anomaly detection task has its own **problem characteristics**: The nature of the input data, the type of anomaly, the labels available for the learning phase and the output format. We will discuss these features in details next. Finally, the same problem can be faced using techniques that come from different **research areas**. This chapter will focus on techniques emerging from recent advances on AI research. We will cover it in details on Section 11.3.

11.2.3 Problem Characteristics

Nature of data refers to the input data at hand. Anomaly detection aims at finding anomalous *data instances*. The applicability of an algorithm depends both on the **data type** of instances and the **relationship** among instances.

Instances are represented by attributes of different types, such as text, integer numbers, or images/video. Anomaly detection may be performed over single or multiple attributes (i.e. multivariate problem). The multiple attributes that characterize instances can eventually be of different types.

Data instances may be related to each other according to some criteria. *Point data* refers to instances with no relationship, e.g. a dataset composed of multiple images or a collection of server log files. Anomaly detection could be used to detect instances deviating from the "usual" ones. *Time-series* are instances recorded over time. The anomaly detection task could be to find outliers in the series. Instances connected based on any other generic relation are called *graph-based*, e.g. the graph of Autonomous System peering. Anomaly detection could be applied to find anomalous connections between instances.[2]

Anomaly types refer to anomaly macro-categories. Anomalies are classified as pointwise, collective, or contextual [2, 5, 6], regardless of the nature of data. Examples are provided in Figure 11.2 considering a numeric attribute forming a time series.

Pointwise anomalies are individual data instances that diverge in a dataset. We see an example in Figure 11.2a, in which a single spike deviates from the regular behavior of the series. Examples of pointwise anomalies in network *security* and *performance* domains are (i) an abrupt rise in the number of packets reaching a server during a DDoS attack and (ii) increases in the round trip time (RTT) between two networks due to a temporary congestion.

Contextual anomalies are cases where an instance becomes anomalous thanks to its context, even if it would not be anomalous in isolation. Figure 11.2b provides

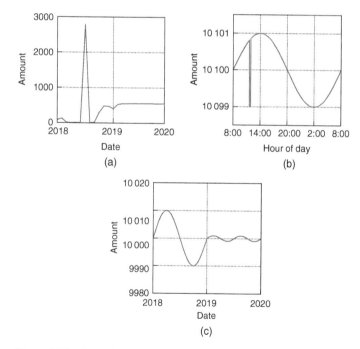

Figure 11.2 Examples of anomaly macro-categories. (a) Pointwise; (b) Contextual; (c) Collective.

---

2 Other categories are found in the literature, such as *spatial* and *spatio-temporal* [1] – those are not discussed here for brevity.

an example. A single low value appears while the series is reporting (smooth) high values for the attribute. However, low values are expected to be seen in other contexts. In network *accounting*, such behavior could represent the bytes per hour in a backbone link during a short daytime outage. Yet, low values would still be expected during night periods, thus characterizing the contextual anomaly in the former case.

Collective anomalies are cases in which various data instances, in conjunction, form the anomaly. Figure 11.2c provides an example. Here a time series that periodically oscillates suddenly changes trend. Whereas the newer values remain in the same range, they collectively change the series behavior. This is a classic example sometimes considered as *novelty*, as the changed behavior may become the new normal after the anomalous transition. A permanent decrease or increase in traffic volume in a link caused by a *fault* on peering links could produce a similar anomaly.

Labels refer to the presence or not of ground truth that confirms that a particular data instance is anomalous. Having ground truth allows us to rely on *supervised* techniques, i.e. algorithms that learn the anomalous patterns from labeled datasets. When only the normal behavior is known, the problem is defined as *semi-supervised*. When neither normal nor anomalous instances are known, the problem is considered *unsupervised*.

Finally, **output** defines the format of the prediction provided by the anomaly detection algorithm. Most commonly, algorithms output either a *discrete* label (e.g. anomalous and normal) or a *score*, defined according to the problem at hand. For example, the anomaly score can represent the deviation of a particular instance from parameters of a probability distribution computed over normal instances, or it can report an arbitrary distance metric between new instances and the expected value for normal instances.

11.2.4 Classic Approaches

Section 11.3 will explore anomaly detection approaches based on recent AI developments. To position them, we here make a brief summary of classic anomaly detection approaches.

Anomaly detection has been faced with multiple **machine learning** and **data mining** algorithms. In fact, most classic algorithms used for classification and clustering can be applied on anomaly detection too. As for classification algorithms, for example, anomalies can be detected by training a model to recognize the normal or anomalous instances. Clearly, labels are needed for training these supervised algorithms, thus limiting their applicability. In some cases, to partially solve this issue, only normal instances are labeled, forcing any testing instance not assigned to a class to be marked as anomalous. In the case of clustering algorithms, data points are split into clusters based on arbitrary distance measures, which may be problem-specific. Points belonging to small clusters as well as those left unassigned are considered possible anomalies. As we will discuss later, many recent AI algorithms target classification and clustering problems. As such, they can be applied to anomaly detection following the above steps.

Statistical anomaly detection is instead based on the assumption that normal instances can be mapped to a stochastic model. Algorithms in this category mark as anomalies, instances that deviate partially or completely from the model. A classic technique belonging to the category is the so-called *boxplot rule*, which marks data instances as outliers considering a reference probability distribution.

Many algorithms have been derived from the **information theory** research. These techniques exploit different measures to quantify the information in a dataset, with the entropy being the most well-known alternative. For example, when considering entropy, some algorithms assume that normal instances would present attributes with a relatively low entropy, whereas the introduction of anomalies would cause an increase in entropy.

Several other categories of anomaly detection algorithms have been documented in the literature, and readers are invited to refer to [2, 4–9] for a deeper discussion on them.

11.3 AI and Anomaly Detection

Recent advancements in AI and deep learning in particular have also contributed to anomaly detection research. In this section, we discuss some of the most relevant developments and new methodologies that can be applied for this purpose.

11.3.1 Methodology

We have performed a literature survey to identify the *research areas* that drive novel trends in anomaly detection. First, we have used a research portal[3] to select top conferences and journals (based on conference H-indexes or journal impact factors) that cover *AI, machine learning*, and *data mining*. Among a wide set of topics, we have picked five broad, yet relatively recent, research directions that had contributions to anomaly detection research. While doing this process, we have collected articles that apply the chosen techniques to the anomaly detection problem, even if not related to computer networks. Finally, we complemented these articles with others by performing a targeted search on google scholar using

---

3 http://www.guide2research.com

anomaly and novelty detection as keywords together with the identified research areas (e.g. anomaly detection representation learning).

In the remainder, we illustrate the applicability of the identified approaches by listing only some of the most relevant articles in each area, published in recent years. By doing so, we intend to give the reader the big picture and the intuition behind each approach. The reader shall thus consider these references as entry pointers to further explore the topic if needed.

11.3.2 Deep Neural Networks

Deep Neural Networks (DNNs) are neural networks (NNs) that have many hidden layers between the input and the output layers (see example in Figure 11.3). Research on DNNs has gained momentum in the last decade, thanks to the increase on computing capabilities and on data availability, and had led to breakthroughs in many machine learning tasks [10]. DNNs have been indeed useful for multiple problems, such as classifying images and voice, as well as time series prediction. As such, they can be used for anomaly detection too, similarly to the classic approaches described on Section 11.2.4. DNNs are praised for their capability to generalize well and to work on complex input data without complex feature engineering [10], achieving high performance, e.g. high precision in classification problems.

DNNs can be built based on multiple architectures that suit best different problems. For example, in Recurrent Neural Networks [11] each node in the hidden layers forwards its result not only to the next layer, but also to itself. This scheme allows the network to *remember* patterns of previous data instances, e.g. helping the network to learn temporal patterns. Long Short-Term Memory [12] networks generalize the idea introducing an architecture able to remember information

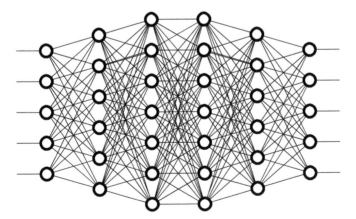

Figure 11.3 Example of deep neural network with four hidden layers.

about long-term sequences. In Convolutional Neural Networks (CNNs), nodes rely on convolutional matrices [13] to compute outputs. This scheme works as a filter, allowing the network to extract complex features from the input data. CNNs have been used successfully for image processing, e.g. due to their capacity to identify image borders.

Applied to anomaly detection, DNNs can be used similarly to how classic machine learning was used, e.g. using supervised learning in case the anomalous labels are present, or using approaches such as one-class classification where the goal is to learn to identify membership to one class (e.g. normal) only from past examples of this class.

The work presented in [14, 15] illustrates the applicability of CNNs to the detection of anomalies in images and videos. Authors of [14] deploy a cascade of DNNs to identify anomalies in crowded scenes. Their solution achieves state-of-the-art performance, but requiring shorter identification time. Authors of [15] improve the method by transferring a pre-trained CNN classifier into a fully convolutional network. The obtained model further reduces the computational time, thus being suitable for real-time applications, such as video surveillance.

Authors of [16] study a hybrid solution to find anomalies in multivariate time-series, which can be applied to networking problems too. It is based on a one class classifier built upon a convolutional long-short term memory network. The solution can identify anomalies as well as report their severity and root-causes, e.g. sensors creating the anomalous series.

Specifically considering security and intrusion detection, authors of [17] deploy a Channel Boosted and Residual learning classifier based on Deep CNNs. A one-class classifier is trained to identify normal instances of the KDD-NSL dataset. Similarly, in [18], authors develop a supervised DNN framework to identify anomalies focusing on interpretability. The framework provides prediction confidence, textual description of anomalies, and the most important features used for prediction. Authors of [19] provide a comparison of different DNN architectures for intrusion detection. Using public datasets (again KDD-NSL), they show that DNNs achieve better performance than state-of-art classifiers supervised classifiers.

As a final example, authors of [20] propose DeepLog, a DNN based on long-short term memory (LSTM) that models system logs as natural language. By learning normal patterns from the logs (i.e. in a semi-supervised anomaly detection setup), the network detects when log patterns deviate from the trained model. The model then evolves based on users' feedback.

Takeaway: DNNs are revolutionizing supervised learning on different problems and, as for classic approaches, are used for anomaly detection. The surveyed works have tried various NN topologies, suitable for different types of data. Similar path could be followed for networking problems.

11.3.3 Representation Learning

The advent of deep learning and its automated feature learning abilities has opened the way to advances on representation learning. This latter includes the vast collection of techniques that directly or indirectly allow to learn rich features or representations from unstructured data [21].

Applied to anomaly detection, the idea would be to *constrain* the learned representations to produce a latent space where normal and anomalous (or novel) samples can be easily separated. Known examples that use such a trick are the auto-encoders, which will be further developed in Section 11.3.4: They learn small latent vectors from which it is possible to reconstruct the original input data. Anomalous instances can be in this case detected by measuring the reconstruction errors [22, 23].

A number of recent work follows the representation learning ideas. Authors of [24] propose to augment the above reconstruction error approach with an additional *surprisal* metric, which assesses how likely a representation should occur under the learned model. The authors argue that detecting anomalies can leverage two approaches: (i) the ability to *remember* what has been seen and (ii) the ability to spot novelties, i.e. *surprisal*. They propose a novelty score that incorporates both. For the first, they leverage the reconstruction error. For the second, they learn an autoregressive model on the latent vectors of the autoencoder and use the resulting likelihood of the latent vector as a proxy for surprisal.

On the same line, authors of [25] leverage the learning of latent representations of normal instances in multiple domains, and the learned boundaries between normal and anomalous in some specific domains (for which they have a ground-truth) to *transfer* anomaly detectors from source domains (supervised, known) to target domains (unknown).

A somewhat similar intuition has been used for multi-view anomaly detection [26] where data instances can have multiple views – e.g. a video represented by audio, video, and subtitles; a face that has multiple views; pages that have versions in different languages. The intuition is to have multiple views of normal data instances generated from the same latent vector, while data instances that are anomalous shall have multiple latent vectors.

Another related approach has been proposed by the authors of [27] for anomaly detection in images. They train a classifier to distinguish between a set of geometric transformations applied to images. The learned representations in this auxiliary task are useful to detect anomalies at testing phase, by analyzing the output of the model when applied on transformed images.

Authors of [28] couple a similar approach with a *student–teacher* framework for unsupervised anomaly detection and pixel-precise anomaly segmentation in images. While the teacher network learns latent features from a set of images, an ensemble of student networks is trained to regress the teacher's output on anomaly-free input. When fed with data with anomalous parts during the testing phase, the student networks will exhibit higher regression errors and lower predictive certainties in areas involving anomalies.

Takeaway*: Representation learning groups a set of techniques that can learn a latent feature space derived from the input variables. An anomaly is any instance whose latent features are significantly distinguishable from others. Representation learning has been applied to different data types and scenarios with multiple algorithms.*

11.3.4 Autoencoders

A famous representation learning technique which we further detail now is autoencoders. These are neural networks that compress the input data into a latent space and, then, reconstruct the input based on the latent variables. Figure 11.4 depicts the basic idea: considering X as input, the encoder (left) compresses the input to a latent space $h = f(X)$. The decoder (right) reconstructs the input based on h, with $X' = g(h)$. The quality of the reconstruction is then evaluated by means of the *reconstruction error*.

The autoencoder is trained with normal instances for anomaly detection. Then, the reconstruction error grows when the autoencoder is fed with anomalous instances during the testing phase, pointing to anomalies. Autoencoders are largely used to detect anomalies in images, videos, and text, but have found applications in several other scenarios too. For example, authors of [29] propose a deep autoencoder – called robust deep autoencoder – that not only discovers high-quality, nonlinear features from input instances, but also eliminates outliers and noise without access to clean training data. The model takes inspiration form robust principal component analysis, as defined in [30]. It aims at splitting the input instances into two parts: a low-dimensional representation of the input

Figure 11.4 Basic autoencoder structure.

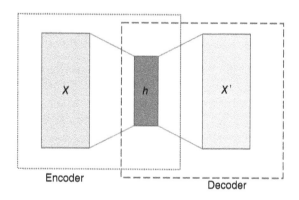

Encoder

Decoder

data, that can be effectively reconstructed by a deep autoencoder, and another one that contains element-wise outliers.

Authors of [31] highlight challenges for the application of autoencoders on anomaly detection, in particular the sensitiveness of the technique to noise and the need for large training sets. The authors then propose RandNet, an ensemble of autoencoders relying on different NN architectures for outlier detection. In a somehow similar direction, authors of [32] present the deep autoencoding Gaussian mixture model, which combines a compression network with an estimation network. The joint optimization of the two networks is claimed to improve performance of unsupervised anomaly detection on multivariate, high-dimensional data.

The assumption that autoencoders produce large reconstruction errors for every anomaly is questioned in [33] and others [22, 32, 34]. Some anomalies may be subtle, and the autoencoders may generalize normal instances to the point of overlooking anomalies. Authors propose the memory-augmented autoencoder, i.e. MemAE, which memorizes prototypes of normal instances. In testing phase, the network will always use one of the prototypes in memory for reconstruction, hopefully increasing the reconstruction error in case of anomalies.

Finally, a similar problem is targeted in [35], emerging when the dataset used for training the autoencoders is contaminated with anomalies (e.g. noise). The autoencoders could learn how to reconstruct the anomalous instances, reducing the reconstruction error for other anomalies. To counter this problem, authors employ an adversarial autoencoder: A generative adversarial networks (GANs) is trained using the encoder output and an arbitrary prior, so to identify and remove possible anomalies already during training phase. This leads us to the next family, GANs, which we discuss in details next.

Takeaway: Autoencoders are among the most well-known representation learning techniques. They compress the input to a latent space. By decompressing the latent features, the reconstruction error is used to spot anomalies. Most surveyed works apply the technique to tabular, video, or image datasets.

11.3.5 Generative Adversarial Networks

Statistical anomaly detection relies upon models, e.g. density functions learned from the normal data. GANs [36] are a recent alternative to learn density functions using two neural networks in an adversarial setting. As such, they can be used to learn density functions for anomaly detection too.

Figure 11.5 summarizes the GANs architecture. The two networks compete against each other. The generator G takes noise as input, e.g. independent samples from a Gaussian distribution. G has the goal to generate instances $G(z)$ that resemble the real data ($x \sim P_r$ in Figure 11.5). The discriminator D acts as a

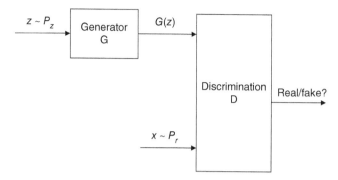

Figure 11.5 Basic generative adversarial network architecture.

binary classifier whose aim is to distinguish between real (x) and generated ($G(z)$) samples. The loss function used on training takes into account the similarity between the data generated by G and the distribution of real data. The two networks thus have opposing objectives: While G must learn how to generate realistic samples, D should distinguish real and synthetic samples. GANs have been used for detecting anomalies on images and high-dimensional tabular data. Thus, the approach can be applied on networking problems too.

One possible way to apply GANs for anomaly detection consists on training a GANs with the normal data (i.e. semi-supervised). The model learned by the generator G is then used to decide whether new instances are anomalous. Authors of [37] develop *f-AnoGAN* that performs such steps using tomography images. AnoGAN is based on an anomaly score. When applying the system to unknown images, these images are mapped back to the latent space z. The score is computed based on the fact that the latent space has smooth transitions – i.e. neighbors in z produce similar images. As such, any normal image should be mapped nearby previously known normal images.

Authors of [38] combine GANs and autoencoders to learn latent representations of in-class examples. The discriminator D is used for refining the latent space of an auto-encoder improving the detection of images diverging from a given class. The approach adopted in [39] combines the generator and the discriminator (both LSTM networks) to devise an anomaly score for high-dimensional, multivariate time-series coming from sensor networks.

Authors of [40] propose a similar approach, called *Adversarially Learned Anomaly Detection* (ALAD). It is based on bidirectional GANs, a concept proposed in [41], which learns a network to perform the inverse map back to the latent space simultaneously to the training of the generator. A similar strategy is adopted also in [42]. All the mentioned approaches are semi-supervised, as they require prior knowledge of the normal image samples to train the models.

Takeaway: *GANs are another example of representation learning algorithms that can be used to spot anomalies. The application of GANs to anomaly detection is however less straightforward, usually requiring additional steps to obtain a latent space that highlights anomalies. Alternatively, the GANs discriminator is often used as classic supervised machine learning algorithms to learn normal/anomalous patterns.*

11.3.6 Reinforcement Learning

Reinforcement learning (RL), in the context of AI, is a type of dynamic programming that trains algorithms using a system of reward and punishment. RL algorithm learns by interacting with its environment. The agent of the algorithm, e.g. a self-driving car, interacts with its environment and receives rewards depending on how it performs, e.g. driving safely. Conversely, the agent receives a penalty for performing incorrectly, e.g. crashing with another car. The agent learns without intervention from a human by maximizing its reward and minimizing its penalty. RL proved to be a great solution to problems that require decisions which influence and are influenced by the environment.

RL main contribution for anomaly detection is on helping finding good data points for training other techniques. This goal is complementary to the previously discussed algorithms. For example, in domains such as cyber-security, attack scenarios change continuously. As such, it is important to have a continuous learning system. This could be achieved using online learning where a supervised signal is fed back to the system to update models with novel data. Alternatively, one could formulate anomaly detection as a RL problem. In a nutshell, the RL model is trained by giving it reward in accordance with a metric of the quality of detected anomalies. Thanks to the exploration algorithms in RL, the model will find more anomalies and thus get rewards. Then, the system maximizes the reward by improving the quality metric, becoming better in finding anomalies over time. Note that the RL techniques would change the data distribution of samples with respect to other techniques, e.g. random sampling.

The first attempts to use RL for anomaly detection are not recent. The work in [43] is one of the first to combine anomaly detection and RL. The author applies adaptive neural networks to detect intrusion on networked systems. The method is capable of autonomously learning new attacks using feedback from the protected system, autonomously improving performance over time. Authors of [44] focus on fraud detection, combining probabilistic techniques with RL. The methodology computes deviations from the expected Benford's Law distributions as an indicator of anomalous behavior. Authors of [45] study networks attack as a traffic anomaly problem using RL for detection. RL agents analyze different parameters of traffic data to distinguish legitimate and DDoS traffic.

Many techniques that use RL for anomaly detection together with other recent developments on AI systems have been proven effective. Authors of [46] propose a time-series anomaly detector powered by RL and a recurrent neural network. The technique (i) makes no assumption about the underlying mechanism of anomaly patterns, (ii) works without any threshold setting, and (iii) keeps evolving with anomaly detection experience. Authors of [47] present a deep RL for anomaly detection targeting surveillance videos. They consider normal and abnormal videos as bags and the selection of videos clips as actions. The network then computes probabilities for each video segment in both anomalous and normal bags indicating how likely a clip contains an anomaly.

In the context of cyber security, authors of [48] faces the problem of sequential anomaly detection, which consists in modeling and predicting a series of temporally related patterns. RL helps to detect sequential behaviors by estimating the value functions of a Markov reward process. Sequential anomaly detection is also studied in [49], focusing on streaming data gathered from sensors. The authors solve the problem by using inverse RL, where the goal is to detect inherent functions triggering the behavior of decision-making agents.

Authors of [50] propose an RL approach to detect cyberattacks in smart grids. The online attack detection is formulated as a partially observable Markov decision process and solved with RL. Authors of [51] use a distributed reinforcement network for DDoS attack detection. Multiple RL agents are deployed on routers to throttle or rate-limit traffic toward victims. Finally, authors of [52] focus on anomaly detection on motors of unmanned aerial vehicles. Using RL, the motor is judged to be operating abnormally or not, dynamically changing the threshold on the environment conditions.

Takeaway: RL complements the deep learning techniques by helping finding good data points for training the algorithms. RL has been applied to networking problems, in particular in security scenarios.

11.3.7 Summary and Takeaways

Table 11.1 summarizes the research areas positioning the selected papers according to the characteristics of problems faced by each reference. Table 11.1 includes the nature of data, type of anomaly, label and output format handled by algorithms discussed in the references.

We can see heterogeneous interests regarding the applications faced by the surveyed works, with several papers applying AI-based anomaly detection on images, videos, tabular and textual data, etc. We have found some works that apply the techniques on networking problems too. However, only a couple of the application domains identified on Figure 11.1 have been covered so far.

Table 11.1 Summary of reviewed papers.

References	Nature of data		Anomaly type			Labels			Output	
	Data type[a]	Relationship[b]	Pointwise	Contextual	Collective	Supervised	Semi-supervised	Unsupervised	Discrete	Score
[14]	VD	TS	✓				✓		✓	
[15]	VD	TS	✓				✓		✓	
[16]	NM	TS	✓	✓				✓		✓
[17]	TAB	PD	✓				✓		✓	
[18]	TAB	PD	✓			✓				✓
[19]	TAB	PD	✓			✓			✓	
[20]	TXT	TS		✓			✓		✓	
[25]	IMG/TAB	PD/TS	✓		✓	✓				✓
[28]	IMG	PD	✓					✓		✓
[24]	IMG/VD	PD/TS	✓					✓		✓
[27]	IMG	PD	✓				✓		✓	
[26]	TAB	PD	✓		✓	✓			✓	
[53]	IMG	PD	✓	✓	✓	✓			✓	
[29]	IMG	PD/GR	✓	✓	✓			✓		✓
[31]	TAB	PD	✓					✓		✓
[32]	TAB	PD	✓					✓		✓
[33]	IMG	PD/GR	✓	✓	✓			✓		✓
[35]	TAB	PD	✓				✓			✓
[37]	IMG	PD	✓				✓			✓
[40]	IMG/TAB	PD	✓				✓			✓
[42]	IMG	PD	✓				✓			✓
[39]	NM	TS	✓					✓		✓
[38]	IMG	PD	✓				✓		✓	
[43]	TAB	TS	✓	✓			✓		✓	
[44]	TAB	PD	✓					✓	✓	
[45]	TAB	TS/GR	✓				✓		✓	
[46]	NM	TS	✓	✓			✓		✓	
[47]	IMG/VD	TS		✓			✓			✓

Table 11.1 (Continued)

References	Nature of data		Anomaly type			Labels			Output	
	Data type[a]	Relationship[b]	Pointwise	Contextual	Collective	Supervised	Semi-supervised	Unsupervised	Discrete	Score
[48]	TAB	TS/GR		✓	✓		✓		✓	
[49]	TAB	TS		✓	✓		✓			✓
[50]	TAB	PD	✓	✓			✓		✓	
[51]	TAB	PD	✓	✓			✓		✓	
[52]	TAB	TS	✓	✓	✓		✓		✓	

a) NM (Numeric), IMG (Image), VD (Video), TXT (Text), TAB (Tabular).
b) PD (Point Data), TS (Time-series), GR (Graph).

Most works focus on point data and time series, with a couple of initial options facing anomalies on generic graphs. Semi-supervised and unsupervised approaches dominate, proving it is still a difficult task to have datasets with labeled anomalies. Finally, we observe a balanced picture regarding the output aspect: Table 11.1 shows almost as many algorithms yielding a discrete output as those returning an anomaly score.

11.4 Technology Overview

We now summarize recent tools that perform anomaly detection. We first focus on alternatives maintained by top Internet players and available on popular programming frameworks. Most of such options include classic techniques only (i.e. as on Section 11.2). Noting the absence of mature alternatives based on the novel AI methodologies, in particular available as open source, we close the section listing libraries proposed in research papers.

11.4.1 Production-Ready Tools

In this first part, we survey tools actively used (and maintained) by large Internet players or available as libraries on popular programming frameworks. Our goal is to map what one can obtain in stable and active off-the-shelf tools, potentially ready for production environments.

Prophet is an open source library maintained by Facebook for time series forecasting [54]. It works based on a modular/addictive regression model that allows users to represent nonlinear trends with different seasonality, e.g. yearly, weekly, daily, etc. The prediction system is coupled with a module to spot and report anomalies in the series, i.e. points falling outside predictions with predetermined confidence levels. Anomalies can be used to extend the models, increasing the system precision. No particular novel AI algorithms are employed for prediction or anomaly detection. Prophet is available both in CRAN (for R) and PyPI (for Python).

Authors of [55] introduce Yahoo's **EGADS**, an open source Java library that implements a collection of time series prediction models. The former includes algorithms such as Kalman filters, ARIMA, and moving averages. The time series prediction is coupled with an anomaly detection module that computes an anomaly score, e.g. using kernel-based or density-based change-point detection. No particular novel AI algorithms are employed.

Microsoft offers its **Anomaly Detector** on the Azure platform. AI algorithms are employed, but the source code and models are not open. Authors of [56] describe some of the used algorithms, which target the detection of anomalies in time series. They combine spectral residual and CNNs, borrowing ideas of saliency detection in images, to increase quality of anomaly detection on time series.

Similarly, Google offers anomaly detection in the cloud with its **streaming analytics and AI**. Here again, hints of the used algorithms can be obtained from research papers [57], where different types of DNNs are trained with TensorFlow to predict future values of time series. Anomaly detection rules focus on collective anomalies making use of thresholds and properties of statistical distributions of data points.

Luminol is an open source python library developed by LinkedIn.[4] It supports anomaly detection and time series correlation. In the former case, Luminol allows one to select the detection algorithm, e.g. based on time series bitmap representations or based on exponential smoothing. Luminol then provides an anomaly score and a correlation module to search for correlated anomalies in different series.

Twitter offers its **AnomalyDetection** technology as an open source R package. It can be used to detect anomalies in time series as well as on vectors of numerical values. The algorithms, described in [58], are built on classic statistical methods and, in particular, on a Seasonal Hybrid Extreme Studentized Deviate (S-H-ESD) test. The test employs time series decomposition and robust statistical metrics (e.g. median absolute deviation) for detecting anomalies in the presence of seasonality.

---

4 https://github.com/linkedin/luminol

In terms of frameworks, Scikit-learn [59] is a prominent example for Python. It includes the **novelty and outlier detection** library. The Scikit-learn project implements many different machine learning algorithms that can be used to make predictions that are passed on to the anomaly detection library. A vast range of classic outlier and novelty detection algorithms are available, e.g. to calculate anomaly scores. Similar frameworks exist for R, Matlab, and Java in different maturity levels. For example, **ELKI data mining framework** is an open source package that implements data mining algorithms in Java [60]. It includes methods for unsupervised data clustering as well as methods for outlier detection, e.g. distance-based and clustering-based.

11.4.2 Research Alternatives

Scikit-learn, StatModels,[5] and other well-established alternatives focusing on anomaly detection do not include algorithms described in Section 11.3. Libraries and tools such Keras,[6] PyTorch,[7] and TensorFlow[8] provide these AI algorithms, but without explicit APIs for anomaly detection. This last step is however covered by tools recently proposed in research works. We briefly provide some examples in the following.

Proposed in [61], **Python Outlier Detection (Pyod)** is a Python toolkit focusing on multivariate data. The toolkit implements more than 30 anomaly detection algorithms, including a vast range of classic approaches, outlier ensembles and different types of DNNs (e.g. autoencoders, using Keras). The same author maintains **SUOD: A Scalable Unsupervised Outlier Detection Framework** [62], which focuses on accelerating training and prediction when lots of detectors are available, e.g. to perform anomaly detection with ensembles.

Some works discussed in Section 11.3 contribute open source implementations of the proposed techniques. Authors of [42] rely on PyTorch to train GANs for anomaly detection on images, delivering **GANnomaly** to the community. Authors of [37] release **f-AnonGAN** that relies on TensorFlow for training GANs on a similar scenario. Relying on TensorFlow, authors of [39] release **MAD-GANs** for anomaly detection on time-series.

ARAE-AnoGAN focuses on text anomaly detection using a combination of GANs and autoencoders, implemented with TensorFlow. Authors of [63] contribute **telemanom**, which performs anomaly detection on multivariate time series using the LSTM neural networks implemented with Keras/TensorFlow.

---

5 https://www.statsmodels.org/
6 https://keras.io/
7 https://pytorch.org/
8 https://www.tensorflow.org/

Finally, considering anomalies in generic graphs, very few public tools can be found, and virtually nothing implements recent AI algorithms. Authors of [64] contribute with **MIDAS**, which finds anomalies on time-evolving graphs using statistical tests. MIDAS searches for *microcluster anomalies*, defined as suspicious edges arriving in bursts. Similarly, **StreamSpot** [65] reports anomalies on evolving graphs (arriving as edge streams) using an algorithm based on graph sketches and statistical tests.

11.4.3 Summary and Takeaways

Table 11.2 summarizes the discussed tools, putting them in perspective of the taxonomy in Figure 11.1. Interesting remarks emerge.

Table 11.2 Summary of the reviewed tools.

	Nature of data		Anomaly type			Labels			Output	
	Data type[a]	Relationship[b]	Pointwise	Contextual	Collective	Supervised	Semi-supervised	Unsupervised	Discrete	Score
Facebook prophet[c]	NM	TS	✓	✓	✓			✓	✓	✓
Yahoo! EGADS[d]	NM	TS	✓	✓	✓			✓	✓	✓
Microsoft anomaly detector[e]	NM	TS	✓					✓	✓	
Google streaming analytics & AI[f]	TAB	PD	✓					✓	✓	
LinkedIn luminol[g]	NM	TS	✓	✓	✓			✓	✓	✓
Twitter's anomaly detection[h]	NM/TAB	PD/TS	✓	✓	✓			✓	✓	
Scikit-learn novelty and outlier detection[i]	TAB	PD	✓		✓		✓	✓	✓	✓
ELKI data mining framework[j]	TAB	PD	✓					✓		✓
PyOD[k]	TAB	PD	✓		✓	✓	✓	✓	✓	✓
SUOD[l]	TAB	PD	✓					✓	✓	
telemanom[m]	TAB	TS	✓					✓		✓
GANomaly[n]	IMG	PD	✓				✓			✓
f-AnoGAN[o]	IMG	PD	✓				✓			✓

Table 11.2 (Continued)

	Nature of data		Anomaly type			Labels			Output	
	Data type[a]	Relationship[b]	Pointwise	Contextual	Collective	Supervised	Semi-supervised	Unsupervised	Discrete	Score
ARAE-AnoGAN[p]	TXT	PD	✓				✓			✓
MAD-GAN[q]	NM	TS	✓					✓		✓
MIDAS[r]	TAB	GR	✓	✓				✓		✓
StreamSpot[s]	TAB	GR	✓		✓			✓		✓

a) NM (Numeric), IMG (Image), VD (Video), TXT (Text), TAB (Tabular).
b) PD (Point Data), TS (Time-series), GR (Graph).
c) https://facebook.github.io/prophet
d) https://github.com/yahoo/egads
e) https://azure.microsoft.com/en-us/services/cognitive-services/anomaly-detector
f) https://cloud.google.com/blog/products/data-analytics/anomaly-detection-using-streaming-analytics-and-ai
g) https://github.com/linkedin/luminol
h) https://github.com/twitter/AnomalyDetection
i) https://scikit-learn.org/stable/modules/outlier_detection.html
j) https://elki-project.github.io
k) https://github.com/yzhao062/pyod
l) https://github.com/yzhao062/suod
m) https://github.com/khundman/telemanom
n) https://github.com/samet-akcay/ganomaly
o) https://github.com/tSchlegl/f-AnoGAN
p) https://github.com/tedyap/ARAE-AnoGAN
q) https://github.com/LiDan456/MAD-GANs
r) https://github.com/ritesh99rakesh/pyMIDAS
s) https://sbustreamspot.github.io

First, production-ready tools (upper part of Table 11.2) are strongly concentrated around *unsupervised* techniques targeting point data and *time series*. This concentration can be explained by the vast availability of such data at the involved Internet players, e.g. from telemetry of production systems. Interestingly, the lack of *supervised* tools reconfirms the well-known problem with the lack of ground-truth for building supervised models [66].

Second, tools found in research works (bottom part) focus on more elaborate datasets, e.g. multivariate time series, tabular data, graphs, and images. Supervised and semi-supervised approaches are used, which can be explained by the need for validation in such typical research settings. Recent AI algorithms are employed, suggesting that the research community identifies these algorithms as prominent alternatives to face anomaly detection on complex datasets. However, in almost all cases, only the simplest anomaly type is faced, i.e. *pointwise* anomalies. This fact suggests that more research work is needed to face complex anomalies, e.g. collective anomalies on multivariate data.

11.5 Conclusions and Future Directions

Tables 11.1 and 11.2 confirm a lively landscape around the use of recent AI advances for anomaly detection. New algorithms such as GANs and autoencoders have proven effective data-driven alternatives for a variety of problems. Yet, several research challenges remain clearly ahead.

To name an example, the transition of algorithms from origin fields (e.g. computer vision and speech synthesis) to network problems is not straightforward. The latter is characterized by multiple and diverse data sources, forming inherently complex relations, i.e. a multimodal graph-based problem.

Explicitly representing network monitoring datastreams with their relations is a prominent way to face anomaly detection – i.e. a *graph-based* problem. For example, one could couple logs of network devices and traffic telemetry with network topological information to search for complex network anomalies. However, we see in Table 11.1 that only a few papers have applied AI-based anomaly detection to graph-based problems so far.

Authors of [6] provide a survey of anomaly detection research on dynamic graphs. An intrinsic problem, which illustrates the challenge, already emerges from basic definitions: As graphs are used to represent complex and arbitrary relations, the definitions of anomalies on a graph change widely according to the problem at hand.

For static graphs, the previous work lists (i) *anomalous vertices*, i.e. data instances with too many or too few connections; (ii) *anomalous edges*, i.e. connections whose weights deviate from expectations; (iii) *anomalous communities*, i.e. densely connected subgraphs whose aggregation deviates from expectations. Other anomalies emerge if one considers evolving graphs, such as (iv) an *event*, i.e. a pointwise change in the graph in a time instant; or (v) a *change-point*, i.e. a permanent change in the graph structure. The dynamic graph case has been faced by some recent works [64, 67, 68], but using classic anomaly detection approaches only.

Exploiting graph relationships is particularly useful when it comes to studying network anomalies. Take a graph representation for flows observed on a network link, with vertices representing hosts and edge weights representing the number of packets exchanged between a pair of hosts. Searching for groups of vertices – e.g. hosts with strong communication patterns – helps to isolate events on the network, eventually pointing to anomalous and coordinated behaviors that would have been otherwise hard to spot. We profit from such an approach to detect coordinated activities in darknet traffic in [69].

Darknets are sets of IP addresses advertised without hosting any services. Darknets are deployed with the purpose of collecting unsolicited packets reaching a network. They are used to monitor events such as the spreading of malware and network scans. Without hosting services, darknets still receive substantial amount of traffic, e.g. traffic from bots participating in a botnet. Anomalies in darknet traffic (e.g. novel behaviors) can help to shed light on emerging botnets or incipient remote attacks. In [69], we focus on darknet sensors and model darknet activity as a graph, capturing how remote machines contact ports at the darknet addresses. Using community detection algorithms, we found groups of hosts that perform similar activity – at this stage, without identifying anomalies or novelties yet.

That work has proven instrumental to summarize the darknet traffic, but it suffers from some limitations to fully realize the potential of darknet monitoring for security applications. First, the used algorithms suffer from scalability issues, making the generalization of the approach hard. Second, the approach currently deals with few variables only (packets, protocols, etc.), which are mapped to a static graph. Other aspects are ignored, such as the dynamic nature of the graph. Moreover, together with darknet traffic, other sensors (e.g. honeypots and information about production traffic) can provide a rich source to identify malicious activity and attacks.

When multiple sensors and variables are considered, the graph supporting the security activities becomes a multilayer network. Sophisticated approaches are needed. Established techniques coming from complex network analysis can help to filter out uninteresting parts of the graph, extracting a network backbone. Yet, AI-based algorithms can play an important role too. We plan to test promising techniques relying on representation learning ideas to search for latent variables that can summarize the graph and its communities, thus acting to reduce the problem dimension and contributing to a better scalability.

Bibliography

1 Chandola, V., Banerjee, A., and Kumar, V. (2009). Anomaly detection: a survey. *ACM Computing Surveys (CSUR)* 41 (3): 1–58.
2 Bhuyan, M.H., Bhattacharyya, D.K., and Kalita, J.K. (2013). Network anomaly detection: methods, systems and tools. *IEEE communication surveys and Tutorials* 16 (1): 303–336.

3 Landauer, M., Skopik, F., Wurzenberger, M., and Rauber, A. (2020). System log clustering approaches for cyber security applications: a survey. *Computers & Security* 92: 101739.

4 Kwon, D., Kim, H., Kim, J. et al. (2019). A survey of deep learning-based network anomaly detection. *Cluster Computing* 22 (1): 949–961.

5 Ahmed, M., Mahmood, A.N., and Hu, J. (2016). A survey of network anomaly detection techniques. *Journal of Network and Computer Applications* 60: 19–31.

6 Ranshous, S., Shen, S., Koutra, D. et al. (2015). Anomaly detection in dynamic networks: a survey. *Wiley Interdisciplinary Reviews: Computational Statistics* 7 (3): 223–247.

7 Agrawal, S. and Agrawal, J. (2015). Survey on anomaly detection using data mining techniques. *Procedia Computer Science* 60: 708–713.

8 Akoglu, L., Tong, H., and Koutra, D. (2015). Graph based anomaly detection and description: a survey. *Data Mining and Knowledge Discovery* 29 (3): 626–688.

9 Kumar, M., Patel, N.R., and Woo, J. (2002). Clustering seasonality patterns in the presence of errors. *Proceedings of the 8th ACM SIGKDD International Conference on Knowledge Discovery and Data Mining*, ACM.

10 LeCun, Y., Bengio, Y., and Hinton, G. (2015). Deep learning. *Nature* 521 (7553): 436–444.

11 Williams, R.J. and Zipser, D. (1989). A learning algorithm for continually running fully recurrent neural networks. *Neural Computation* 1 (2): 270–280.

12 Hochreiter, S. and Schmidhuber, J. (1997). LSTM can solve hard long time lag problems. *Proceedings of the Advances in Neural Information Processing Systems*, NIPS.

13 Krizhevsky, A., Sutskever, I., and Hinton, G.E. (2012). ImageNet classification with deep convolutional neural networks. *Proceedings of the Advances in Neural Information Processing Systems*, NIPS.

14 Sabokrou, M., Fayyaz, M., Fathy, M., and Klette, R. (2017). Deep-cascade: cascading 3D deep neural networks for fast anomaly detection and localization in crowded scenes. *IEEE Transactions on Image Processing* 26 (4): 1992–2004.

15 Sabokrou, M., Fayyaz, M., Fathy, M. et al. (2018). Deep-anomaly: fully convolutional neural network for fast anomaly detection in crowded scenes. *Computer Vision and Image Understanding* 172: 88–97.

16 Zhang, C., Song, D., Chen, Y. et al. (2019). A deep neural network for unsupervised anomaly detection and diagnosis in multivariate time series data. *Proceedings of the AAAI Conference on Artificial Intelligence*, AAAI.

17 Chouhan, N. and Khan, A. (2019). Network anomaly detection using channel boosted and residual learning based deep convolutional neural network. *Applied Soft Computing* 83: 105612.

18 Amarasinghe, K., Kenney, K., and Manic, M. (2018). Toward explainable deep neural network based anomaly detection. *Proceedings of the 11th International Conference on Human System Interaction*, IEEE.

19 Naseer, S., Saleem, Y., Khalid, S. et al. (2018). Enhanced network anomaly detection based on deep neural networks. *IEEE Access* 6: 48231–48246.

20 Du, M., Li, F., Zheng, G., and Srikumar, V. (2017). Deeplog: anomaly detection and diagnosis from system logs through deep learning. *Proceedings of the ACM SIGSAC Conference on Computer and Communications Security*, ACM.

21 Bengio, Y., Courville, A., and Vincent, P. (2013). Representation learning: a review and new perspectives. *IEEE Transactions on Pattern Analysis and Machine Intelligence* 35 (8): 1798–1828.

22 Hasan, M., Choi, J., Neumann, J. et al. (2016). Learning temporal regularity in video sequences. *Proceedings of the IEEE Conference on Computer Vision and Pattern Recognition*, IEEE.

23 Liu, W., Luo, W., Lian, D., and Gao, S. (2018). Future frame prediction for anomaly detection–a new baseline. *Proceedings of the IEEE Conference on Computer Vision and Pattern Recognition*, IEEE.

24 Abati, D., Porrello, A., Calderara, S., and Cucchiara, R. (2019). Latent space autoregression for novelty detection. *Proceedings of the IEEE Conference on Computer Vision and Pattern Recognition*, IEEE

25 Kumagai, A., Iwata, T., and Fujiwara, Y. (2019). Transfer anomaly detection by inferring latent domain representations. *Proceedings of the Advances in Neural Information Processing Systems*, NIPS.

26 Iwata, T. and Yamada, M. (2016). Multi-view anomaly detection via robust probabilistic latent variable models. *Proceedings of the Advances in Neural Information Processing Systems*, NIPS.

27 Golan, I. and El-Yaniv, R. (2018). Deep anomaly detection using geometric transformations. *Proceedings of the Advances in Neural Information Processing Systems*, NIPS.

28 Bergmann, P., Fauser, M., Sattlegger, D., and Steger, C. (2020). Uninformed students: student-teacher anomaly detection with discriminative latent embeddings. *Proceedings of the IEEE/CVF Conference on Computer Vision and Pattern Recognition*, IEEE.

29 Zhou, C. and Paffenroth, R.C. (2017). Anomaly detection with robust deep autoencoders. *Proceedings of the 23rd ACM SIGKDD International Conference on Knowledge Discovery and Data Mining*, ACM.

30 Candès, E.J., Li, X., Ma, Y., and Wright, J. (2011). Robust principal component analysis? *Journal of the ACM (JACM)* 58 (3): 1–37.

31 Chen, J., Sathe, S., Aggarwal, C., and Turaga, D. (2017). Outlier detection with autoencoder ensembles. *Proceedings of the SIAM International Conference on Data Mining*, SIAM.

32 Zong, B., Song, Q., Min, M.R. et al. (2018). Deep autoencoding Gaussian mixture model for unsupervised anomaly detection. *Proceedings of the International Conference on Learning Representations*.

33 Gong, D., Liu, L., Le, V. et al. (2019). Memorizing normality to detect anomaly: memory-augmented deep autoencoder for unsupervised anomaly detection. *Proceedings of the IEEE International Conference on Computer Vision*, IEEE.

34 Zhao, Y., Deng, B., Shen, C. et al. (2017). Spatio-temporal autoencoder for video anomaly detection. *Proceedings of the 25th ACM International Conference on Multimedia*, ACM.

35 Beggel, L., Pfeiffer, M., and Bischl, B. (2019). Robust anomaly detection in images using adversarial autoencoders. *Proceedings of the Joint European Conference on Machine Learning and Knowledge Discovery in Databases*, Springer.

36 Goodfellow, I., Pouget-Abadie, J., Mirza, M. et al. (2014). Generative adversarial nets. *Proceedings of the Advances in Neural Information Processing Systems*, NIPS.

37 Schlegl, T., Seeböck, P., Waldstein, S.M. et al. (2019). f-AnoGAN: Fast unsupervised anomaly detection with generative adversarial networks. *Medical Image Analysis* 54: 30–44.

38 Perera, P., Nallapati, R., and Xiang, B. (2019). OCGAN: one-class novelty detection using gans with constrained latent representations. *Proceedings of the IEEE Conference on Computer Vision and Pattern Recognition*, IEEE.

39 Li, D., Chen, D., Jin, B. et al. (2019). MAD-GAN: multivariate anomaly detection for time series data with generative adversarial networks. *Proceedings of the International Conference on Artificial Neural Networks*, Springer.

40 Zenati, H., Romain, M., Foo, C.-S. et al. (2018). Adversarially learned anomaly detection. *Proceedings of the IEEE International Conference on Data Mining*, IEEE.

41 Donahue, J., Krähenbühl, P., and Darrell, T. (2016). Adversarial feature learning. *CoRR*, abs/1605.09782.

42 Akcay, S., Atapour-Abarghouei, A., and Breckon, T.P. (2018). GANomaly: semi-supervised anomaly detection via adversarial training. *Proceedings of the Asian Conference on Computer Vision*, Springer.

43 Georgia, J.C. (2000). Next generation intrusion detection: autonomous reinforcement learning of network attacks. *Proceedings of the 23rd National Information Systems Security Conference*, NIST.

44 Lu, F., Boritz, J.E., and Covvey, D. (2006). Adaptive fraud detection using Benford's law. *Proceedings of the Conference of the Canadian Society for Computational Studies of Intelligence*, Springer.

45 Servin, A. and Kudenko, D. (2008). Multi-agent reinforcement learning for intrusion detection: a case study and evaluation. *Proceedings of the German Conference on Multiagent System Technologies*, Springer.

46 Huang, C., Wu, Y., Zuo, Y. et al. (2018). Towards experienced anomaly detector through reinforcement learning. *Proceedings of the AAAI Conference on Artificial Intelligence*, AAAI.

47 Aberkane, S. and Elarbi, M. (2019). Deep reinforcement learning for real-world anomaly detection in surveillance videos. *Proceedings of the 6th International Conference on Image and Signal Processing and their Applications*, IEEE.

48 Xu, X. (2010). Sequential anomaly detection based on temporal-difference learning: principles, models and case studies. *Applied Soft Computing* 10 (3): 859–867.

49 Oh, M.-h. and Iyengar, G. (2019). Sequential anomaly detection using inverse reinforcement learning. *Proceedings of the 25th ACM SIGKDD International Conference on Knowledge Discovery and Data Mining*, ACM.

50 Kurt, M.N., Ogundijo, O., Li, C., and Wang, X. (2019). Online cyber-attack detection in smart grid: a reinforcement learning approach. *IEEE Transactions on Smart Grid* 10 (5): 5174–5185.

51 Malialis, K. and Kudenko, D. (2015). Distributed response to network intrusions using multiagent reinforcement learning. *Engineering Applications of Artificial Intelligence* 41: 270–284.

52 Lu, H., Li, Y., Mu, S. et al. (2018). Motor anomaly detection for unmanned aerial vehicles using reinforcement learning. *IEEE Internet of Things Journal* 5 (4): 2315–2322.

53 Andrews, J., Tanay, T., Morton, E.J., and Griffin, L.D. (2016). Transfer representation-learning for anomaly detection. *Proceedings of the 33rd International Conference on Machine Learning*, JMLR.

54 Taylor, S.J. and Letham, B. (2018). Forecasting at scale. *The American Statistician* 72 (1): 37–45.

55 Laptev, N., Amizadeh, S., and Flint, I. (2015). Generic and scalable framework for automated time-series anomaly detection. *Proceedings of the 21th ACM SIGKDD International Conference on Knowledge Discovery and Data Mining*, ACM.

56 Ren, H., Xu, B., Yi, C. et al. (2019). Time-series anomaly detection service at microsoft. *Proceedings of the ACM SIGKDD International Conference on Knowledge Discovery and Data Mining*, ACM.

57 Shipmon, D.T., Gurevitch, J.M., Piselli, P.M., and Edwards, S.T. (2017). Time series anomaly detection; detection of anomalous drops with limited features and sparse examples in noisy highly periodic data. *CoRR*, abs/1708.03665.

58 Hochenbaum, J., Vallis, O.S., and Kejariwal, A. (2017). Automatic anomaly detection in the cloud via statistical learning. *CoRR*, abs/1704.07706.

59 Pedregosa, F., Varoquaux, G., Gramfort, A. et al. (2011). Scikit-learn: machine learning in Python. *Journal of Machine Learning Research* 12: 2825–2830.

60 Schubert, E. and Zimek, A. (2019). ELKI: A large open-source library for data analysis - ELKI release 0.7.5 "heidelberg". *CoRR*, abs/1902.03616.

61 Zhao, Y., Nasrullah, Z., and Li, Z. (2019). PyOD: a python toolbox for scalable outlier detection. *Journal of Machine Learning Research* 20 (96): 1–7.

62 Zhao, Y., Ding, X., Yang, J., and Bai, H. (2020). SUOD: toward scalable unsupervised outlier detection. *Proceedings of the Workshops at the 34th AAAI Conference on Artificial Intelligence.*

63 Hundman, K., Constantinou, V., Laporte, C. et al. (2018). Detecting spacecraft anomalies using LSTMs and nonparametric dynamic thresholding. *Proceedings of the 24th ACM SIGKDD International Conference on Knowledge Discovery & Data Mining*, ACM.

64 Bhatia, S., Hooi, B., Yoon, M. et al. (2019). MIDAS: microcluster-based detector of anomalies in edge streams. *CoRR*, abs/1911.04464.

65 Manzoor, E., Milajerdi, S.M., and Akoglu, L. (2016). Fast memory-efficient anomaly detection in streaming heterogeneous graphs. *Proceedings of the 22nd ACM SIGKDD International Conference on Knowledge Discovery and Data Mining*, ACM.

66 D'Alconzo, A., Drago, I., Morichetta, A. et al. (2019). A survey on big data for network traffic monitoring and analysis. *IEEE Transactions on Network and Service Management* 16 (3): 800–813.

67 Hooi, B., Shin, K., Song, H.A. et al. (2017). Graph-based fraud detection in the face of camouflage. *ACM Transactions on Knowledge Discovery from Data (TKDD)* 11 (4): 1–26.

68 Jiang, M., Cui, P., Beutel, A. et al. (2016). Catching synchronized behaviors in large networks: a graph mining approach. *ACM Transactions on Knowledge Discovery from Data (TKDD)* 10 (4): 1–27.

69 Soro, F., Allegretta, M., Mellia, M. et al. (2020). Sensing the noise: uncovering communities in darknet traffic. *Proceedings of the 18th Mediterranean Communication and Computer Networking Conference (MedComNet)*, IEEE.

12

Automated Orchestration of Security Chains Driven by Process Learning*

Nicolas Schnepf¹, Rémi Badonnel², Abdelkader Lahmadi², and Stephan Merz²

¹ Department of Computer Science, Aalborg University, Aalborg, Denmark
² Université de Lorraine, CNRS, Loria, Inria, Nancy, France

12.1 Introduction

The relentless growth in the number of connected smart devices such as smartphones and tablets has attracted the attention of malicious actors who exploit these devices as both targets and vectors of attacks against user data and the network infrastructure. For example, 4 million malicious applications were detected on the Google Play Store in 2019 [1]. Although necessary, preventive screening of applications by the store operators is not sufficient for detecting all malicious applications. Moreover, limited resources in terms of CPU and battery makes it difficult to develop and deploy sophisticated on-device security mechanisms: this certainly applies to IoT applications, but can be true even for smartphones or tablets, depending on the nature of expected processing. Finally, end users may be overwhelmed by the technical details and unaware of unintended functionality that such applications exhibit.

With the development of software-defined networking (SDN), it is attractive to deploy chains of security functions – including firewalls (FWs), intrusion detection systems (IDS), deep packet inspection (DPI) or data leakage prevention (DLP) mechanisms – into cloud infrastructures for a network-based protection. SDN relies on decoupling the network into the data plane, realized by SDN switches, that forward traffic according to configuration rules, and the control plane, commonly realized by a single controller, that reconfigures the switches. The standard OpenFlow protocol may typically support communications between

*Partially supported by the Concordia project that has received funding from the EU's Horizon 2020 research and innovation programme under grant agreement No. 830927.

Communication Networks and Service Management in the Era of Artificial Intelligence and Machine Learning,
First Edition. Edited by Nur Zincir-Heywood, Marco Mellia, and Yixin Diao.

the controller and switches. Although we do not rely on it here, network function virtualization (NFV) provides an elegant abstraction for implementing such services.

However, deploying security chains in practice can be challenging. Their complexity and dynamics is prone to inconsistencies and misconfigurations, resulting in security breaches that could be exploited by attackers. As we discuss in Section 12.2, there has been interesting work on applying formal methods in view of verification or synthesis of chains. However, formal verification techniques tend to exhibit exponential complexity in terms of response time, limiting their applicability in practice, in particular for dynamic deployment in the network. Moreover, mobile applications have heterogeneous networking behavior and vulnerabilities and therefore require tailored security chains for protecting end users and the network infrastructure itself.

This chapter proposes a method for automatically generating security chains for deployment in SDN infrastructures. This method is driven by the security requirements based on the networking behavior of individual applications that we infer using process learning methods. We classify potential attacks using logical predicates and then infer security rules that are grouped into security functions. The resulting application-specific security chains are merged and optimized in view of deploying them in the network.

The remainder of this chapter is organized as follows: Section 12.2 presents related work, Section 12.3 introduces background notions, Section 12.4 gives an overview on the proposed method, whereas Sections 12.5–12.8 describe in more detail the different steps for learning networking behavior, generating security chains, formal verification of correctness properties, and optimization. Section 12.9 presents results of performance evaluations and Section 12.10 concludes the chapter and points out future research perspectives.

12.2 Related Work

This section is centered on existing work related to chains of security functions and formal verification techniques, in the context of the protection of smart devices and their applications. Different methods have already been designed to mitigate attacks targeting smart devices. These environments are exposed to a large variety of attacks, such as denial of service attacks, port scans, worms and botnets [2]. Android systems are particularly concerned with a growing number of malicious applications, as discussed in [3]. In addition, their resources are often limited, making it impractical to deploy advanced security mechanisms on the devices [2]. The permission system of Android is an important element contributing to security [4]. Permissions grant applications the right for using different resources of the Android device, such as for instance the Internet connection or the cameras. Nevertheless, this system may be the source of misconfigurations [5]. It is possible to monitor the method calls of the applications and compare them to the declared permissions to detect misbehaviors, such as in [6]. However, this approach shows limitations with respect to attacks targeting the network infrastructure, as the Internet permission does not provide a fine granularity. Establishing network profiles of applications has also been explored by several authors. In [4], a security monitoring framework was proposed to combine permissions, user interactions, system calls, and network traffic, whereas the solution developed in [7] permits to learn the communication behaviors of Android applications from their binary files. These approaches are mainly intended for screening new applications rather than protecting end users from malicious behaviors of already installed applications. The lack of reactive methods for protecting devices from installed malicious applications, together with the constrained resources of these environments, goes in favor of exploring learning techniques for detecting specific misbehaviors and developing protective measures that can be outsourced from the devices.

12.2.1 Chains of Security Functions

The development of SDN as well as NFV has contributed to the deployment of security chains [8]. In particular, Virtual Network Functions (VNF) can be deployed to enforce security mechanisms. They correspond to network functions that are virtually deployed on commodity hardware [9], and may implement different security functions. These functions can then be chained using the facilities offered by software-defined networks.

There exists a large body of literature addressing the challenge of formally modelling security functions implemented as VNFs. In [10], a generic model is proposed, based on a dedicated language to express security functions. This language relies on pre-conditions and post-conditions regarding the network traffic accepted by a security function. While it provides a very precise and explicit specification of security functions, it suffers from the lack of concrete implementations that would enable the practical deployment of the specified functions. In addition, the high diversity of security functions available on the market makes it difficult to design a generic language without losing semantic properties. In [11], a method is introduced for resolving conflicts that may occur when combining several security functions. While the work has been implemented, it is only focused on the case of firewalls, and does not cover any other security functions. A major issue in the field of security function modelling is therefore to build a model that is general enough to ensure a large coverage, and that can be exploited in practice to support an automated deployment of security functions.

A second challenge concerns more specifically the deployment and configuration of security functions. The use of NFV jointly with the programmability provided by SDN enables a more flexible deployment of security policies. For instance in [12], a framework is described for chaining network and security functions implemented as middleboxes based on a high-level specification of composition. This latter is then translated into low-level rules that are interpreted by SDN infrastructures, in order to deploy security chains. Several research efforts analyze the deployment of security functions as a resource allocation problem: indeed the problem is known to be NP-hard in the general case. In [13], a three-stage formalization is considered: service function chain (SFC) composition, SFC embedding and SFC scheduling. The first stage of the problem is solved by characterizing the service requests in terms of network functions, and optimally building the SFCs using an integer linear programming (ILP) approach. The allocation takes into account that multiple SFCs may exploit the same virtualized network functions, but also that there may exist dependencies amongst some of the VNFs, requiring to place them in a specific order. In the same manner, the authors of [14] explore automated security function allocation to support reactive security in 5G infrastructures. The proposed framework relies on SDN supervisory control and data acquisition honeypots. It follows the standardization efforts from the IETF SFC working group, and exploits OpenDayLight controllers to configure the infrastructure. It permits a continuous monitoring of industrial networks and a fine-grained analysis of potential attacks that then serves to isolate attackers and evaluate their level of sophistication.

Another important challenge is to exploit security patterns to drive the configuration of security chains. In particular, network security patterns have been introduced for leveraging the best practices from the security experts, and capturing different security constraints that enable the efficient selection of adequate security functions [15]. That paper also introduces a scalable networking and computing resources-aware optimization framework to properly provision different chains based on an open-source cloud environment. For Android environments, the system developed in [16] can be used to analyze the behavior of applications and to build behavioral patterns. These are then exploited to select pre-configured security functions when some deviations from the behavioral patterns are observed. This approach is extended by Hurel et al. [17], which integrates in the decision process the permissions initially declared by the applications. However, the security functions are not automatically chained in these scenarios.

12.2.2 Formal Verification of Networking Policies

Formal verification techniques are a key enabler for automating the orchestration of security chains. Model checking [18] designates a collection of techniques for evaluating if a property (typically expressed as a formula of temporal logic) is true in a structure, such as a transition system. These techniques were originally applied to the verification of concurrent and distributed systems, and their main limitation is the exponential growth of the number of reachable states in terms of the number of system components. Satisfiability modulo theory (SMT) [19] is also relevant to our work. SMT extends the satisfiability problem of propositional logic (SAT) by considering decidable theories such as fragments of arithmetic, the theory of binary words or strings. Again, SMT solving is at least NP-hard, but works surprisingly well in many practical cases.

These techniques have received much interest in recent years in the context of SDN, in order to check the consistency of network policies before their deployment. The programmability of networks may introduce misconfigurations, and even configuration vulnerabilities that can then be exploited by attackers. For instance, techniques developed in [20] target the verification of the control plane of network infrastructures. Considering a collection of router configurations and a high-level specification of the network behaviors, the approach checks that these configurations correctly enforce the specification for all possible network behaviors. Alternatively, the solution can be exploited to synthesize correct configurations from the high-level specification, to be implemented by network routers. In a similar manner, the Vericon framework [21] is designed to verify that a SDN program (north-bound interface) is correct for all admissible topologies and for all possible sequences of network events. It exploits first-order logic to specify admissible network topologies and desired network-wide invariants, that are then implemented using deductive verification with the Z3 prover as an automatic backend. However, the approach does not take into account temporal logics, which may restrict its overall coverage for preventing some security attacks, such as Denial of Service (DoS) attacks.

Another important challenge is to verify the correctness of updates that are applied to the network configuration at runtime. For instance, Foster et al. [22] proposes a solution based on model checking for the verification of network updates. In this approach, each state of the built automaton corresponds to a state of the network, transitions capture the events affecting the network, such as the sending of packets or the deployment of new rules. This approach mainly focuses on rule updates and is designed to verify that the configuration of the network remains correct after updates are applied. Due to the considered level of granularity, it seems difficult to be applied in a fully dynamic context, and provides better performance with an offline usage. An alternative solution [23] aims at verifying network-wide invariants that are checked at runtime. The objective is to tame the complexity of the models by considering incremental rather than overall verification. The solution targets the verification of new rules, with respect to the remainder of the network policy. However, it requires the specification of invariants that are defined manually by network experts.

Formal verification has been largely used for checking firewall policies. For instance, SMT solving methods have been used to detect anomalies in large and distributed firewall policies [24]. The insertion or modification of filtering rules may impact the security of the infrastructure and its services. The solution aims at detecting conflicts and redundancies that may occur amongst firewall rules. Two rules are in conflict when they correspond to contradictory decisions, while two rules are redundant when they partially or fully overlap. The verification is performed both in an intra-firewall manner (concerning the rules of the same firewall), and in an inter-firewall manner (concerning the rules that are distributed over several firewalls). Process algebra has also been exploited [25] to support formal verification for SDN-based firewalls. The verification is performed at each update of the network configuration.

12.3 Background

The method that we propose requires some background elements with respect to the flow-based detection of attacks and to programming SDN controllers.

12.3.1 Flow-Based Detection of Attacks

According to RFC 5101 [26], network flows can be defined as collections of IP packets observed at a certain point in the network during a certain time interval. They are generally described by different attributes such as source and destination IP addresses and port numbers (*srcaddr*, *dstaddr*, *srcport*, and *dstport*), their network protocol (*protocol*) and the numbers of packets or bytes they contain (*packets* and *bytes*). We assume that flows are collected on-device [27], and extended with a timestamp (*timestamp*) and the name of the application that produced them (*appname*). Although the network flows do not represent the payload transmitted during a communication, their analysis can indicate certain kinds of security attacks [28]. Combining the *appname* attribute with the permission system of Android enables furthermore gaining some insight into the kind of data that may be transmitted in a flow.

Denial of service (DoS) attacks target a victim in order to prevent it from providing a service [29]. We consider DoS attacks that can be observed from a networking point of view in that they produce abnormal quantities of traffic from or to a certain equipment. For example, in a SYN flood attack a large number of SYN packets are sent to a host in order to overload the TCP stack with connections that will never be closed.

In port scanning attacks, an application initiates connections with multiple port numbers in order to detect open ports. For example, the port scanner nmap available on standard Linux platforms gives rise to characteristic patterns in network flows.

A worm is a program that can execute independently while consuming the resources of its host and that can replicate a fully executable version of itself to other devices [30]. Worms replicate by exploiting vulnerabilities of applications and operating systems, or by methods of social engineering. We consider worms that scan certain ports on devices.

A potentially malicious bot is a program installed on a system in order to execute tasks, typically under the control of a remote administrator, called bot master [31]. The detection of botnets has been extensively studied. In particular, certain botnets communicate using HTTP requests that are hard to identify from a networking point of view, and some are based on a peer-to-peer architecture in order to transmit messages of the bot master. We consider botnets that can be detected based on the large amount of traffic that they exchange with their controller or by the use of network protocols that are abnormal in a certain context.

The objective here is to detect such attacks by profiling the behavior of an application, based on methods of process learning. Modeling the interactions of an application as a Markov automaton, we leverage methods designed to infer the automaton structure such as the K-tail algorithm [32] or its extensions Synoptic [33] or Invarimint [34]. These methods sometimes result in overly complicated models, and we introduce techniques for reducing this complexity in order to make them applicable for dynamically orchestrating security chains for smart devices.

12.3.2 Programming SDN Controllers

Whereas SDN controllers typically use the OpenFlow protocol to communicate with programmable switches, several higher-level languages have been designed for programming them. Our method is based on the Pyretic language [35], part of the Frenetic [36] family of programming languages developed by Foster, Rexford et al. This programming language, implemented in Python, describes the behavior of the data plane for any kind of traffic accepted by the network. Pyretic provides some basic policies as well as operators for combining policies. The basic policies include:

- *identity* to forward all incoming packets,
- *drop* to remove all incoming packets,
- *match*$(x_1 = y_1, \ldots, x_n = y_n)$ to forward packets whose header fields x_i equal y_i,
- *modify*$(x_1 = y_1, \ldots, x_n = y_n)$ to forward all packets and changes the header fields x_i to y_i,

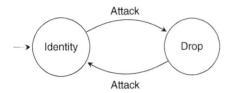

Figure 12.1 Example of a Kinetic control plane automaton.

- *query* to send packets to the controller for deeper analysis,
- *countPackets*($x_1 = y_1, \ldots, x_n = y_n$) to count the number of packets whose header fields x_i contain the values y_i,
- *limitFilters*($k, x_1 = y_1, \ldots, x_n = y_n$) to forward at most k packets whose header fields x_i contain the values y_i,
- *regexpQuery*(*pattern*) to forward packets whose payload matches the given regular expression.

Operators for combining policies include sequential composition, parallel composition, and complement. The sequential composition $p_1 \gg p_2$ forwards all packets accepted by both policies p_1 and p_2 (where p_2 receives packets accepted and potentially modified by p_1). The parallel composition $p_1 + p_2$ forwards all packets accepted by p_1 or by p_2, whereas the complement $\sim p_1$ forwards all packets rejected by p_1 and vice versa.

The Kinetic [37] extension of the Pyretic language enables the verification of the control plane described as a finite state automaton. As a simple example, Figure 12.1 illustrates an automaton that switches between the identity and the drop policies on the basis of the detection of an attack. The idea is that the traffic is normally forwarded without any further control unless an attack is detected, which would cause the traffic to be dropped. Kinetic users can provide properties expressed in the computation tree logic (CTL) temporal logic and verify the control plane automaton against this property. However, verification is restricted to the control plane in Kinetic, and properties of the data plane cannot be verified.

12.4 Orchestration of Security Chains

We now describe a collection of techniques for orchestrating chains of security functions that are deployed in SDN environments. The illustrative use case corresponds to the protection of smart devices with limited CPU and battery capacities such as presented in [38], in particular Android devices, but the methodology is also applicable to SDN infrastructures in general. We give here an overview of the security chain orchestrator, while the different steps related to the orchestration methodology will be described in more detail in Sections 12.5–12.8.

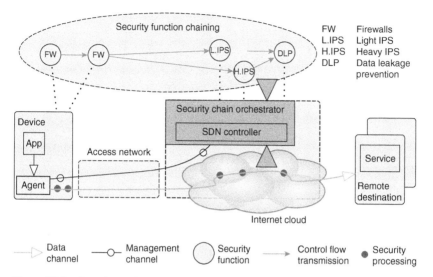

Figure 12.2 Security chain orchestrator integrated into a SDN infrastructure.

A high-level picture of the orchestrator is provided in Figure 12.2. An agent installed on the device shown on the bottom left registers the security requirements of the installed applications. As discussed in Section 12.2, the actual security functions such as firewalls or IDS are deployed in a cloud infrastructure, symbolized by dark gray points in the cloud in the bottom part. These security functions are orchestrated by the security orchestrator which exploits different techniques to build, verify and optimize the chains of security functions. These are then compiled into low-level configuration rules and transmitted to the controller in order to be deployed. The purpose of the chains is to filter the traffic between the device and the remote destinations that it contacts, represented on the right. Devices transmit to the orchestrator the list of applications that connect to the network. Security requirements related to an application are inferred based on a model of its networking interactions in terms of network flows as well as on the permissions requested by the application in its manifest file. While network flows do not represent the data transmitted in messages, the permissions declared by the application are used in order to over-approximate the data that may be exchanged.

The four main problems that we address are the following, and correspond to the different steps of the proposed method (as shown in Figure 12.3):

1. Build a model of the security requirements of the applications to be protected;
2. Synthesize automatically the chains of security functions;
3. Verify that the generated security chains meet the requirements;
4. Optimize their deployment in order to minimize the impact on the network.

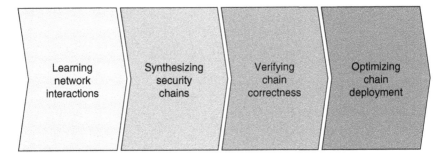

Figure 12.3 Different steps of the proposed methodology.

The first step includes constructing concise and accurate models of the networking behavior of an application; it is addressed by applying process learning techniques. This step results in a finite automaton (more precisely, a Markov chain) that represents the networking interactions of an application based on flow traces collected during its execution. This model is analyzed in order to detect anomalies that may indicate some malicious behavior of an application. These anomalies are represented as predicates that will be used by the subsequent steps for generating abstract representations of chains of security functions that will then be optimized before being compiled into a concrete implementation and deployed.

Concretely, the predicates inferred from the behavior model permit to generate functional representations of single chains (second step). Chains corresponding to individual applications can be combined in order to factor common parts and minimize the overall number of rules to be deployed. They are also formally verified to check their consistency and user-specified correctness properties (third step). Finally, an optimization step computes the optimal placement of security rules according to the topology of the network and criteria specified by the network operator (fourth step).

12.5 Learning Network Interactions

Process learning techniques are applied for modeling the networking behavior of applications, as presented in more detail in [39]. In preparation to the construction of a security chain, network flows for an application are collected by the Flowoid agent [27] that is deployed on the device, and they are then collected as a dataset. For our experimental evaluation, we consider a pre-existing dataset of flows of multiple Android applications.

These learning techniques are of limited use when applied to strongly heterogeneous datasets. Network flows typically contain many different IP addresses that correspond to a single service provider. The flows collected by Flowoid are therefore enriched by a field representing the owner of an IP address. This piece of information, abbreviated as *orgname*, can be retrieved using the well-known whois tool, which also provides the *netname*, i.e. the name of the network in which the IP address is deployed. Usually, the netname is more specific than the orgname, and we decide on which of the two fields to use based on a threshold for the number of occurrences.

Although whois is still the most widely used tool for querying the owner of an IP address, it is also quite common that this information is not available or outdated, motivating the interest for possible alternatives. A first good candidate is the registration data access protocol (RDAP) protocol [40] proposed as a successor to whois. This protocol is based on HTTPS and provides its answer in the JSON format. A second possible alternative is the reverse DNS protocol (RDNS) used to retrieve the domain name associated with an IP address. This solution is actually used by most mail servers to filter out IP addresses that do not belong to any domain name, which could also be used in our approach to identify unsafe IP addresses.

After collecting and enriching the flows of an application we use them to build its behavioral model. A representation in the form of a finite automaton with probabilistic transitions appears particularly appropriate, and we examined existing techniques for learning automaton structures such as the K-tail algorithm [32] or its Synoptic [33] extension, as well as Invarimint [34]. These three methods receive a list of the logs of a system and output an automaton describing the behavior that can be derived from the input logs. Both K-tail and Synoptic learn a Markovian automaton whose transitions are labeled by probabilities, the limit of these approaches is nevertheless the high level of complexity of their outputs. In contrast, Invarimint produces a simpler automaton without probabilities that qualitatively describes the behavior observed in the input logs. We found that on our datasets, Synoptic produced overly complicated automata while Invarimint produced simpler automata, but it does not take into account probabilities.

We therefore designed an algorithm that produces a Markov chain (similar to Synoptic) while producing a compact representation (similar to the automata generated by Invarimint). Algorithm 12.1 represents the automaton using the tables *States* and *Transitions*. It takes as input a list of size N of orgnames, obtained from the flows in the dataset by splitting them into chunks with identical orgname attribute. Automaton states correspond to orgnames, while transitions indicate the probability of succession between orgnames.

The algorithm creates an automaton with as many states as the input contains orgnames, and the weight of a state corresponds to how often it appears. For every pair of successive states, a transition is created and its weight is computed similarly. At the end of the algorithm, transition probabilities are assigned by dividing the

Algorithm 12.1 Learning a Markov chain.

Input: $flow$, a list of size $N + 1$ of orgnames (or netnames)

$States := \emptyset$

$Transitions := \emptyset$

$orgname := flow[0]$

$States[orgname] := 1$ ▷ Count the occurrences of states and of transitions

for $i \in 1..N$ **do**

    $transition := (orgname, flow[i])$

    $orgname := flow[i]$

    if $orgname \in States$ **then**

        $States[orgname] += 1$

    else

        $States[orgname] := 1$

    end if

    if $transition \in Transitions$ **then**

        $Transitions[transition] += 1$

    else

        $Transitions[transition] := 1$

    end if

end for ▷ Compute the probability of each transition

for $transition \in Transitions$ **do**

    $Transitions[transition] := Transitions[transition]/States[transition_0]$

end for

weight of a transition by the weight of its source state. The states of the automaton can be enriched in order to express more information contained in the original flows. Concretely we compute the following standard network metrics from the flows directed to addresses corresponding to each state l of the automaton; these will be used in the following for generating chains of security functions:

- $l.ports$: the set of ports appearing in flows for l;
- $l.protocols$: the protocols used;
- $l.count(x)$: the highest number of occurrences of the address or port x;
- $l.avg\_size$: the average number of packets;
- $l.avg\_interval$: the average distance between communications based on timestamps.

Moreover, $bgp\_ranking(ip)$ denotes a metric corresponding to a value of trust of the IP address ip. In practice, this value is obtained by contacting a remote service relying on various data sources to compute the trust ranking of an IP address.

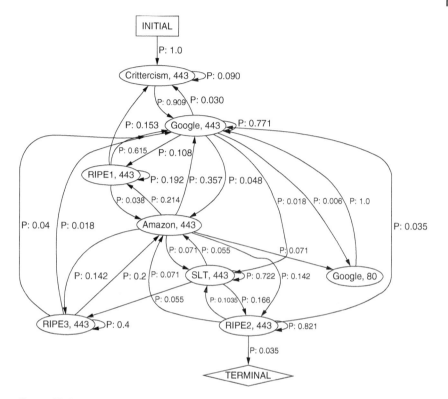

Figure 12.4 Inferred Markov chain of the Pokemon Go Android application.

As a concrete example, Figure 12.4 shows the automaton computed for the dataset corresponding to the Pokemon Go application.[1] Compared to similar automata computed by existing algorithms, our automata have 29.6 states and 141.5 transitions on average against 27.6 states and 142.5 transitions on average for Invarimint. The automata sizes are therefore comparable, but our automata include transition probabilities, and they are much more compact than the automata computed by Synoptic (55 states and 150 transitions on average).

12.6 Synthesizing Security Chains

From the Markov chain representing the network interactions of an application and some thresholds set by the network operator, the next step in our method is

---

1 Probabilities have been rounded and may not add up to 1.

to synthesize a high-level representation of a chain of security functions designed to protect the application against the types of attacks mentioned in Section 12.3. Indeed, once a user inadvertently installed a malicious application on a smart device it is then necessary to protect the user as well as the network against potential attacks. The Markov model of the behavior of an application helps detect suspicious application behavior and prevent unfortunate consequences for the user or the network.

We use a rule-based approach for generating security chains in order to make the algorithm easy to understand and easy to adapt. We assume the following thresholds corresponding to the metrics introduced previously; concrete values for each of these will be set by network operators.

- *attack_limit*: maximal probability of transitions looping on a single state,
- *min_interval*: minimal interval between flow arrivals,
- *min_size*: minimal number of packets in a flow,
- *ip_limit*: maximal number of occurrences for an IP address,
- *port_limit*: maximal number of occurrences for a port number,
- *port_scan_limit*: maximal number of ports in a flow,
- *unsafe_threshold*: maximal value of *bgp_ranking*.

We also assume given a set D_{danger} of Android permissions considered as potentially dangerous. Given a Markov chain with states L_{app} and transitions T_{app}, each of the form (l, p, l') for states $l, l' \in L_{app}$ and a probability $p \in [0; 1]$, as well as trace t_{app}, observe that every flow record $f \in t_{app}$ corresponds to precisely one state $l \in L_{app}$, corresponding to $f.orgname$; we denote this state as l_f.

The core of the detection corresponds to an algorithm for classifying destination addresses a appearing in flows of t_{app}. Instead of hardwiring a fixed classification algorithm, we represent each class of attack as a logical predicate and associate with it a rule that characterizes flows that exhibit the respective attack. In our work, we used the rules shown below; however, they can be modified based on the domain knowledge of the network operator. Use of a declarative programming framework (Prolog in our implementation) helps making these definitions readable and easy to change.

$$dos(a) \leftarrow \exists f, p : f \in t_{app} \land a = f.dstaddr \land (l_f, p, l_f) \in T_{app} \land p \geq$$
$$attack\_limit \land l_f.count(a) \geq ip\_limit \land l_f.avg\_interval \leq$$
$$min\_interval \land l_f.avg\_size \leq min\_size$$

$$port\_scan(a) \leftarrow \exists f, p : f \in t_{app} \land a = f.dstaddr \land (l_f, p, l_f) \in T_{app} \land p \geq$$
$$attack\_limit \land l_f.count(a) \geq ip\_limit \land l_f.avg\_interval \leq$$
$$min\_interval \land l_f.avg\_size \leq min\_size \land | l_f.ports | \geq$$
$$port\_scan\_limit$$

$$worm(a, pt) \leftarrow \exists f, p : f \in t_{app} \wedge a = f.dstaddr \wedge pt = f.dstport \wedge$$
$$(l_f, p, l_f) \in T_{app} \wedge$$
$$p \geq attack\_limit \wedge l_f.count(pt) \geq port\_limit$$

$$botnet(a, pt) \leftarrow \exists f : f \in t_{app} \wedge a = f.dstaddr \wedge$$
$$l_f.count(a) \geq ip\_limit \wedge pt = f.dstport \quad \vee$$
$$l_f.protocols \cap \{tcp, udp\} \neq \emptyset \wedge l_f.avg\_interval \leq$$
$$min\_interval$$

$$unsafe(a) \leftarrow \exists f : f \in t_{app} \wedge a = f.dstaddr \wedge bgp\_ranking(a) \geq$$
$$unsafe\_threshold$$

$$safe(a) \leftarrow \neg dos(a) \wedge \neg port\_scan(a) \wedge \neg unsafe(a) \wedge$$
$$\neg \exists pt : (worm(a, pt) \vee botnet(a, pt))$$

$$danger(pm) \leftarrow pm \in P_{f.appname} \cap \mathcal{D}_{danger}$$

Based on these classification rules, we associate elementary security rules with IP addresses that appear in the trace. These rules are then composed in parallel, yielding security functions such as firewalls or IDS that are in turn composed in sequence for building chains of security functions. We continue to describe our methodology using declarative rules, and later explain how to translate these into a Pyretic program.

We represent network traffic as a sequence $t \in \mathcal{P}^*$ where $\mathcal{P}$ denotes the set of network packets. A security function $f: \mathcal{P}^* \to \mathcal{P}^*$ transforms network traffic. For an integer $n \in \mathbb{N}$, the function $cut(t, n)$ returns the prefix of t of length (at most) n. Given a predicate $pred(p)$ on packets, the function $restrict(t, pred)$ returns the subsequence of t of those packets satisfying $pred$.

Given two traces t_1 and t_2, their *merge* $t_1 \oplus t_2$ corresponds to the unique trace formed by the elements of t_1 and t_2 in increasing order of time stamps, with the proviso that whenever t_1 and t_2 contain flows f_1 and f_2 with $f_1.timestamp = f_2.timestamp$, then f_1 appears in $t_1 \oplus t_2$ while f_2 is dropped. Security functions can be composed in sequence ($\circ_{\gg}$) or in parallel ($\circ_+$):

$$(f \circ_{\gg} g)(t) = g(f(t)) \qquad (f \circ_+ g)(t) = f(t) \oplus g(t)$$

and these operators generalize to n-ary compositions $\bigcirc_{\gg}$ and $\bigcirc_+$.

Elementary security rules make use of the following predicates that can be implemented directly in Pyretic or using VNF rules if we were using NFV:

- *regexp*(s, pm): true if the string s, representing the payload of the packet, satisfies the regular expression associated with the permission pm;
- *tcp_check*(t): true if the traffic t respects the standards of a TCP connection;

- *http_check*(s): true if the string s, representing the payload of the packet, is a valid HTTP request;
- *inspect_payload*(s): true if the string s, representing the payload of the packet, complies with the underlying DPI policy.

We now define elementary security rules:

$$forward(a, t) = restrict(t, \lambda pk : pk.dstaddr = a)$$

$$block(a, pt, t) = restrict(t, \lambda pk : pk.dstaddr \neq a \wedge pk.dstport \neq pt)$$

$$limit(a, n, t) = cut(forward(a, t), n)$$

$$filter(a, pm, t) = restrict(t, \lambda pk : pk.dstaddr = a \wedge regexp(pk.payload, pm))$$

$$inspect(a, t) = restrict(t, \lambda pk : pk.dstaddr = a \wedge inspect\_payload$$
$$(pk.payload))$$

$$tcp(a, pt, t) = \begin{cases} restrict(t, \lambda pk : pk.dstaddr = a \wedge pk.dstport = pt) \\ \quad \text{if } tcp\_check(t) \\ \langle \rangle \quad \text{otherwise} \end{cases}$$

$$udp(a, pt, t) = restrict(t, \lambda pk : pk.dstaddr = a \wedge pk.dstport = pt)$$

$$http(a, pt, t) = restrict(t, \lambda pk : pk.dstaddr = a \wedge pk.dstport = pt \wedge$$
$$http\_check(pk.payload))$$

The following rules infer which security rules should be associated with addresses classified according to the predicates presented above:

$$deploy_{block}(a, pt) \leftarrow worm(a, pt)$$

$$deploy_{block}(a, pt) \leftarrow botnet(a, pt)$$

$$deploy_{forward}(a) \leftarrow \neg \exists pt : worm(a, pt) \vee botnet(a, pt)$$

$$deploy_{limit}(a, ip\_limit) \leftarrow dos(a)$$

$$deploy_{limit}(a, ip\_limit) \leftarrow port\_scan(a)$$

$$deploy_{tcp}(a, pt) \leftarrow \begin{array}{l} f \in t_{app} \wedge a = f.dstaddr \wedge pt = f.dstport \wedge \\ f.protocol = tcp \end{array}$$

$$deploy_{udp}(a, pt) \leftarrow \begin{array}{l} f \in t_{app} \wedge a = f.dstaddr \wedge pt = f.dstport \wedge \\ pt \neq 80 \wedge pt \neq 443 \wedge f.protocol = udp \end{array}$$

$$deploy_{http}(a, 80) \leftarrow f \in t_{app} \wedge a = f.dstaddr \wedge f.dstport = 80$$

$$deploy_{http}(a, 443) \leftarrow f \in t_{app} \wedge a = f.dstaddr \wedge f.dstport = 443$$

$$deploy_{filter}(a, pm) \leftarrow unsafe(a) \wedge danger(pm)$$

$$deploy_{inspect}(a) \leftarrow unsafe(a)$$

Using the predicates *deploy* derived based on the flows, we now construct security functions by composing elementary actions in parallel:

$$stateless\_firewall(t) = \bigcirc_+ \{ \, forward(a, t) : deploy_{forward}(a), a \in \text{Addr} \, \}$$
$$\circ_+ \bigcirc_+ \{ \, block(a, pt, t) : deploy_{block}(a, pt), a \in \text{Addr}, pt \in \text{Port} \, \}$$
$$ids(t) = \bigcirc_+ \{ \, limit(a, n, t) : deploy_{limit}(a, n), a \in \text{Addr}, \, n \in \mathbb{N} \, \}$$
$$stateful\_firewall(t) = \bigcirc_+ \{ \, tcp(a, pt, t) : deploy_{tcp}(a, pt), a \in \text{Addr}, \, pt \in \text{Port} \, \}$$
$$\circ_+ \bigcirc_+ \{ \, udp(a, pt, t) : deploy_{udp}(a, pt), a \in \text{Addr}, \, pt \in \text{Port} \, \}$$
$$\circ_+ \bigcirc_+ \{ \, http(a, pt, t) : deploy_{http}(a, pt), a \in \text{Addr}, \, pt \in \text{Port} \, \}$$
$$dpi(t) = \bigcirc_+ \{ \, inspect(a, t) : deploy_{inspect}(a), a \in \text{Addr} \, \}$$
$$dlp(t) = \bigcirc_+ \{ filter(a, pm, t) : deploy_{filter}(a, pm), a \in \text{Addr}, pm \in D \, \}$$

On the basis of these security functions we now define the chains to be deployed for filtering traffic generated by the target application by associating addresses to those chains corresponding to the classes to which the address belongs:

$$safe\_chain = stateless\_firewall \circ_\gg stateful\_firewall$$
$$unsafe\_chain = stateless\_firewall \circ_\gg stateful\_firewall \circ_\gg dpi \circ_\gg dlp$$
$$dos\_chain = stateless\_firewall \circ_\gg ids \circ_\gg stateful\_firewall$$
$$port\_scan\_chain = dos\_chain$$
$$worm\_chain = stateless\_firewall$$
$$botnet\_chain = stateless\_firewall$$

Finally, we provide rewriting rules for converting security functions into Pyretic code. The argument *t* representing network traffic becomes implicit in Pyretic, which applies the transformation to concrete incoming traffic. The functions *DPIQuery*, *TCPFilter*, *UDPFilter*, and *HTTPFilter* exploit the rules of dynamic query that Pyretic provides. The overall security functions are obtained from the elementary ones by using the combinators $\gg$ and $+$ of Pyretic that correspond to $\circ_\gg$ and $\circ_+$:

$$forward(a, t) \rightsquigarrow match(dstaddr = a)$$
$$block(a, pt, t) \rightsquigarrow \, \sim match(dstaddr = a, \, dstport = pt)$$
$$limit(a, n, t) \rightsquigarrow LimitFilters(n, dstaddr = a)$$
$$filter(a, pm, t) \rightsquigarrow match(dstaddr = a) \gg RegexpQuery(regexp(pm))$$
$$inspect(a, t) \rightsquigarrow match(dstaddr = a) \gg DPIQuery$$
$$tcp(a, pt, t) \rightsquigarrow match(dstaddr = a, dstport = pt) \gg TCPFilter$$

$$udp(a, pt, t) \;\rightharpoonup\; match(dstaddr = a, dstport = pt) \gg UDPFilter$$
$$http(a, pt, t) \;\rightharpoonup\; match(dstaddr = a, dstport = pt) \gg HTTPFilter$$

To sum up, our approach consists in synthesising a program that encods the chain of security functions to be deployed in the network. To this end, we first learn the security properties to be guaranteed by the chain from the Markov automaton encoding the behavior of the application. These predicates are then used to derive the abstract specification of the chain to be deployed based on a constraint programming method. Finally this high level specification is used to generate the actual code of the concrete chain of security functions that can be either directly deployed in the network or be used for further optimizations.

12.7 Verifying Correctness of Chains

The next step consists in verifying correctness properties of security chains. As explained below, the chains generated by our method satisfy certain properties by construction.

12.7.1 Packet Routing

Two desirable properties for packet routing are the absence of black holes and of loops. A black hole occurs when traffic is directed to a link where no security function is installed. A loop is a cycle in the connections between security functions, such that network packets will be transmitted to a security function that they are already cleared.

Proposition 12.1 *The synthesis of security chains described in Section 12.6 avoids black holes and loops.*

Proof: Our security functions and chains are constructed from elementary rules by parallel and sequential composition. In particular, each component of the chain is completely defined before being used, and there is no fixpoint construction or similar cyclic construct. This ensures that no black holes or cycles can exist at the high level of chain construction. We rely on the correctness of the translation to Pyretic to ensure that this property is preserved at the implementation level. □

12.7.2 Shadowing Freedom and Consistency

A security function is shadowing free if for any packet it contains at most one applicable rule.

Proposition 12.2 *Security functions generated by the algorithm of Section 12.6 guarantee shadowing freedom.*

Proof: In the definition of *stateless_firewall*, shadowing would arise if for some address a and port pt, both rules *forward*(a, t) and *block*(a, pt, t) were composed in parallel. However, this is impossible because by definition the corresponding *deploy* predicates are mutually exclusive. Similarly, the different *deploy* predicates used in the definition of *stateful_firewall* are incompatible for any given address and port. □

We now show that our chains of security functions are consistent with the security properties determined on the basis of the traces t_{app} used for their generation.

Proposition 12.3 *Given a trace t_{app} characterizing the network traffic generated by an application, the chain generated by the algorithm of Section 12.6 forwards traffic classified as safe to the corresponding destinations but blocks or limits malicious traffic.*

Proof: An address is considered as malicious if its t_{app} contains flows associated with the orgname of the address that are classified as worm, botnet, DoS, port scan or unsafe. Traffic directed to addresses considered as worm or botnet will immediately be blocked by the stateless firewall. Traffic towards addresses belonging to flows classified as DoS or port scan is transmitted to the IDS, which imposes a limit on the number of packets that will be allowed to pass.

Addresses associated with unsafe flows, i.e. network traffic that potentially compromises the confidentiality of private data, are handled by the DPI and DLP security functions that check for packet payload, according to the predicates *regexp* (associated with Android permissions) and *inspect_payload*. Encrypted traffic would have to be handled by specific inspection methods [41]. Traffic directed to IP addresses considered as safe is only subject to the stateless and stateful firewalls, which forward it and simply check conformance with the declared protocol. □

Beyond these structural correctness properties, we implemented techniques for verifying user-specified properties of both the control and the data planes of security chains [42]. These techniques build upon the Kinetic extension [35] of the Pyretic language that includes model checking capabilities for properties of the control plane, but they enable the verification of properties of the data plane as well.

The first technique is based on constraint solving. We encode elementary Pyretic actions as formulas in SMT-LIB, the input language of SMT solvers. For example,

F1 = match (srcip = IP ("198.122.37.15")) + match (srcip = IP ("253.182.3.14"))
F2 = match (srcport = 100) + match (srcport = 200) + match (srcport = 300)
F3 = match (srcport = 400) + match (srcport = 500) + match (srcport = 600)
F4 = match (dstport = 700) + match (dstport = 800) + match (dstport = 900)
chain = ((F1 >> F2) + (~F1 >> F3)) >> F4

$$allowed \equiv \bigwedge \bigvee \bigwedge srcip = ip0 \bigvee srcip = ip1$$
$$\bigwedge srcpt = pt1 \bigvee srcpt = pt2 \bigvee srcpt = pt3$$
$$\bigvee \bigwedge \neg(srcip = ip0 \bigvee srcip = ip1)$$
$$\bigwedge srcpt = pt4 \bigvee srcpt = pt5 \bigvee srcpt = pt6$$
$$\bigwedge dstpt = pt7 \bigvee dstpt = pt8 \bigvee dstpt = pt9$$

Figure 12.5   A toy security chain in pyretic and its encoding as a constraint.

identity and *drop* are represented as *true* and *false*, and *match* and *modify* give rise to equational constraints on packet headers, where concrete IP addresses and port numbers are mapped to symbolic constants. Sequential and parallel composition correspond to conjunction and disjunction, and complement to negation. For example, Figure 12.5 shows the encoding of a simple security chain as a logical formula. Data plane properties, such as whether certain packets are allowed to proceed or blocked, can then be verified by querying the constraint representing the chain.

The second technique is implemented based on symbolic model checking. In this case, Pyretic chains are represented as finite state machines. For this purpose, we extract strictly sequential subchains (such as $F1 \gg F2$ in the example of Figure 12.5), and these give rise to state transitions that are guarded with conditions on header fields. A packet is accepted by the chain if there exists a path to the final state of the state machine all of whose transition conditions are satisfied, and this can be expressed using formulas of the CTL temporal logic and verified by the symbolic infinite-state model checker nuXmv [43]. This technique integrates well with the verification capabilities that exist in Kinetic, but extend them to encompass the data plane.

12.8   Optimizing Security Chains

When applying the techniques for generating security chains described in Section 12.6 for several applications, we obtain multiple chains that must be deployed in the network. However, many applications share certain services, such as for serving advertisements or for performing analytics, and the security chains corresponding to these applications are likely to contain similarities. Instead of simply combining chains using Pyretic's operator for parallel composition, or of deploying several chains using independent control planes (which could increase the overall vulnerability of the architecture), we aim at transforming several chains into a single one in a way that combines similar elements in different chains, minimizing the number of security functions and rules. A security chain corresponds to a graph of security functions of different types such as firewalls, intrusion detection or DLP systems [17]. In turn, a security function consists of a set of security rules applied in parallel, where a rule is described by a guard and an action.

Our transformation, presented in [44], is based on two procedures. The procedure *merge_functions* takes two security functions (assumed to be of the same type) as inputs and merges them. Rules of either of the two functions whose guards are disjoint from the rules of the other function cannot be conflicting and are simply added to the merged function. For guards that appear in both security functions, if the associated action is the same, the rule is again added to the result. In case of different actions, we rely on priorities provided by the network operator in order to determine which rule to include in the result. The procedure *merge_chains* composes two chains. It first identifies security functions of the same type that appear in the input chains and merges them using *merge_functions*, while functions that have no equivalent in the other chain are simply added to the resulting chain. The edges of the output chain mirror those of the input chains. A quantitative evaluation is provided in Section 12.9.

The transformations *merge_functions* and *merge_chains* do not involve structural modifications of the chains and therefore preserve the structural properties stated in Propositions 12.1 and 12.2. The consistency with classification (Proposition 12.3) may not be preserved when an address is classified differently by the flows collected for two different applications. In our experiments based on existing benchmarks, we have never observed this happening.

The second aspect of optimization concerns the deployment of security chains in the SDN network. The placement of security functions in a network has to satisfy certain constraints: the order in which functions appear in the chain has to be respected, the number of rules deployed on any given switch must not exceed the capacity of that switch, and the capacity of channels connecting switches must be respected. Within these constraints, we aim at optimizing metrics such as the number of required switches, the congestion of the service, and its probability of availability.

In order to make the optimization problem feasible using standard solvers, we aggregate destination addresses, as well as network resources in our model. In our context, we can consider collections of IP addresses that will be associated with SDN switches and then assign the rules for these collections of destinations to the corresponding network equipment. We call these collections of IP addresses *destination aggregates*. It is important for the optimality of the placement that destination aggregates represent comparable traffic load. Thus, we compute the destination aggregates as the result of a knapsack problem. The number of knapsacks is computed as the ratio between the overall traffic load and the capacity of the smallest channel in order to guarantee that we will be able to place every destination aggregate on every channel.

We also aggregate switches as network paths, i.e. as sequences of switches connected in line without branching. The properties of network paths are computed depending on the properties of their internal switches and channels. We will consider the following properties in the remainder of this chapter:

- *length*: the number of switches connected in sequence,
- *rule_capacity*: the minimal rule capacity in the path,
- *load_capacity*: the minimal load capacity in the path, and
- *path_probability*: the availability probability of the path.

The information describing the chains and the network are provided as input for the placement. Destination aggregates are represented by the set *dests*. The number of flows to handle per destination aggregate is represented by a dictionary *dest_load* indexed by the set *dests*. In a similar manner, we introduce a dictionary *dest_weight* that associates with each destination the number of aggregated IP addresses. The dictionary *function_weight* associates with each security function its number of rules per destination. For each security function, our synthesis algorithm guarantees that a destination will be protected by exactly two rules, one for incoming trafic and one for outgoing trafic. Network paths are represented by the set *paths*. The dictionary *path_length* provides information about the length of each path, *rule_capacity* associates with each path the smallest rule capacity among the switches on that path, *load_capacity* stores the load capacity of each path, and *path_probability* indicates the availability probability of each path. The relation *path_connection* indicates if a path is the successor of another path. Finally, we derive two sets *incomings* and *outgoings* which represent incoming and outgoing paths of the network. Namely, $i \in$ *incomings* if and only if $\forall p \in$ *paths*, *path_connection*$_{(p,i)} = 0$ and $o \in$ *outgoings* if and only if $\forall p \in$ *paths*, *path_connection*$_{(o,p)} = 0$.

We represent the placement of rules by the variables *dest_placement*, a matrix of binary variables indexed by *paths* and *dests* that indicate whether the rules concerning a destination d are placed on a path p and the array *used_path* of binary variables that identify used network paths. The following constraints must be respected for a placement to be valid:

1. Constraints on path usage:
    (a) A path is used if the rules for at least one aggregate of destinations are placed on it.
    $$\forall p \in paths, \ |dests| \times used\_path_p \geq \sum\nolimits_{d \in dests} dest\_placement_{(p,d)}$$

(b) A path can be used only if at least one of its successors is used.

$\forall p \in paths, \ |dests| \times used\_path_p$

$\geq \sum_{suc \in paths} path\_connection_{(p,suc)} \times used\_path_{suc}$

(c) The symmetric constraint requiring that a path can be used only if at least one of its predecessors is used.

2. Constraints on destination placement:

(a) Each destination must be placed on at least one incoming path.

$\forall d \in dests, \ \sum_{p \in incomings} dest\_placement_{(p,d)} \geq 1$

(b) The symmetric constraint requiring that each destination must be placed on at least one outgoing path.

3. Capacity constraints:

(a) Constraints on the rule capacity of each path in the network.

$\forall p \in paths, \ rule\_capacity_p \geq$

$function\_weight \times \sum_{d \in dests} dest\_weight_d \times dest\_placement_{(p,d)}$

(b) Constraints in terms of traffic load of each path in the network.

$\forall p \in paths, \ load\_capacity_p \geq \sum_{d \in dests} dest\_load_d \times dest\_placement_{(p,d)}$

We want to optimize several objectives while ensuring the above constraints: (i) network utilization, i.e. the number of switches needed for deploying the security chains, (ii) service congestion due to the concentration of traffic load on a few channels and (iii) probability of availability, i.e. the probability for the service to be available and not affected by network downtimes. In our case, these three criteria are combined in a single objective function to minimize. Results of experiments with non-linear solvers, linear approximations, and optimizing SMT solvers are described in Section 12.9.

12.9 Performance Evaluation

We implemented the techniques described in this chapter in a prototype consisting of 13 457 lines of Python 2.7 and 111 lines of SWI-Prolog (v7.6.4) and evaluated them on a Macbook Pro (13-in., 2017) with an Intel® core i7 processor (2.5 GHz) and 16 GB RAM. The back-end solvers used for verification (Section 12.7) are the model checker nuXmv (v1.0.1) and the SMT solvers cvc4 (v1.5) and veriT (v201506). For optimization (Section 12.8) we employed the simplex solver glpsol (v4.64), the mixed integer nonlinear programming (MINLP) solver couenne (v0.5.6) and optimization module of the SMT solver z3 (v4.8.0). During our experiments we considered 10 Android applications given in Table 12.1. For each application we indicate the number of recorded flows, the corresponding number of IP addresses, the presence of a manifest file and the number of requested permissions.

In our experiments we evaluated the following criteria: the complexity of the chains (numbers of security functions and rules), the response times for synthesis, factorization and verification, the accuracy with which security chains detect attacks, and the overhead incurred by deploying chains in a network.

12.9.1 Complexity of Security Chains

Table 12.1 shows the numbers of security functions and rules of the chains generated for the different applications. Each chain contains either four or five security functions, depending on the presence of the manifest file, which causes DLP rules to be generated. The number of rules clearly illustrates the high disparity of network behavior observed for the applications.

Table 12.2 shows the number of rules corresponding to a chain obtained by successively combining chains for individual applications. We compare three different approaches: parallel composition simply composes individual chains using Pyretic's + operator, combined generation generates a single chain from the concatenation of the flows corresponding to the applications, and chain merging implements the algorithm presented in Section 12.8. The results show that parallel composition and merging produce significantly fewer rules than a combined generation. In contrast to parallel composition where the number of overall functions corresponds to the sum of the numbers of functions per application, merging preserves the number of security functions to be deployed, reducing overhead and attack surface.

Table 12.1 The set of Android applications considered for evaluation.

Applications	Flows	Addresses	Manifest	Permissions	Functions	Rules
Disneyland	282	5	No	—	4	44
Dropbox	1000	17	Yes	5	5	311
Faceswitch	151	30	Yes	3	5	425
Lequipe	1000	151	No	—	4	1640
Meteo	1000	80	No	—	4	716
Ninegag	1000	88	No	—	4	930
Pokemongo	275	24	Yes	6	5	485
Ratp	779	3	No	—	4	28
Skype	1000	161	Yes	11	5	6529
Viber	1000	78	Yes	15	5	4163

Table 12.2 Number of rules for combined chains.

Nb. of apps	Parallel composition	Combined generation	Chain merging
1	311	311	311
2	1 951	3 987	1 947
3	2 376	6 033	2 367
4	2 420	6 153	2 407
5	3 136	8 289	3 119
6	3 164	8 361	3 143
7	9 693	25 949	9 667
8	13 856	51 041	13 825
9	14 341	61 181	14 305
10	15 271	71 147	15 231

12.9.2 Response Times

In our experiments, the time needed for learning the behavior of an application from a recorded trace is on the order of minutes, whereas generating and merging chains takes at most a few seconds. For example, merging the security chains for the 10 applications in our benchmark set takes five seconds. These numbers clearly illustrate the fact that learning the Markov automaton representing an application is not feasible at runtime. However, assuming that applications are relatively stable, learning can be done offline, and the cost of a learning session can be amortized over time. In our overall architecture, we suggest that security chains corresponding to applications be stored in a database. Given the applications to protect, we can at deployment time load the corresponding chains, merge them, and install the result through the SDN controller.

In order to evaluate the performance of formally verifying properties of chains, we artificially generated chains whose numbers of rules varied between 1000 and 10 000 [42]. The SMT-based verification technique results in linear growth with about 15 seconds for the largest chains, whereas nuXmv exhibits super-linear growth and requires more than 40 seconds for the largest chains. However, nuXmv performed better for long chains with many security functions composed sequentially. In both cases, these numbers indicate that formal verification is feasible as an off-line task.

12.9.3 Accuracy of Security Chains

In order to evaluate the accuracy of the generated chains, we used 70% of the recorded flows for an application for generating the chain and then used the

Table 12.3 Accuracy of chains generated for protecting applications.

Applications	Average Accuracy	Minimum Accuracy	Maximum Accuracy
Viber	0.683	0.502	0.997
Faceswitch	0.812	0.518	0.990
Dropbox	0.997	0.993	1.000
Ninegag	0.509	0.498	0.526
Disneyland	0.992	0.986	1.000
Pokemongo	0.743	0.512	0.994
Skype	0.998	0.998	0.998
Lequipe	0.518	0.496	0.537
Meteo	0.837	0.510	0.998
Ratp	0.940	0.692	0.999

remaining 30%, into which we injected a simple port scan, for evaluating its accuracy. We measured accuracy as the ratio between the sum of true positives and true negatives by the total number of flows. We also fixed a threshold, varying between 0 and 10, corresponding to the number of attack flows that must be analyzed before blocking the traffic. Table 12.3 shows the minimal, maximal and average accuracy observed for each chain of security functions. We also computed the corresponding results for the combined chain for all 10 applications in order to observe a potential loss of accuracy, but obtained identical values.

The results are mixed, depending on the considered application. For certain applications, the 30% of logged flows used for the evaluation only contain flows that were already encountered during the learning phase, and we obtain an accuracy close to 100% while for other applications the recorded flows have stronger disparity. These results indicate that the quality of the data used for learning is important. We believe that our approach is acceptably stable, since the orgnames of servers contacted by an application should not change in between major updates. We also believe that the definitions of the predicates that we use for classifying attacks are probably quite naive. Since our approach makes it easy to plug different definitions into our algorithm for chain synthesis, one can experiment different rules without modifying the overall architecture.

12.9.4 Overhead Incurred by Deploying Security Chains

In order to evaluate the cost in terms of bandwidth related to deploying our security chains, we simulated the traffic generated by each application with and

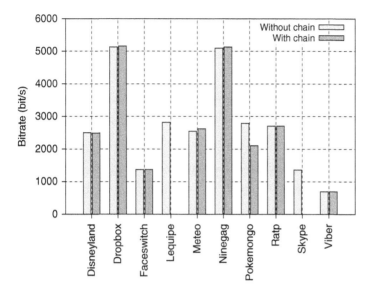

Figure 12.6 Overhead in terms of bandwidth introduced by security chain deployment.

without the corresponding chains and measured the resulting bit rate. The results of these experiments are presented in Figure 12.6.

In contrast to the other evaluations described here, we used a Frenetic implementation of security chains because the Pyretic language is no longer supported by modern SDN controllers. For most applications, the overhead is negligible. The observed differences are minor, probably due to the underlying Open vSwitch (OVS) switches and their dictionary-based flow tables. However, we were unable to deploy the chains for two applications (lequipe and skype) because the Frenetic controller generated too many rules by compiling our chains into OpenFlow. Because our approach is agnostic to implementation languages, it could easily be extended by new implementations based on P4 or involving NFV.

12.10 Conclusions

This chapter introduces a method for automating the orchestration of security functions driven by process learning, and illustrates how it could be used for protecting Android devices by relying on software-defined networks. It contributes to bridging the gap between learning and verification techniques.

The method that we propose addresses four main problems: (i) modeling the specific security needs of applications through process learning techniques, (ii) generating corresponding chains of security functions based on methods of formal synthesis, (iii) verifying the correctness properties of these chains, and (iv) optimizing their deployment by merging chains and adapting them to the network infrastructure. We evaluated the performance of the method through extensive series of experiments.

The flexibility of SDN infrastructures enables synthesizing and deploying security chains that are specific to the networking behavior of individual applications running on smart devices. By construction, the obtained chains ensure certain correctness properties, and specific properties can be formally verified based on SMT solving and model checking. Finally, by applying appropriate optimization methods, the impact of deploying security chains on network performance can be substantially reduced.

This work opens several directions for future research. A closer coupling of network and system aspects could be investigated, beyond the generation of regular expressions based on the permissions declared in manifest files of applications. Emerging methods from explainable artificial intelligence could also be considered for facilitating the interpretation of automation results, together with the use of more elaborated detection techniques. Finally, it could interesting to explore complementary synthesis techniques for taking into account the dynamics of attacks, for instance with more sophisticated models expressed in temporal logic, by following a similar overall methodology.

Bibliography

1 GData (2019). Mobile Malware Report. http://www.gdatasoftware.com (accessed 21 April 2021).

2 La Polla, M., Martinelli, F., and Sgandurra, D. (2012). A survey on security for mobile devices. *IEEE Communication Surveys and Tutorials* 15: 446–471.

3 Faruki, P., Bharmal, A., Laxmi, V. et al. (2015). Android security, a survey of issues, malware penetrations and defenses. *IEEE Communication Surveys and Tutorials* 17: 998–1022.

4 Wei, X. (2012). ProfileDroid: multi-layer profiling of android applications. *Proceedings of the 18th Annual International Conference on Mobile Computing and Networking (MOBICOM 2012)*.

5 Android Permissions System. https://developer.android.com/guide/topics/security/permissions.html (accessed 21 April 2021).

6 Backes, M., Bugiel, S., Derr, E. et al. (2016). On demystifying the android application framework: re-visiting android permission specification analysis. *Proceedings of the 25th USENIX Security Symposium (NSDI 2016)*.

7 Kim, J., Choi, H., Namkung, H. et al. (2016). Enabling automatic protocol behavior analysis for android applications. *Proceedings of the 12th International*

Conference on Emerging Networking EXperiments and Technologies (CONEXT 2016), New York, NY, USA, pp. 281–295.

8 Feamster, N., Rexford, J., and Zegura, E. (2014). The road to SDN, an intellectual history of programmable networks. *SIGCOMM Computer Communication Review* 44 (2): 87–98.

9 Bernardos, C.J., Rahman, A., Zúñiga, J.-C. et al. (2019). Network Virtualization Research Challenges. RFC 8568. https://rfc-editor.org/rfc/rfc8568.txt (accessed 21 April 2021).

10 Dilip, J. and Ion, S. (2008). Modeling middle boxes. *IEEE Network: The Magazine of Global Internetworking Archive.*

11 Hu, H., Han, W., Ahn, G.J., and Zhao, Z. (2014). FlowGard: building Robust firewalls for software-defined networks. *Proceedings of the 3rd Workshop on Hot Topics in Software Defined Networking (SIGCOMM 2014).*

12 Ayyub, Z. and Miao, R. (2013). Simple-fying middlebox policy enforcement using SDN. *ACM SIGCOMM Computer Communication Review.*

13 Ocampo, A.F., Gil-Herrera, J., Isolani, P.H. et al. (2017). Optimal service function chain composition in network functions virtualization. In: *Proceedings of the IFIP International Conference on Autonomous Infrastructure, Management and Security (IFIP AIMS'17)* (ed. A. Biere and R. Bloem), 62–76. Springer International Publishing.

14 Petroulakis, N., Fysarakis, K., Askoxylakis, I., and Spanoudakis, G. (2017). Reactive security for SDN/NFV-enabled industrial networks leveraging service function chaining. *Transactions on Emerging Telecommunications Technologies* 12. https://doi.org/10.1002/ett.3269.

15 Shameli-Sendi, A., Jarraya, Y., Pourzandi, M., and Cheriet, M. (2016). Efficient provisioning of security service function chaining using network security defense patterns. *IEEE Transactions on Services Computing* 10. https://doi.org/10.1109/TSC.2016.2616867.

16 Hurel, G., Badonnel, R., Lahmadi, A., and Festor, O. (2015). Towards cloud based compositions of security functions for mobile devices. *IFIP/IEEE International Symposium on Integrated Network Management (IM 2015).*

17 Hurel, G., Badonnel, R., Lahmadi, A., and Festor, O. (2015). Behavioral and dynamic security functions chaining for android devices. *Proceedings of the 11th IFIP/IEEE/ACM SIGCOMM International Conference on Network and Service Management (CNSM 2015).*

18 Clarke, E.M., Henzinger, Clarke, E.M., Henzinger, and Bloem, R. (eds) (2016). *Handbook of Model Checking.* Springer.

19 Biere, A., Heule, M., van Maaren, H., and Walsch, T. (2008). *Handbook of Satisfiability.* IO press.

20 Beckett, R. (2018). Network control plane synthesis and verification. PhD thesis. University of Princeton.

21 Ball, T., Bjørner, N., Gember, A. et al. (2014). VeriCon: towards verifying controller programs in software-defined networks. Proceedings of the 35th ACM SIGPLAN International Conference on Programming Language Design (PLDI'14), Edinburgh, UK, pp. 282–293.

22 Foster, N., McClurg, J., Hojjat, H., and Cerny, P. (2015). Efficient synthesis of network updates. *Proceedings of the 36th ACM SIGPLAN Conference on Programming Language Design and Implementation (PLDI'15).*

23 Khurshid, A., Zou, X., Zhou, W., Caesar, M., and Brighten, P. (2012). VeriFlow: verifying network-wide invariants in real time. *Proceedings of the 1st Workshop on Hot Topics in Software-Defined Networks (HotSDN'12).*

24 Al-Shaer, E.S. and Hamed, H.H. (2004). Discovery of policy anomalies in distributed firewalls. *Proceedings of the 23rd Annual Joint Conference of the IEEE Computer and Communications (INFOCOM'04).*

25 Kang, M.-Y., Choi, J.-Y., Kang, I. et al. (2016). A verification method of SDN firewall applications. *IEICE Transactions on Communications* E99.B (7): 1408–1415.

26 Claise, B. (2008). Specification of the IP Flow Information Export (IPFIX) Protocol for the Exchange of IP Traffic Flow Information. RFC 5101.

27 Lahmadi, A., Beck, F., Finickel, E., and Festor, O. (2015). A platform for the analysis and visualization of network flow data of android environments. *IFIP/IEEE International Symposium on Integrated Network Management (IM 2015).*

28 Sperotto, A. (2010). Flow-based intrusion detection. PhD thesis. University of Twente.

29 Handley, M.J. and Rescorla, E. (2006). Internet Architecture Board. Internet Denial-of-Service Considerations. RFC 4732.

30 Malkin, G.S. and Parker, T.L.Q. (1993). Internet Users' Glossary. RFC 1392.

31 Barthel, D., Vasseur, J.P., Pister, K. et al. (2012). Routing Metrics Used for Path Calculation in Low-Power and Lossy Networks. RFC 6551.

32 Biermann, A.W. and Feldman, J.A. (1972). On the synthesis of finite-state machines from samples of their behavior. *IEEE Transactions on Computers* C-21: 592–597.

33 Beschastnikh, I., Abrahamson, J., Brun, Y., and Ernst, M.D. (2011). Synoptic: studying logged behavior with inferred models. *Proceedings of the 19th ACM SIGSOFT Symposium and the 13th European Conference on Foundations of Software Engineering*, ESEC/FSE '11. New York, NY, USA: ACM, pp. 448–451.

34 Beschastnikh, I., Brun, Y., Abrahamson, J. et al. (2015). Using declarative specification to improve the understanding, extensibility, and comparison of model-inference algorithms. *IEEE Transactions on Software Engineering* 41: 408–428.

35 Foster, N., Freedman, M.J., Guha, A. et al. (2016). Languages for software-defined networks. *Software Technology Group.*

36 Foster, N., Freedman, M.J., Harrison, R. et al. (2011). Frenetic, a network programming language. *Proceedings of the 16th ACM SIGPLAN International Conference on Functional Programming (ICFP 2011).*

37 Kim, H., Reich, J., Gupta, A. et al. (2015). Kinetic: verifiable dynamic network control. *Proceedings of the 12th USENIX Conference on Networked Systems Design and Implementation (NSDI 2015).*

38 Schnepf, N. (2019). Orchestration et Vérification de Fonctions de Sécurité pour des Environnements Intelligents. PhD thesis. University of Lorraine.

39 Schnepf, N., Merz, S., Badonnel, R., and Lahmadi, A. (2018). Towards generation of SDN policies for protecting android environments based on automata learning. *Proceedings of the 16th Network Operations and Management Symposium (IEEE/IFIP NOMS 2018).*

40 Newton, A., Ellacott, B., and Kong, N. (2015). HTTP Usage in the Registration Data Access Protocol (RDAP). RFC 7480. https://rfc-editor.org/rfc/rfc7480.txt (accessed 21 April 2021).

41 Sherry, J., Lan, C., Popa, R.A., and Ratnasamy, S. (2015). BlindBox: deep packet inspection over encrypted traffic. *Proceedings of the ACM Conference on Special Interest Group on Data Communication (SIGCOMM 2015).* New York, NY, USA: ACM, pp. 213–226.

42 Schnepf, N., Merz, S., Badonnel, R., and Lahmadi, A. (2017). Automated verification of security chains in software-defined networks with synaptic. *Proceedings of the 3rd IEEE Conference on Network Softwarization (IEEE NetSoft 2017).*

43 Cavada, R., Cimatti, A., Dorigatti, M. et al. (2014). The nuXmv symbolic model checker. *Proceedings of the 26th International Conference on Computer Aided Verification (CAV 2014),* Vienna, Austria, pp. 334–342. https://doi.org/10.1007/978-3-319-08867-9_22.

44 Schnepf, N., Merz, S., Badonnel, R., and Lahmadi, A. (2019). Automated factorization of security chains in software-defined networks. *Proceedings of the 16th IFIP/IEEE Symposium on Integrated Network and Service Management (IM 2019).*

13

Architectures for Blockchain-IoT Integration[1]

Sina Rafati Niya, Eryk Schiller, and Burkhard Stiller

Communication Systems Group CSG, Department of Informatics IfI, University of Zürich UZH, Zürich, Switzerland

13.1 Introduction

Internet-of-Things (IoT) as a key technology for distributed data collection and operations encompasses a myriad of use cases. The expansion of IoT happens at an impressive pace and it is foreseen that by 2030 the number of IoT devices will reach 125 billion (cf. Figure 13.1 [1]). Although typical IoT-based applications are based on the centralized client-server (C/S) paradigm, which connects these devices to cloud servers through the Internet, the C/S model may no longer be sustainable, since the enormous growth of different IoT-based architectures creates the need for exploring manageable paradigms, such as decentralized management approaches for many IoT devices. Additionally, centralized architectures do not guarantee automatically data transparency, since clients do not know how and where their data is being used or modified. An interesting solution to circumvent side effects of centralization became available by the fully decentralized paradigm of Blockchains (BC).

In general, BCs are decentralized and distributed data storage systems, which operate on the basis of (i) a Peer-to-Peer (P2P) network protocol, (ii) a full copy of the Distributed Ledger (DL) per node, (the data storage), and (iii) miners, validators, and BC clients. The immutability of BCs is enabled via *consensus mechanisms*, which defines a decentralized and trusted *principle-set* providing a global agreement. An example mechanisms is the Proof-of-Work (PoW) consensus used by Bitcoin (BTC). PoW requires the validation of transactions (Txs), the re-validation

1 This work was supported partially by (i) the University of Zürich UZH, Switzerland, and (ii) the European Union Horizon 2020 Research and Innovation Program under Grant Agreement No. 830927, namely the CONCORDIA Project.

Communication Networks and Service Management in the Era of Artificial Intelligence and Machine Learning, First Edition. Edited by Nur Zincir-Heywood, Marco Mellia, and Yixin Diao.

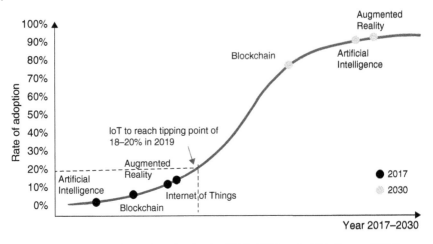

Figure 13.1 IoT adoption estimation for the 2017–2030 period [1]. Source: DBS Group Research [1].

of previously mined blocks, and selecting the "winning" miner in a decentralized manner. Furthermore, PoW requires miners to check issued and open, but not yet mined Txs against the balance of Tx issuers' account balance (BC clients). In case of cryptocurrencies based on a PoW BC, this avoids double spending of funds by BC-clients, i.e. account owners. The Txs checked will be added to a block, for which a hash will be computed. Additionally, PoW mechanisms include the solving of a cryptographic puzzle, which is computed by all "competing" nodes (miners) planning to mine a block. Since this step is computationally intensive, the miner providing the solution to this puzzle will be offered a reward from the BC network and his block will be persisted in the ledger, if he offers that solution first.

While an extensive overview of incentives and potentials of employing BC and IoT integration (BIoT) within integrated architectures is presented in Section 13.2.1, a selected set of important BIoT reasons were already collected by Fernández-Caramés and Fraga-Lamas [2]. Due to the decentralization of BCs, the integration with IoT can (i) optimize costs of centralized management solutions, (ii) offer trust and access rights management, (iii) overcome potential gaps of transparent operations, and (iv) support synchronization. Costs stemming from the deployment and maintenance of IoT solutions in a traditionally centralized fashion, e.g. with centralized clouds and server farms, have been largely due to networking, storage, computational, and identity management resources' operations. Additionally, trust concerns may exist for IoT adapters, when Trusted Third Parties (TTP), e.g. service providers, manufacturers, or governments, may manage device access rights. While in selected cases the TTP may collect and analyze user data, the privacy of data owners, whose data are collected may be tampered with.

In recent years centralized architectures, like Azure IoT solution accelerators [3], Amazon Web Services (AWS) IoT [4], or IBM Watson [5], have contributed to cloud-based IoT platforms. However, if data transmitted to these is not encrypted before their transmission, reliability and the transparency of these data may be compromised. Hence, centralized IoT architectures are more prone to security flaws. Additionally, synchronization concerns with centralized IoT architectures emerge due to the increase of devices. At the end, these platforms are vulnerable as Single Point-of-Failure (SPoF).

Driven by key BCs characteristics, however, a solution for most of these drawbacks seems at hand. While initially, BCs were associated with FinTech and cryptocurrencies, like BTC, they are exploited by now within a wider area of use cases, such as secure data storage provisioning, supply chain tracing (SCT), and IoT-oriented applications [6]. BCs are distinct from other distributed systems by being public or private, permissioned or permissionless, and immutable. By adding a BC to a technical system, trust and transparency are being added by design, since data is persisted within transactions (Tx) via backward-linked and cryptographically strengthened blocks. Thus, the trust in BCs is due to these cryptographic algorithms of hashing and signing. They "govern" the operation of participating decentralized nodes, i.e. miners and BC clients. With BCs equal rights are provided to all participating BC clients, since anyone can persist a Tx on a BC. A Tx, once sent and validated, contains data that cannot be reverted or censored anymore [7] without being detected.

Thus, these BC characteristics and the demand for decentralized IoT management are discussed within this chapter to explore BIoT advantages and deficits, leading to the analysis of the efficiency of BIoT architectures in general and a new BIoT architecture combining major advances.

13.1.1 Blockchain Basics

BCs are distinguished in terms of public vs. private and permissioned vs. permissionless characteristics. Public BCs are accessible for all users to read and write, thus, to participate within the consensus mechanisms. In "private" BCs (termed Distributed Ledgers, DLs), however, only a limited set of users can have write access to the BC and the data persisted. A permissioned BC only allows for pre-defined users to access the BC, whereas in permission-less ones, everyone has the same privilege and access at no explicit permissions needed for anyone. BTC is known to be the first public BC. Ethereum [8] offers both options depending on how an application or system utilizes it.

While only the public, permissionless BC determines a real BC, the three other combinations define DLs, which are characterized by restricted read, write, or consensus participation options and, thus, very different trust models and assumptions. Note that the "distributed ledger" is used in two different notions, one as just described (with limited access characteristics) and another one in the notion of a decentralized database, the (distributed) ledger only.

For instance, since BTC exploits PoW, it is considered to be secure and trusted (i.e. a public and permissionless BC). However, this security trades off the BC's scalability and energy efficiency. BTC is configured with a block time of ~10 minutes. The block time is the time of a BC in which new Tx are collected to be mined into the next new block and added to the chain after a successful mining. Due to the maximum block size and varying Tx sizes, BTC's Tx rate is at about 7 Tx per second (TPS) [9]. Such a low Tx rate leads to a scalability bottleneck for use cases, which need thousands of Txs being persisted, such as for many IoT devices at hand. Since PoW-based mining is computationally expensive, IoT devices offer a very limited capability to perform the full mining process. However, once a Tx is persisted into a BC, the immutability of this BC's data storage is achieved, determining a major incentive for integrating BCs with an IoT architecture. Thus, the question remains: which BC and in which scenario can be utilized to avoid computational bottlenecks, while maintaining an IoT-compatible Tx rate?

13.1.2 Internet-of-Things (IoT) Basics

An IoT system is characterized as a set of interrelated computing devices, each of which identified with a unique identifier, where the transfer of data happens over an attached network without requiring human or human-to-computer interactions. An attached wide-area communication network establishes the connection of these devices to the Internet, which can be instantiated as a Local Area Network (LAN) approach like Ethernet, or based on wireless technologies, such as WiFi, Bluetooth, Long-range Wide Area Networks (LoRaWAN), Low-Power Wide Area Networks (LP-WAN), or cellular networks.

IoT plays a key role in monitoring, measuring, sensing, and collecting data from different devices within an environment (e.g. a house, a factory, or nature). IoT can be deployed in supply chain management, health care, or smart cities (cf. Section 13.2.2). Moreover, the utilization of IoT devices is crucial, where human presence is either physically impossible, economically demanding, or risky (e.g. measuring toxic gas emissions of a volcano). Due to their distributed nature, IoT protocols and architectures are often compared against the three metrics: (i) scalability, (ii) energy efficiency, and (iii) security.

Due to frequent technical developments in IoT, presently seen in smaller device sizes, higher computational and storage capacity, and stronger network connection features, IoT applications and architectures need to be flexible to adopt these improvements. For instance, the Industry 4.0 era without IoT devices could not have progressed the industrial automation evolution that quickly. Therefore, respective IoT architectures have evolved over time, shifting from closed and centralized architectures to open-access, cloud-centered architectures. Future IoT architectures need integrate cloud functionality among multiple BC nodes as the next technology adaptation and evolution path [2].

In case of fully centralized approaches, IoT application platforms must be connected to the Internet, to retrieve raw data from IoT networks, process these data, and return information processed to IoT infrastructure owners. These significant amounts of data stored at central IoT architectures can develop toward a SPoF. Currently, many platforms follow this centralized paradigm, such as "The Things Network (TTN)" [10], and they facilitate easy to establish data collection and monitoring services with proprietary levels of IoT compatibility. For instance, TTN offers limited services to LoRaWAN communications, but provides worldwide communications through its gateways, connecting LoRa nodes to TTN servers. Additionally, centralized IoT platforms suffer from security concerns with respect to Denial-of-Service (DoS) attacks on cloud nodes, eavesdropping of user data, and nodes controlling attacks, where the adversary takes control of an entity, e.g. an IoT device, besides forming an SPoF already.

13.2 Blockchain-IoT Integration (BIoT)

Data integrity and transmission reliability is the key for BIoT, here comprising of the IoT infrastructure in use and the BC deployed. Since IoT devices are the initiators of IoT-to-BC communications, they are in the front line to support data integrity and thus its reliability. Therefore, IoT devices need to operate as BC nodes, either as BC clients or BC miners. Generally, BC clients using BC wallet applications and BC miners utilize cryptographic functions to issue a Tx or mine a block. Since hashing is a major task of BCs, the performance of IoT devices in running cryptographic functions, such as (i) SHA-256 for sealing Txs in a block upon mining or (ii) the light-weight Elliptic Curve Digital Signature Algorithm (ECDSA) Ed25519 with SHA-512 [11] for signing Txs, have to be studied on different computing architectures. Their current performance levels reached [12] are summarized in Table 13.1.

Here, the performance of selected cryptographic functions is evaluated on a TelosB (MSP430), an ATmega 2560 (Arduino Mega) node, a Raspberry Pi 3 (RPI), and a regular PC (Personal Computer) with an Intel i7 CPU (Central Processing Unit) of 2.4 GHz [12]. The evaluation was performed for SHA-256 and SHA-512, measured in Hashes per Second (HPS), and for Ed25519, measured in signatures

Table 13.1 Performance of selected cryptographic functions on different IoT hardware in hashes per second (HPS) and signatures per second [12].

Cryptographic function	TelosB	Arduino Mega 2560	RPI 3	Intel-i7
SHA-256 (HPS)	34.22	79.8	31 989	182 216
SHA-512 (HPS)	4.79	10.46	12 194	97 655
Ed25519 (signatures/s)	0.0036	0.0179	30.1	84.2

Source: Source: Niya et al. [12].

per second. These determine those cryptographic functions as used by BCs. Consequently, once IoT devices would be required to operate as a BC node, thousands of TelosB and Arduinos would be needed to perform BC operations as efficient as a single Application-specific Integrated Circuit (ASIC)-based BC miner could do it.

It has been proven that only signing operation for a single Tx with Ed25519 can be considered heavy for constrained IoT devices. Thus, IoT-based mining of PoW-based consensus mechanisms using SHA-256 is neither realistic nor practically achievable in general. However, if the computational power of IoT devices is high, such as with an Arduino and an RPI, they can operate as BC clients. The IoT device signs a Tx with its Secret Key (SK) and attaches the corresponding Public Key (PK) to the signed data, and/or the hash of the data. Such a signed Tx, consisting of the raw data, its hash, and the IoT device's PK, will ensure data integrity and origin of a Tx. Therefore, even operating as a BC client, IoT devices in BIoT applications have to be powerful for Tx signing, to enable data integrity for the full BIoT path.

Further requirements, such as scalability and energy-efficiency, are required by BIoT applications, too. Hence, it is crucial to determine which functions and configurations can be supportive. For instance (cf. Section 13.2.3), de-fragmentation, pre-processing, time stamping, event handling, en-decryption, and compression/decompression of data collected before any IoT-to-BC transmission are functions, impact the scalability and energy-efficiency of IoT-to-BC communications. To configure and manage IoT devices to perform these functions, an efficient and flexibly managed BIoT architecture is necessary. Thus, BIoT architectures are expected to consider BCs and IoT protocols and according to specific characteristics of the employed IoT protocol, corresponding settings for the IoT configuration and the type of a BC or DL selected are needed.

13.2.1 BIoT Potentials

Studies have shown, a well-designed and preconfigured IoT setup can benefit from BCs in many ways [2, 13, 14]. BIoT potentials encompass the driving force for application developers as well as platform providers to explore BIoT operational advances. Six key incentives and potentials in BIoT exist:

Decentralized security provisioning: BCs do not require a TTP, which means that a secure and trusted "decentralization" created by BCs prevents any individual or authority to control or tamper with particular data persisted in the BC. Thus, depending on application requirements the full range of BCs to DLs can be deployed.

Reliability and resiliency: Once a Tx is validated and appended to a BC, the content of this Tx is immutable and distributed across all BC nodes. The Tx stored will be accessible for all BC clients at a high, uninterrupted availability, since full BC nodes store an entire copy of the DL locally. Thus, the establishment of a reliable and resilient data storage is possible.

Traceability: Since all Txs are stored within a BC in chronological order, all BC clients can trace any content and order of these Txs. Thus, data persisted once in a BC will be accessible with respect to their time, possibly geo-location, and content. Note that different data storage alternatives exist, e.g. to store only a hash of data on-chain and storing the full data off-chain. This is often adopted in BIoT use cases, such as SCT (cf. Section 13.2.2), since private data can be maintained effectively.

Autonomic interactions: BCs grant IoT devices the ability to interact with autonomous processing entities, i.e. Smart Contracts (SCs), which define immutable programs, e.g. running in the Ethereum BC [15]. Autonomic communications of IoT nodes without a TTP are possible, since contract clauses embedded in SCs are executed within a self-governed fashion, when a certain condition is satisfied (e.g. the user breaching a contract will be fined automatically) or when a threshold is violated (e.g. IoT-connected environment monitoring sensors sense Carbon Monoxide [CO] levels reaching to a threshold amount; consequently, the corresponding SC will execute an alarm function) [16]. Thus, a BIoT system can autonomously process and stimulated actions.

Identity and Access Management (IAM): IoT devices are uniquely identified, however, only IAM provides authentication and authorization. Since the maintenance of credentials, such as keys and certificates for IoT devices, are important, trusted and distributed authentication and authorization services for IoT devices are required. While the authentication refers to the process of verifying the identity of a user or process, which, once successfully performed, can lead to authorization by providing access to resources. An Authentication and Authorization Infrastructure (AAI) can be extended to operate with and for IoT device identities, too.

Interoperability is key for heterogeneous IoT systems, across communication protocols and with alternative BCs. Thus, machine-to-machine, device-to-device, and IoT-to-BC node communications have to be harmonized.

13.2.2 BIoT Use Cases

The emerging BC and IoT domains lead to newly integrated BIoT use cases. Especially BIoT potentials as listed in Section 13.2.1 enabled manifold use case areas, from which the five most important ones are discussed:

A **Smart Grid** operates as a electricity grid combined with information technology to enable the mutual exchange of control information and to allow for the management and monitoring of the distribution of electricity from various sources. Smart Grids need communications among a large number of devices to be tracked, monitored, analyzed, and controlled within a network. Therefore, distributed automation is essential and it can be achieved through the application of IoT-based sensors, smart actuators, meters in production, transmission and distribution, storage, and the management of electricity consumption. With Smart Grids, the role of consumers is reshaped into so-called *prosumers*, who can generate and consume energy. For instance, prosumers holding unused energy can sell their electricity produced. The respective trading process defines a P2P energy trading [17]. Such energy trading can benefit from secure and trusted BIoT platforms, since the storage of volumes of consumed/produced or traded energy can be stored in decentralized ledger.

A **Smart City** shows various interconnected infrastructure components, mobility and traffic management, citizen relations, and environmental resource monitoring [17]. Thus, IoT-based approaches for interoperations are prone to potential attack targets. To protect against security attacks and related risks, a Smart City ecosystem must support data encryption, anonymization, and pseudonymization. This can be supported by BIoT architectures with the employment of SC-based autonomous interactions and an execution of automated actions on top of immutable data handling and persistence.

Smart Home: A traditional house being equipped with a human-intractable set of IoT sensors and processing is termed Smart Home, since respective units of the infrastructure can monitor resources and the environment. Configuration settings of energy, light, or shutters within a given home could be updated – preferably automatically – based on data sensed. Such an approach can enhance the comfort and safety of residents by achieving a higher energy efficiency, higher sustainability, and surveillance, while BCs offer the traceability of decisions taken based on these data monitored and persisted immutably.

Healthcare: In general, healthcare platforms interact with multiple stakeholders, such as patients, healthcare providers (e.g. hospitals), insurance companies, government agencies, clinical researchers, and pharmaceutical suppliers. Given their role on the well-being of people, IoT-integrated healthcare systems support the sharing of data generated by IoT sensors. However, IoT actuators need to be stringently managed, since only authorized and correct commands shall be processed. Thus, with a suitable IAM integration for data collection, storage, and control decisions, health platforms need to guarantee the transparency, immutability, and decentralization inherently.

Therefore, the use of SCs residing within the BC can restrict data accessibility on a "need-to-know" basis, which meets privacy requirements. Privacy preservation challenges and the security of healthcare data can be achieved by storing the proof of data integrity on the BC. By relying on these features, data collected by medical sensors can be automatically sent to the central control by triggering a SC, thus, supporting real-time monitoring of patients. BIoT-enabled methods support the privacy of healthcare users and can verify their authenticity and identity. As of today, a BIoT-based healthcare system can monitor a pandemic outbreak, like for COVID-19 as of late 2019, such that infected people with IoT sensors could be tracked and countermeasures could be applied, while protecting the patient's privacy [7, 18].

Supply Chain Tracing: In SCT, not only the sequence of fund Tx is needed, but also the sequence of actions performed on a particular product during its life cycle is valuable. Thus, the traceability of funds and actions rely on accurate time stamping, which is enabled by BIoT-integrated applications. To circumvent boundaries between different stakeholders, a large number of use cases integrate BCs and IoT SCT. IoT devices are used for sensing, collecting, and monitoring data automatically in distributed settings. BCs attach to these data time stamps, geo-location information, and an IoT device identifier. Thus, actions and conditions monitored throughout a product's life-cycle, are persisted and traceable afterwards. Quantitative and qualitative data from such processes can be mapped to resources and actions, which control the full life-cycle in a transparent manner. Recent studies elaborate on several aspects employing BCs in SCT, such as [19–21].

13.2.3 BIoT Challenges

BC IoT integration (BIoT) is an interdisciplinary approach. Due to the very different BC and IoT characteristics several risks are experienced in providing support for practically efficient applications. Those risks are categorized, as shown in Figure 13.2, into the social, operational, performance, technical, functional, and architectural fields [7, 22].

Figure 13.2 Categories of BIoT risks.

Social risks refer to the social acceptance risks of a solution. A purely "solution-oriented" design of an application may offer an output that is not seen in the same importance and usability by end-users who are mainly facing a product from a "problem-oriented" perspective. In the BIoT case, a clear understanding of BC benefits by end-users is key for any application to be socially adopted. Otherwise, there is a high risk that even a technically perfect application ends up with no costumers nor consumers.

Technical risks can be caused by the algorithms and protocols employed in each technology, i.e. BC and IoT. For instance, using Ethereum as an underlying BC infrastructure of many platforms has shown a reliable security level in its latest versions, however, there have been a few security breaches experienced with its SCs in the past. Furthermore, technical risks can be caused by the way programmers develop BC-based applications including how they use the programming languages, e.g. solidity when developing Ethereum-based SCs. Technical risks have resulted in data and financial loss several times in the past both for platform owners and users.

Functional risks exist where an application cannot deliver a specific functionality due to technical problems or updates. For instance, a functional issue can be caused due to the incompatibility of software or hardware deployed.

Operational risks refer to the inability in delivering expected operations either due to a weak design of an application or users lacking knowledge in interacting with the application. Both functional and operational risks potentially cause dissatisfaction of users and their reluctance in using a platform.

Performance and Architecture risks are correlated, hence, intentionally placed adjacently in Figure 13.2. As mentioned in Section 13.2, an efficient BIoT is not achievable unless corresponding components and functions are enabled. Therefore, the set of important metrics directly impacting a BIoT architecture

Figure 13.3 BC-IoT integration (BIoT) metrics.

and consequently the application performance include (i) scalability, (ii) security, and (iii) energy efficiency (cf. Figure 13.3).

Each of these categories has a group of directly relevant and impacting parameters and reasons. For instance, studies and experiences in BIoT show that the scalability of BIoT applications is highly impacted by BC consensus, BC type, BC client actions and location, processing nodes, and the computational complexity of the mining and signing processes. The scalability of BCs is measured by different metrics such as Tx validations per time, measured in TPS. Section 13.2.3.1 discusses these concerns in depth.

BIoT security as shown in Figure 13.3, is a representation of data security and integrity provision throughout data collection, transmission, and persistence. The more the security of a BIoT architecture, the higher will be the user trust in the data/information shown by the applications based on that architecture. In order to offer a higher trustability, BIoT systems shall offer highly sophisticated encryption and cryptographic algorithms while providing user privacy and transparency. Even if not applicable in all countries today, regulation and governments influence on the verification of the validity of the data stored in a BC is highly important and impacts the trust to that application or system. Section 13.2.3.2 discusses the security concerns more in-depth.

As today's world requires sustainable approaches urgently, the energy efficiency of BIoT architectures is gaining high importance [23]. On one hand, processes of an IoT device including all the actions it has to perform for collecting data and the ones performed prior or in parallel to data transmission such as encapsulations, and aggregation of data, are all affecting the energy efficiency of a BIoT approach. On the other hand, the high energy demand of BC mining processes has raised serious concerns that must be considered in its selection for a BIoT application. Section 13.2.3.3 discusses these concerns more in-depth. These metrics are mostly collected based on a set of experiences and literature review by authors of this chapter [12, 24, 25].

13.2.3.1 Scalability

The scalability of a BIoT depends mainly on the scalability of the underlying BC, the IoT protocol, and the architecture design. Thus, the BC consensus mechanism impacts the Tx rate, which is used in general to evaluate the scalability of BCs. Since the consensus mechanism and a mining process are tightly-coupled, these algorithms are affected by the underlying networking layer. Since the networking layer enables P2P communications between all BC nodes, the overall latency and Tx rate are impacted (cf. Figure 13.3).

Firstly, BC miners' communications for (i) synchronization and state transmissions, (ii) broadcasting a newly mined block, (iii) asking for lost Txs, or (iv) block/Tx validations cause delays and affect the scalability in case of networking instability, which can cause packet losses. Thus, a suitable BIoT has to take into account the number of consensus findings and mining-related Txs [26]. For example, un-successful miners in BTC affect the divergence of the latency of PoW miners in the BTC network, discriminated by the consensus mechanism. As a consequence, miners may be delayed in synchronizing themselves with the BC and their effort in mining new blocks might be lost due to the weak networking situation, i.e. by exceeding the blocktime or inability to broadcast their mined block to a greater portion of the BTC network on time.

Secondly, the BC type impacts the scalability of BIoT applications, since it defines a user participation level as a client or miner. One of the main reasons for the scalability difference (in terms of the Tx rate) between private, permissioned DLs, and public BCs is the consensus mechanisms and its computational complexity. Since permissioned BCs show a centralized authority deciding on mining participation, they do not need computationally expensive consensus mechanisms. Thus, usually a Proof-of-Authority (PoA), a Proof-of-Stake (POS), or a Byzantine Fault Tolerant (BFT)-like consensus mechanisms is used. A recent study performed on the detailed effects of consensus mechanisms on the IoT-based use cases can be found in [13].

Thirdly, the BC size and growth need to be considered for a scalable BIoT. BC size is the accumulation of all the data records stored and maintained on the main chain of a BC by its miners. When a new miner intends to join a BC it has to receive all these records which means that the new miner has to dedicate an equivalent storage space according to the size of that BC. For instance, the BTC size reaches almost 300 GB and the Ethereum BC size is over 1 TB as of June 2020. Considering that IoT devices can generate gigabytes of data in real-time, corresponding techniques need to be employed to limit a BC size growth, while still providing a trusted and tamper-proof data storage [27]. Moreover, the design of public BCs does not provide fast and cheap storage of large amounts of data. Hence, different approaches have to be considered to filter, normalize, and compress IoT data to reduce the data size [22]. In this context, the removal of older Txs from the "older" blocks or aggregating them have been recently proposed, too [14, 27].

Fourthly, the scalability of BIoT is related to the (i) employed IoT technology, (ii) transmission scheme, and (iii) processing tasks on IoT nodes. The Maximum Transmission Unit (MTU), available bandwidth, available air time, the Medium Access Control (MAC) protocol, the location of in-/outdoor Gateways, the computational complexity of cryptographic operations, and en-/decryptions determine IoT-related characteristics, which affect the scalability, i.e. the data size transmitted from IoT sensors to the BC.

Finally, the BIoT architecture itself affects the overall scalability depending on its design since edge, fog, or cloud oriented approaches facilitate different scalability levels. BC-related tasks operated by IoT devices, such as the signing of packets by IoT devices, impact computationally demanding, resource-constrained IoT devices [12, 24]. Thus, the location of BC clients and the complexity of cryptographic algorithms for hashing or signing Txs are important. In this regard, a BIoT architecture needs to take into account required computational and storage resources on different layers.

13.2.3.2 Security

With the increasing number of attacks on IoT networks, security measures and functionality are needed for BIoT architectures and include (i) Data Integrity, (ii) Trust, (iii) Regulators and Governments' Influence, (iv) Transparency, (v) Privacy, and (vi) Encryption and Cryptographic signatures (cf. Figure 13.3).

Several reasons can cause security concerns for IoT protocols, e.g. failure of devices, vandalism, and users (cf. Section 13.1.2). Thus, it is important to perform a health check of IoT devices before and while they are integrated with BCs. Thus, the implementation of automated security alerts based on Hardware Security Modules (HSM) or Physically Unclonable Functions (PUF)-based periodical IoT device identity verification determines an essential element of secure BIoT architectures.

Trust and reliability of IoT-generated data determines a vital security aspect of BIoT, since if data had already been corrupted before persisted into the BC, it will remain secure, but wrong. Thus, data integrity, immutability, and the identification of changes have to be provided by BIoT architectures, such that data reliability can be ensured throughout data collection, transmission, and storage. Furthermore, trusted distributed authentication and authorization services for IoT devices can be provided by BC-preserved SC-based IAM. For example, storing the hash of a device firmware and state will create a permanent record on the BC that can be used to verify the identity of the device and its settings, making sure they have not been manipulated [14].

A safe preserving of the private key or SK of IoT devices acting as BC nodes is crucial, since SK losses causes users to be unable to access his/her account and their funds. If the SK is stolen, the user can even lose all digital assets in the form of cryptocurrencies or tokens [20]. Thus, for security reasons special mechanisms are needed for storing the SK within the users' hardware, particularly not in the BIoT's fog or cloud. The responsibility of maintaining the SK's privacy at full is with IoT owners.

Data privacy in BIoT architectures can be required at the time, when data are being stored in the BC – by a careful decision on what shall be stored within the BC, especially in public BCs – but also within the entire IoT-to-BC path involving IoT data collection, communications, and the application domain. Taking into account the General Data Protection Regulation (GDPR) user privacy is crucial to be integrated into BIoT architectures.

The risk of using a public BC is related to the data privacy of information being stored within this BC. BCs being immutable by design cannot be deleted and being public, all data is persisted in clear text. Thus, any non-trusted participant may join the BIoT application as a user, e.g. SCT stakeholders. Diversely, private DLs can partially manipulate the data stored or delete the data. In such a case, their privacy is preserved by different policies, nonetheless, only at the cost of making very different premises on the underlying trust model, especially the centralization of the consensus mechanism.

13.2.3.3 Energy Efficiency

Energy efficiency and scalability of BIoT architectures are closely intertwined [4]. For instance, networking and communication layer characteristics and configurations, such as the MAC protocol of the IoT infrastructure deployed, affect energy efficiency. Specifically, the IoT transmission scheme plays a crucial role in providing energy-efficient communication. E.g. the lack or presence of Automatic Repeat reQuest in LoRaWAN affects the throughput and packet loss in the IoT network [24]. Energy efficiency is particularly relevant for the transmission of signature packets, since a lost signature affects Txs integrity and consequently the reliability of the architecture.

It is recommended to sign IoT packets inside an IoT node with a BC client Private Key, i.e. SK, which is the IoT device owner SK. Also, it is recommended to use the same cryptographic algorithm while hashing and signing, that is used by BC and BC clients to provide data integrity [12]. When data packets signed by the IoT device owners' SK, the signature indicates the owner and origin of that packets. However, if the IoT MAC protocol – due to the small MTU settings (e.g. in LoRaWAN with 55–200 Bytes) – requires data fragmentation on the IoT nodes, such that the collected data can fit in the packets, data aggregation will be needed on the BIoT edge or fog, or the BC client. Hence, the energy consumption of the encryption and cryptographic functions is a decisive element on the design of BIoT architectures, given that not all IoT devices are capable of performing computationally expensive operations.

The energy deficiency of consensus mechanisms determines a drawback of important BCs, like BTC. Typically, BTC miners need to try a large number of nonce values in case of PoW, where a nonce is the number a miner has to select such that the block hash will meet the block difficulty (number of 0's in the left most bits of a hash). In a block containing a Tx-set, miners look for a *golden nonce* that satisfies the target. BTC mining uses an estimated 61.76 TWh of electricity per year – more than many countries such as Switzerland and Czech Republic and ~0.28% of total global electricity consumption in 2019 [28]. If BTC was a country, it would be the 41st most-energy-demanding nation in the world. This information on BTC energy consumption is a clear indication of how a consensus mechanism affects energy consumption. Thus, a BIoT architecture must consider this metric with a high priority.

13.2.3.4 Manageability

The flexiblity of a BIoT architecture is essential to adapt its settings with the changes of the underlying infrastructure. BIoT applications benefit from adopting communication settings of IoT devices based on network congestion, such that IoT devices can connect to a new gateway, thus, providing a more reliable IoT-to-BC path. Such a modification demands software-defined management of the BIoT architecture. Thus, it is vital to provide controller units and adjustable components to facilitate flexible and configurable BIoT architectures.

13.3 BIoT Architectures

Based on the decentralized and distributed BIoT applications and protocols developed [16, 24, 25, 29–33], these practical demands of BIoT use cases

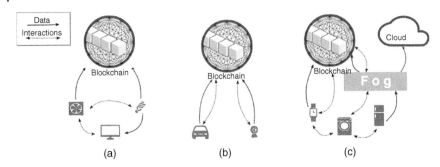

Figure 13.4 BC and IoT integration models [22]. (a) IoT–IoT, (b) IoT–Blockchain, and (c) hybrid approach. Source: Reyna et al. [22]. Licensed under CC BY-4.0.

(cf. Section 13.2.3) are now taken into account, to analyze, design, and implement BIoT architectures, which have to encompass necessary functionality to provide security, scalability, and energy-efficiency.

BIoT depends on the communication between the underlying IoT infrastructure and BC. IoT devices can generate massive amounts of data in real-time, and this can cause network congestion, when a large number of IoT devices stream data at the same time [34]. While BIoT can be instantiated in different ways, such as establishing IoT devices via fog computing as a layer between the cloud and edge, main BIoT architectures are relying on the models as of [22] differentiated into (i) IoT–IoT models, (ii) IoT–BC models, and (iii) hybrid models (cf. Figure 13.4). BIoT architectures are designed based on one or a combination of these models.

The **IoT–IoT Model** operates off-chain, storing data within databases, while only a proof of data integrity, i.e. data hashes, are stored in the BC. Thus, data storage demands are minimized and IoT interactions occur without involving the BC. This model is employed, when a reliable data channel between IoT devices exists and it is protected with IoT security measures, and IoT interactions occur with low latency.

The **IoT–BC Model** stores all interactions in the BC. Thus, the autonomy of IoT devices increases, since IoT devices act as BC clients, where IoT-to-BC transactions are triggered. By collecting all IoT data within immutable records of transactions, IoT-BC offers details of all interactions in the BIoT platform.

For the **Hybrid Model** a set of IoT transactions will be stored in the BC, and the remainder will be transmitted throughout the IoT network without involving the BC. The challenging part is to optimally categorize transactions, either in advance or on the fly, such that BIoT applications leverage the benefits of BCs and real-time IoT interactions. Hence, the hybrid model employs fog and cloud-based computing architectures to benefit from these processing units.

13.3.1   Cloud, Fog, and Edge-Based Architectures

In cloud-based architectures data collected by the IoT device layer (i.e. edge layer) is forwarded without further processing directly to the cloud through IoT gateways. Thus, access to cloud servers becomes a SPoF. If an IoT device connected to the cloud or central server is broken [2], IoT devices may be compromised. Moreover, the physical distance between cloud data centers and IoT devices can cause delay in data transfer. Hence, Quality-of-Service (QoS) is negatively impacted for time-critical applications [34]. To support distributed, low latency, and QoS-aware applications in IoT, cloud-based BIoT architectures have evolved with optimizing fog and edge computing approaches.

Since fog and edge computing-based architectures support BIoT by shifting parts of processing tasks from cloud servers to the network edge [34], they depend on cloud servers and services [2]. Edge computing refers to moving computational resources to the edge of the network, where IoT devices are located. However, resource-constrained devices at the edge do not support strong computational operations [7, 14, 17]. Thus, fog computing emerged as a subset of edge computing [2], offering an intermediate layer between the edge and the cloud. Fog-based architectures facilitate computational, storage, or network-intensive BIoT applications. Fog devices are considered distributed computing instances of the architecture deployed across the edge network.

13.3.2   Software-Defined Architectures

Deploying Software-defined Networking (SDN) enhances the performance of BIoT architectures. For instance, the SDN-based BIoT architecture [35] is based on a BC-based, distributed cloud architecture with SDN enabling controller fog devices at the edge of the network (cf. Figure 13.5). SDN architectures expand across three layered structures consisting of the (i) device or edge layer, (ii) fog layer, and (iii) cloud layer.

The device layer lies at the edge of the network. This layer monitors infrastructure environments and transmits unfiltered data to be processed to the fog layer. Hence, the device layer requires listening and transmission services to collect data from a monitored environment and passes it to the fog layer. The fog layer covers a community consisting of end points and carries out data analysis and service delivery promptly. If needed, the results of data processed can be sent to the cloud layer. Here, the fog layer accesses the distributed cloud layer to utilize application service and storage or computational resources. The cloud layer provides monitoring and control on a widespread level, whereas the fog layer provides localization. Cloud and fog layers collaborate to materialize large scale event detection, long-term pattern recognition, and behavioral analysis through distributed computing and

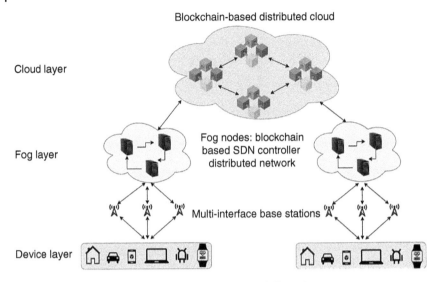

Figure 13.5 An SDN-enabled fog and cloud-based BIoT architecture.

storage. The distributed cloud layer can take up computing workload of fog nodes, if they become incapable of processing local data due to a lack of sufficient computing resources [35].

Sharma et al. [35] defines a distributed cloud-based BC. Moreover, a BC-based distributed SDN controller network operates within the fog layer. Each SDN controller is empowered by an analysis function of the flow rule and a packet routing function. Base Stations (BS) provide security in case of security attacks. Moreover, multi-interfaced BS at the edge of the network enable the adoption of new IoT protocols. A multi-layered BS consists of wireless gateways to collect all raw data coming from local IoT devices. Thus, BSes keep track of the traffic at the data plane and create user sessions. Furthermore, SDN programming interfaces are provided to network management operators.

13.3.3 A Potential Standard BIoT Architecture

A BIoT requires adaptations at BC and IoT infrastructures complemented with a BIoT architecture providing configurable components required for an efficient ecosystem. In this context and considering the challenges of BIoT (cf. Section 13.2.3) the Blockchain and Industrial IoT Integration (BIIT 1.0) architecture (cf. Figure 13.6) based on [12] is proposed as a potential standard BIoT architecture. To provide scalability, security, energy efficiency, and manageability, BIIT considers and supports a wider range of BIoT solutions depending on

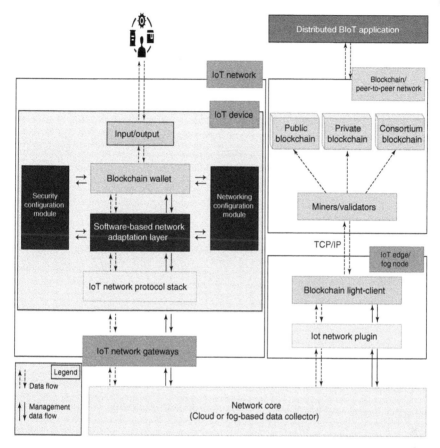

Figure 13.6 BIIT 1.0 architecture.

different BC and IoT specifications. As a proof-of-concept BIIT was partially implemented [12, 25].

For higher scalability, BIIT provides configurable BIoT communication schemes, such as for the IoT communication standards Thread [36], LoRa, or cellular networks. Moreover, Open Application Programming Interfaces (APIs) of BIIT allow for the implementation of logical components compatible to a wide variety of different BCs. BIIT BCs, where miners or validators deploy computing infrastructures, support a broad range of BC types, i.e. public, private, or consortium based, all with application oriented consensus mechanisms, e.g. PoS and BFT. BIIT enriches the edge and fog adaptation by specifying elements to smoothly integrate IoT with BCs. The management components defined in BIIT allow for the specification of efficient transmission schemes.

A BIIT BC Tx adaptation scheme on-the-fly adapts the transmission scheme to underlying networking interfaces, e.g. by employing a Tx fragmentation level in case network interfaces with low MTUs are used, such as for IEEE 802.15.4, to guarantee a high Tx throughout, transmission reliability, and energy efficiency. This adaptation is supported by the Software-based Network Adaptation Layer (cf. Figure 13.6). The fragmentation and aggregating of IoT data packets have been simulated in different scenarios [24] which proves the importance of such technical considerations in IoT-to-BC communications. Furthermore, the generic architecture of BIIT allows for the implementation of communication protocols adjusted to particular BC specifications allowing the programmer to derive BC-compatible protocol data units, e.g. with an appropriate Tx data format. Finally, BIIT considers a flexible edge, fog, and cloud-based service-and-resource management scheme for BIoT.

For a high security, BIIT complaint system has its BC Txs signed on the IoT device itself. Hence, the origin and authenticity of the data submitted to the BC can be verified at the later stage. Moreover, to allow for a required level of protocol security, BIIT appropriately handles security-specific Tx fields rooted in a given BC specification, such as the Ethereum's Tx nonce field, which protects against double-spending. BIIT suggests a set of configurable APIs that play a key role in the IoT-to-BC communication safety. These APIs shall establish configurable security settings via encryption protocols according to the BIoT use case requirements. Therefore, BIIT considers and offers a high security and flexibility at the same time.

BIIT does not define a framework for the implementation of a BIoT application, but it outlines most critical considerations of BIoT system's design and development in support of a realistic application. Thus, BIIT cannot fragment or aggregate IoT data automatically or it cannot run BC clients with specified details, but crucial requirements are derived.

In the security provision context, this generic architecture covers a wide range of functionalities related to authentication, encryption, data integrity, and authenticity. Authentication engine challenges IoT devices by asking for credentials before access to a specific resource may be granted. For example, a permissioned BC is equipped with authentication engine, Access Control Lists, and membership registers that allow authorized devices to read/write to the BC, excluding third party users explicitly. Furthermore, similar authentication engines may be placed at the edge, when only authorized devices may use the edge infrastructure to offload heavy processing toward edge with high processing capabilities. These authentication functions are embedded in the BC wallet on the IoT device and as light BC clients at the edge.

Data integrity is guaranteed by the message digest computed over chunks of data submitted through hashing functions, such as Ed25519, SHA2, and SHA3.

Additionally, data authenticity is established through digital signatures via public cryptography mechanisms using two mutual cryptography keys used, such as with the ECDSA. Therefore, BIIT components securely submit a Tx from an IoT device to the BC. Niya et al. [12] implements BC clients – according to BIIT's specific considerations – on LoRa IoT devices, which transmit data to a BC by first singing them with the user's BC address SK. In a BIIT-compliant approach IoT devices shall connect to the BC client which may be located on IoT infrastructure or even on the fog, and collect the user related information such as account address, balance, and the SK without user interaction, and via an automated mechanism.

The corresponding data integrity and authenticity functions on the IoT device are delivered by the security module. The node perception layer has to be secured as well against node tempering attacks to ensure that data of the environment remain intact, before they are eventually sealed in the BC.

Finally, the energy efficiency via BIIT is handled by the management components which have to be adjusted based on users preferred parameters, such as the (i) maintained security level, (ii) computational complexity, (iii) storage requirement, and (iv) power efficiency. These management decisions are executed in the networking configuration module.

13.4 Summary and Considerations

The use of BC and IoT in combination leads to a challenging, but rewarding Blockchain-IoT integration (BIoT). Based on BC and IoT basics the BIoT approach was outlined. Mainly six integration incentives were discussed in detail and relevant use cases for BIoT were added. Driven by these examples the integration challenges have been refined into specifically the social, operational, performance, technical, functional, and architectural fields, respectively. Since an interdisciplinary approach contains risks, dedicated metrics were introduced for the scalability, security, and energy efficiency to specifically determine those technical components, which impact those.

BCs form the cornerstone of the BIoT integration and have evolved in the last decade to offer a better scalable and more energy efficient approach. In addition, BIoT-integrated use cases drive such optimizations to circumvent the existing scalability concerns with BCs. Furthermore, software-defined BIoT management platforms offer the functionality to administrators to modify technical and security specifications of those BCs and IoT devices in use. Especially the pre-processing of IoT-to-BC traffic by those platforms can offer real-time monitoring and the respective transmission scheme helps.

Choice of a BC and IoT infrastructure goes hand in hand with the user and use case practical demands. Thus, an efficient BIoT architecture shall allow the flexible adaptation and replacement of different BCs and IoT infrastructures. In this regard, the overview presented on BIIT 1.0 shows a BIoT integration architecture, which enables a modular approach for integrating edge, fog, and cloud computing, all connected to a software-defined controlling core.

Bibliography

1 DBS Group Research. https://www.dbs.com.sg/treasures/aics/pdfController .page?pdfpath=/content/article/pdf/AIO/151102_insights_capitalising_on_ internet_of_things.pdf (accessed 21 April 2021).

2 Fernández-Caramés, T.M. and Fraga-Lamas, P. (2018).A review on the use of blockchain for the internet of things. *IEEE Access* 6: 32979–33001. https://doi.org/10.1109/ACCESS.2018.2842685.

3 Azure IoT Solution Accelerators. https://azure.microsoft.com/en-us/features/ iot-accelerators/ (Last visit: 6 November 2020).

4 Sanju, S., Sankaran, S., and Achuthan, K. (2018). Energy comparison of blockchain platforms for internet of things. *IEEE International Symposium on Smart Electronic Systems (iSES 2018)*, Hyderabad, India, pp. 235–238.

5 IBM. The Internet of Things Delivers The Data. AI Powers The Insights. https://www.ibm.com/uk-en/internet-of-things (Last visit: 6 November 2020).

6 Huh, S., Cho, S., and Kim, S. (2017). Managing IoT devices using blockchain platform. *International Conference on Advanced Communication Technology (ICACT 2019)*, Bongpyeong, South Korea, pp. 464–467.

7 Panarello, A., Tapas, N., Merlino, G. et al. (2018). Blockchain and IoT integration: a systematic survey. *Sensors* 18 (8): 2575. https://doi.org/10.3390/ s18082575.

8 Ethereum is a Global. Open-Source Platform for Decentralized Applications. https://www.ethereum.org (Last visit: 6 November 2020).

9 Croman, K., Decker, C., Eyal, I. et al. (2016). On scaling decentralized blockchains (A Position Paper). In: *Workshop on Financial Cryptography and Data Security (FC 2016)*, Lecture Notes in Computer Science (LNCS), vol. 9604 (ed. K. Rohloff, J. Clark, S. Meiklejohn et al.), 106–125. Berlin, Heidelberg: Springer.

10 The Things Network (TTN). https://www.thethingsnetwork.org/ (Last visit: 6 November 2020).

11 Bernstein, D.J., Duif, N., Lange, T. et al. (2012). High-speed high-security signatures. *Journal of Cryptographic Engineering* 2 (2): 77–89.

12 Rafati Niya, S., Schiller, E., Cepilov, I., and Stiller, B. (2020). Standardization of blockchain-based I2oT systems in the I4 era. *IEEE/IFIP Network Operations and Management Symposium (NOMS 2020)*, Budapest, Hungary, pp. 1–9.

13 Salimitari, M., Chatterjee, M., and Fallah, Y. (2020). A survey on consensus methods in blockchain for resource-constrained IoT networks. *Internet of Things*, Volume 11, p. 100212.

14 Maroufi, M., Abdolee, R., and Tazehkand, B.M. (2019). On the convergence of Blockchain and Internet-of-Things (IoT) technologies. *Journal of Strategic Innovation and Sustainability* 14 (1): 101–119.

15 Antonopoulos, A.M. and Wood, G. (2018). *Mastering Ethereum: Building Smart Contracts and DApps*. Sebastopol, CA: O'Reilly Media.

16 Rafati Niya, S., Jha, S.S., Bocek, T., and Stiller, B. (2018). Design and implementation of an automated and decentralized pollution monitoring system with blockchains, smart contracts, and LoRaWAN. In: *IEEE/IFIP Network Operations and Management Symposium (NOMS 2018)*, 1–4. Taipei, Taiwan. https://doi.org/10.1109/NOMS.2018.8406329.

17 Dai, H., Zheng, Z., and Zhang, Y. (2019). Blockchain for internet of things: a survey. *IEEE Internet of Things Journal* 6 (5): 8076–8094.

18 De Carli, A., Franco, M., Gassmann, A. et al. (2020). WeTrace – a privacy-preserving mobile COVID-19 tracing approach and application, arxiv. https://arxiv.org/abs/2004.08812 (Last visit: 6 November 2020).

19 Bocek, T., Rodrigues, B.B., Strasser, T., and Stiller, B. (2017). Blockchains everywhere - a use-case of blockchains in the pharma supply chain. *IFIP/IEEE Symposium on Integrated Network and Service Management (IM 2017)*, Lisbon, Portugal, pp. 772–777. https://doi.org/10.23919/INM.2017.7987376.

20 Stiller, B., Rafati Niya, S., and Grossenbacher, S. (2019). Application of blockchain technology in the swiss food value chain (foodchains project report), Zürich, Switzerland. https://owncloud.csg.uzh.ch/index.php/s/42GBmDrDbXqG27y (accessed 21 April 2021).

21 Banerjee, A. (2019). Chapter nine - blockchain with IoT: applications and use cases for a new paradigm of supply chain driving efficiency and cost. *Elsevier's Advances in Computers* 115: 259–292.

22 Reyna, A., Martín, C., Chen, J. et al. (2018). On blockchain and its integration with IoT. Challenges and opportunities. *Future Generation Computer Systems* 88: 173–190.

23 Sharma, P.K., Kumar, N., and Park, J.H. (2020). Blockchain technology toward green IoT: opportunities and challenges. *IEEE Network* 34 (4): 263–269. https://doi.org/10.1109/MNET.001.1900526.

24 Schiller, E., Rafati Niya, S., Surbeck, T., and Stiller, B. (2019). Scalable transport mechanisms for blockchain IoT applications. *IEEE 44th Conference on Local Computer Networks (LCN 2019)*, Osnabrück, Germany, pp. 34–41.

25 Rafati Niya, S., Schiller, E., Cepilov, Ile. et al. (2019). Adaptation of proof-of-stake-based blockchains for IoT data streams. *IEEE International*

Conference on Blockchain and Cryptocurrency (ICBC 2019), Seoul, South Korea, pp. 15–16.

26 Wan, L., Eyers, D., and Zhang, H. (2019). Evaluating the impact of network latency on the safety of blockchain transactions. *IEEE International Conference on Blockchain (Blockchain)*, Atlanta, Georgia, USA, pp. 194–201.

27 Rafati Niya, S., Maddaloni, F., Bocek, T., and Stiller, B. (2020). Toward scalable blockchains with transaction aggregation. *Symposium on Applied Computing (SAC 2020)*. Brno, Czech Republic: ACM, pp. 308–315. ISBN 9781450368667.

28 McCarthy, N. (2019). Bitcoin Devours More Electricity Than Switzerland. https://www.forbes.com/sites/niallmccarthy/2019/07/08/bitcoin-devours-more-electricity-than-switzerland-infographic/#6f2a0a3321c0 (Last visit: 6 November 2020).

29 Rafati Niya, S., Jeffrey, B., and Stiller, B. (2020). A Blockchain-based Platform for Self-sovereign IoT Identification. *IFI Technical Report No. 2020.04*. https://owncloud.csg.uzh.ch/index.php/s/cp4JrDMYsBATaXX (accessed 21 April 2021).

30 Rafati Niya, S., Schüpfer, F., Bocek, T., and Stiller, B. (2018). A peer-to-peer purchase and rental smart contract-based application (PuRSCA). In: *It-Information Technology* (ed. S. Conrad and P. Molitor), vol. 60, 307–320. Berlin, Boston, MA: De Gruyter Oldenbourg.

31 Schiller, E., Esati, E., Rafati Niya, S., and Stiller, B. (2020). Blockchain on MSP430 with IEEE 802.15.4. *IEEE 45th Conference on Local Computer Networks (LCN 2020)*, Sydney, Australia, pp. 345–348.

32 Rafati Niya, S., Beckmann, R., and Stiller, B. (2020). DLIT: a scalable distributed ledger for IoT data. *Second International Conference on Blockchain Computing and Applications (BCCA 2020)*, Izmir, Turkey, pp. 100–107.

33 Rafati Niya, S., Dordevic, D., Hurschler, M. et al. (2020). A Blockchain-based Supply Chain Tracing for the Swiss Dairy Use Case. *IFI Technical Report No. 2020.07*. https://owncloud.csg.uzh.ch/index.php/s/rH6sA25C9JegEHW (accessed 21 April 2021).

34 Tuli, S., Mahmud, R., Tuli, S., and Buyya, R. (2019). FogBus: a blockchain-based lightweight framework for edge and fog computing. *Journal of Systems and Software* 154: 22–36.

35 Sharma, P.K., Chen, M., and Park, J.H. (2018). A software defined fog node based distributed blockchain cloud architecture for IoT. *IEEE Access* 6: 115–124.

36 Thread Group. https://www.threadgroup.org/ (Last visit: 6 November 2020).

Index

a

AAI. *See* Authentication and authorization infrastructure (AAI)
ABNO. *See* Application based network orchestrator (ABNO)
Action space 157–158
Active queue management (AQM) 85
Admission control
  cross-slice congestion control problem 56
  state-of-the-art 55
  supervised learning (SL) 56
Adversarially learned anomaly detection (ALAD) 273
AE. *See* Auto encoder (AE)
AI. *See* Artificial intelligence (AI)
AI and anomaly detection
  autoencoders 271–272, 282
  darknets 283
  deep learning 267
  DNNs 268–269
  exploiting graph relationships 282–283
  GAN 272–273, 282
  methodology 267–268
  multiple sensors and variables 283
  point data and time series 277
  reinforcement learning 273–274

representation learning 270–271
research areas 275–277
semi-supervised and unsupervised approaches 277
AI-based optimal configuration 206–207
AI methods in data-intensive applications
  data-processing frameworks 200–203
  DRL algorithms 217
  management techniques 200
  mapping of AI models 216
  state-of-the-art (*see* State-of-the-art)
  traditional 200
ALAD. *See* Adversarially learned anomaly detection (ALAD)
Amazon EC2 virtual servers 202, 205
Amazon Web Services (AWS) IoT 323
ANNs. *See* Artificial neural networks (ANNs)
ANNs algorithms 210
ANNs models
  performance 211
Anomaly detection technology 278
Anomaly detector 278
Anomaly types 265–266
Apache Flink 202–203
Apache Mesos 202
Apache Spark 202, 204, 209

Communication Networks and Service Management in the Era of Artificial Intelligence and Machine Learning,
First Edition. Edited by Nur Zincir-Heywood, Marco Mellia, and Yixin Diao.
© 2021 The Institute of Electrical and Electronics Engineers, Inc. Published 2021 by John Wiley & Sons, Inc.

Apache Storm   200–201, 204
API. *See* Application programming
  interface (API)
Application based network orchestrator
  (ABNO)   87
Application domains   262
Application programming interface
  (API)   202
Application-specific Integrated Circuit
  (ASIC)   326
ARAE-AnoGAN   279
Arduino   326
ARIMA   212, 278
Artificial intelligence (AI)
  classical/symbolic AI models   19
  types   19
Artificial neural networks (ANNs)   47,
  205, 207, 209
ATmega 2560 (Arduino Mega) node
  325
Attributed graphs
  graph convolutional networks   186
  graph neural networks   186
Audio
  representation learning   270
Aurora   228
Authentication and authorization
  infrastructure (AAI)   327
AuTO
  Central System   234–235
  comparison targets   244
  current DRL systems   232
  daemon   253
  deep dive   247–249
  formulations and solutions, DRL
    235–239
  four-queue example   235
  grouped by scenarios   242–243
  homogeneous traffic   244–245
  limitations   252
  overview   232–233

peripheral system   233–234
problem identification   231–232
setting   243–244
spatially heterogeneous traffic
  245–246
system overhead   249–251
temporally and spatially
  heterogeneous traffic   246–247
Auto encoder (AE)   134, 271–272
Automated failure management   61
Automatic Repeat reQuest   334
Auto-scaling problem   214
Azure platform   278

b
Baseband units (BBUs)   44, 80
Bayesian optimization (BO)
  205–207
BC clients   321
BC size and growth   333
Beyond-5G (B5G) systems   73
BigDataBench benchmark   204
BIoT architectures
  applications and protocols   335
  cloud-based   337
  edge-based   337
  fog   337
  models   336
  potential standard   338–341
  SDN-based   337–338
BIoT challenges
  categories of risks   330
  energy efficiency   332, 334–335
  functional risks   330
  manageability   335
  metrics   331
  operational risks   330
  performance and architecture risks
    330–331
  scalability of   332–333
  security   331–334

social risks 330
technical risks 330
BIoT security 331
Bitcoin (BTC) 321
Black-box analysis 204
Blockchain and Industrial IoT
   Integration (BIIT 1.0) architecture
   338–340
Blockchain-IoT (BIoT) integration
   architectures 322
   autonomic interaction 327
   BIoT Challenges 329–335
   clients 325
   communications 325
   costs 322
   cryptographic functions 325–326
   decentralized security 327
   healthcare 329
   HPS and signatures per second 325,
      326
   IAM 327
   interoperability 328
   potentials 326–328
   reliability and resiliency 327
   scalability and energy-efficiency 326
   SCT 329
   Smart City 328
   Smart Grid 328
   Smart Home 328
   traceability 327
   TTP 322–323
Blockchains (BC)
   applications and architectures
      321–322
   BIoT (*see* Blockchain-IoT (BIoT)
      integration)
   characteristics 323
   decentralization of 321, 322
   distributed data storage systems 321
   DLs 324
   PoW 324

public 323–324
Tx rate 324
Bluetooth 324
BO. *See* Bayesian optimization (BO)
Byzantine Fault Tolerant (BFT)-like
   consensus mechanisms 332

C
Cassandra 202
Cellular networks 324
centralized client-server (C/S) paradigm
   321
Chains of security functions
   deployment and configuration 292
   formal verification techniques 290
   OpenFlow protocol 289
   SDN supervisory control 292
   security function modelling 291
   VNF 291
Channel Boosted and Residual learning
   classifier 269
Classification and regression tree
   (CART) 206, 210
Closed loop automation management
   platform (CLAMP) 45
Cloud 337
Cognitive Radio Vehicular Adhoc
   Network (CRAVENET) 138
Collective anomalies 265–266
Command line interfaces (CLI) 10
Common open policy service (COPS)
   10
Computational hardness 177
Congestion control 228
Context-based clustering 80–81
Contextual anomalies 265–266
Controlled delay active queue
   management (CODEL) 85
Convolutional neural networks (CNNs)
   209
   deep CNNs 269

Convolutional neural networks (CNNs)
(*contd.*)
  detection of anomalies   269
  hybrid solution   269
  performance anomaly diagnosis   209
Cost function   21
CRAN (for R)   278
Credit assignment   21
Cross-entropy   211
Cyber-security   274
  context of   275
  DDoS attacks   261
  detection of port scans   261
  functions   77

d
Datacenter TOs   229–230
Data collection and monitoring protocols
  IPFIX protocol   6–7
  IPPM   7–8
  routing protocols   8–9
  SNMP protocol family   5
  Syslog protocol   5–6
Data-driven algorithm   178
  networking research work   190
  tackling algorithmic problems
    188–190
Data-driven network optimization
  learning facility (controller)
    placements   191
  ML/AI-based approaches   181
  network reconfigurations   192
  network verification   191
  optimization pipeline   182
Data integrity
  and authenticity functions   341
  BC and BC clients   335
  guaranteed   340–341
  and origin of Tx   326
  proof of   329, 336
  security   333, 334, 340

  standard configuration protocols   9
  and transmission reliability   325
Data-intensive applications
  Apache Flink   202–203
  Apache Spark   202
  Apache Storm   200–201
  characteristics   200, 201
  Hadoop MapReduce   201
Data leakage prevention (DLP)   289
Data mining   267
Data mining algorithms   266
Data plane programmability
  intra-data center scenarios   76
  P4 switch   77
  SDN phase   77
Data plane slicing
  AQM   85
  CODEL   85
  hypervisor-based virtualization   86
  HyperV proposition   86
Data production and collection
  malicious data   193
  monitoring and storing data   193
Data type of instances   264–265
Deep dive
  optimizing long flows using DRL
    248–250
  optimizing MLFQ thresholds
    247–248
Deep learning (DL) algorithms
  128–129
DeepLog   269
Deep neural networks (DNNs)   205
  classifying images and voice   268
  CNNs   269
  framework   269
  input and the output layers   268
  KDD-NSL dataset   269
  LTSM   268–269
  multiple architectures   268
  users' feedback   269

Deep packet inspection (DPI) 289
Deep reinforcement learning (DRL) 85,
   134
  agent 226
  algorithm 217, 231
  AuTO 227
  environment 226
  flow scheduling problem 230
  formulation 230–231
  function approximation 226–227
  machine learning methods 226
  REINFORCE algorithm 227
Denial of Service (DoS) 293, 325
Dense wavelength division multiplexing
  (DWDM) 78
Density-based change-point detection
  278
Design of experiments (DOE) approach
  203
Deterministic policy gradient (DPG)
  236
Differentiated services code point
  (DSCP) 233
Directed acyclic graphs (DAGs) 203
Discrete label 266
Distributed denial-of-service (DDoS)
  attacks 261
Distributed ledger (DL) 321
DNNs 212, 213. *See* Deep neural
  networks (DNNs)
Double deep Q-learning network
  (DDQN) 49
DRL. *See* Deep reinforcement learning
  (DRL)
DRL agents 234
DRL algorithms in CS
  optimizing long flows 239
  optimizing MLFQ thresholds
  235–239
DWDM. *See* Dense wavelength division
  multiplexing (DWDM)

Dynamic Host Configuration Protocol
  (DHCP) 38
Dynamic scenario
  accumulated reward 168
  MILP model execution time 167
Dynamic slice orchestration 61

e

Ed25519 325, 326
Edge layer 337
EGADS 278
Elasticity 211
ELKI data mining framework 279
Elliptic Curve Digital Signature
  Algorithm (ECDSA) 325
ELYSIUM 212
Encryption and Cryptographic
  signatures 333
End-to-end learning
  mathematical formulation-based
    representations 187
  sequence-based models 188
Energy efficiency
  BIoT 334–335
Enforcement module (EM) 233–234,
  240–241
Enhanced Interior Gateway Routing
  Protocol (EIGRP) 8
Ethereum 330
Ethernet 324

f

Facebook Hadoop system 209
f-AnonGAN 279
FC. *See* Fog computing (FC)
Federated learning 141–142
Federation
  centralized approach 88
  distributed approach 89
  multi-domain orchestration 88
  SDOs 88

Federation paradigm   72
Feedforward NN(FNN)   54
5G management state-of-the-art
  data plane slicing and programmable
    traffic management   85–86
  federation   88–89
  RAN resource management (*see* RAN
    resource management)
  service orchestration   83–85
  wavelength allocation   86–88
5G networks
  AI/ML use cases   102
  degree of programmability   117
  diversity of services   70
  end-to-end (E2E) network slicing
    101
  federation   92
  general configuration   112
  high degree of complexity   118
  isolated networks   70
  KPIs   70
  MANO systems   101
  ML-based mechanisms   118
  mobile communications   69
  model training and evaluation   113
  NWDAF   92
  optimal placement calculation   113
  pre-trained AI/ML models   90
  QoS   102
  RAN resources   91
  resource orchestration algorithms   89
  SFC request generation   112
  slicing and prioritization   71
  SLO requirements   102
  training data generation   113
  use cases   91, 106
  vertical slicer   90
  VNFs   103
  VNF scaling decisions   90
Flink applications   203
Flow-based detection of attacks

appname attribute   294
port scanning attacks   295
FlowVisor   40
Fog   337
Fog computing (FC)   149
  chain requirements   148
  dynamic scenario   167–169
  environment implementation   162
  fog–cloud infrastructure   162
  gym-fog environment configuration
    164–165
  hardware configuration   163
  ILP-based methods   170
  infrastructures   170
  micro-service patterns   150–151
  novel paradigm   147
  OpenAi gym environment structure
    163
  reinforcement learning (RL)   151–152
  resource allocation   152–153
  resource allocation domain   170
  resource provisioning   149–150
  SFC   150
  SFC allocation (*see* SFC allocation)
  state-of-the-art   169
  static scenario   165–167
Formal verification techniques
  DoS   293
  firewall policies   294
  orchestration of security chains   292
  SMT   293
F-score   206, 207

g

GAN-DDQN. *See* GAN-powered deep
    distributional Q network
    (GAN-DDQN)
GANnomaly   279
GAN-powered deep distributional Q
    network (GAN-DDQN)   59
Gaussian processes (GPs)   205

GCNs. *See* Graph convolutional
networks (GCNs)
GDPR. *See* General data protection
regulation (GDPR)
General data protection regulation
(GDPR) 334
Generative adversarial networks (GAN)
272–274
Google 278
GPs 206
Graph-based modeling 179
Graph convolutional networks (GCNs)
49, 186
Graph neural networks (GNNs) 179,
186
Graph representations
graph structure 184
latent space models 185
node and graph features 185
spectral methods 184
Graph structuring 187
Grep 204

h
Hadoop distributed file system (HDFS)
201, 202
Hadoop MapReduce 201
Hardware security modules (HSM) 333
Hashes per Second (HPS) 325
HBase 202
Healthcare, IoT sensors 329
Hill climbing algorithm 204
Hive 202
Homogeneous traffic 244–245
Huawei Tecal RH1288 V2 servers 232
Hybrid model, IoT transactions 336

i
IBM 204
IBM Watson 323

Identity and access management (IAM)
327
Imitation learning 140–141
Information theory 267
Integer linear programming (ILP) 84,
148
Integrated platforms
agent modules 11
database modules 11
media gateway 12
monitoring architecture 11
proxy modules 11
Intel i7 CPU (Central Processing Unit)
325
Intelligent data-intensive software
systems 199–
Intelligent monitoring 60
International Organization for
Standardization's Open Systems
Interconnection (ISO/OSI) 3
Internet engineering task force (IETF)
9
Internet-of-Things (IoT)
adapters 322
adoption estimation 321, 322
applications 321
architectures 321
Azure IoT solution accelerators 323
basis 324–325
BIoT (*see* Blockchain-IoT (BIoT)
integration)
data collection and operations 321
devices 321
security flaws 323
Internet Protocol Flow Information
Export (IPFIX) protocol
components 6
NetFlow 7
Internet protocol performance metrics
(IPPM) 7–8

Intrusion detection system (IDS)   36, 289

Intrusion prevention system (IPS)   10

IoT. *See* Internet-of-Things (IoT)

IoT–BC model   336

IoT–IoT model   336

j

JobManager   203

Job scheduling   228

k

Kalman filters   278

Keras/TensorFlow   232

kernel-based change-point detection   278

Key performance indicators (KPIs)   70

K-means clustering   126–127

K-nearest neighbors (KNN)   125

l

Labels   266

Last-hop top of rack (TOR) switches   230

Latent space models   185

Latin hypercube sampling   204

Learning algorithms

  components   20

  cost function   21

  credit assignment   21

  reinforcement learning   23–24

  supervised learning   21–22

  unsupervised learning   22–23

Link virtualization

  higher layers   42

  physical layer partitioning   41–42

  technologies   41

Local area network (LAN)   37, 324

Logical router (LR)   38

Long-range Wide Area Networks (LoRaWAN)   324, 325

Long short-term memory (LSTM)   206, 268–269, 269

LoRa IoT devices   341

LoRaWAN   334. *See* Long-range Wide Area Networks (LoRaWAN)

Low-power wide area networks (LP-WAN)   324

lRLA   242

LSTM. *See* Long short-term memory (LSTM)

LSTM networks   208, 209

Luminol   278

m

Machine learning (ML)   19, 267

  data driven   20

  facility location problems   180

  testing policy compliance   180

  traffic optimizations (TOs)   224

  virtual network embedding   180

MAD-GAN   279

Manageability, BIoT   3355

Management and orchestration (MANO) system   85, 101

  domain MANO   116

  framework   15

  network slicing   44–46

  NFV   15

  VIM   45

MapReduce   209

MapReduce job   204

MapReduce performance   204

Markov decision process (MDP)   51, 132, 214

Massive machine type communications (mMTCs)   123

Mean squared error (MSE)   109

Microsoft   278

Miners   321

Mixed-integer linear programming (MILP)   148

ML. *See* Machine learning (ML)
ML/AI-based algorithms
  certain probability   194
ML-based algorithms
  federated learning   141–142
  imitation learning   140–141
  quantum   142
  transfer learning   140
ML-based QoE
  assessment and management
    104–105
  costs for deploying ML   107
  data quality and granularity   107
  estimation accuracy   110
  estimation and management
    106–107
  feedback control loops   108
  5G architecture   108
  identification of relevant features
    107
  methodology   108–109
  MSE   109
  networking context   103–104
  proactive VNF deployment   110–111
  trade-off between accuracy and costs
    108
  virtualized networks   104
  VNF placement problem   111
ML-enabled resource allocation
  ML techniques   130
ML techniques
  resource allocation   153
Model-based RL   215
Monitoring module (MM)   234, 240
MRTuner   204
Multi-access edge computing (MEC)
  110–111
Multilayer perceptron (MLP)   209
Multi-level feedback queuing, PS   233,
  234

Multiple level feedback queue (MLFQ)
  thresholds
  AuTO   235
  average FCT   247, 248
  capabilities of DNN   236
  DCSP field   235
  DDPG   237
  deep stochastic and deterministic
    policies   236
  DPG   236
  DRL algorithm   238–230
  DRL formulation   237–238
  flows scheduling   235
  p99 FCT   247, 249
  Q-learning   236
  regular PG algorithm   236
  REINFORCE   236
  sRLA *vs* optimal thresholds   247, 248
Multiple machine learning   266
Multiple sensors and variables
Multiprotocol Label Switching (MPLS)
  9
Multi-service, multi-domain
  interconnect
  components   116
  domain MANO   116
  key service features   115
  optimal allocation   117
  QoE manager   117
  QoS parameters   115
  unsupervised learning   115

n

Nature of data   264
NavieBayes workloads   204
Near-RT RAN Intelligent Controller
  (Near-RT RIC)   75
Network address translation (NAT)   38
Network address translator   36
Network algorithms   178

Network and service management
 AI/ML techniques   24
 control loop   4
 fault management   25
 human experts   26
 ISO/OSI   3
 NFV and SDN technologies   25
 NOC   4
 overview, overall process   4
 security management   24
 TCP/IP   3
Network anomalies
 AI and anomaly detection   267–277
 anomaly detection algorithm   261
 application domains   262
 classic approaches   266–267
 complex anomalies   282
 cyber-security   261–262
 definitions   262
 macro-categories   265
 methods   261
 problem characteristics   264–266
 production-ready tools   277–279, 280
 research alternatives   279–280
 research areas   264
 reviewed tools   280–281
 taxonomy   262, 263
 trends   262
Network-as-a-Service (NaaS)   36
Network configuration protocol
 integrated platforms   10–12
 NETCONF protocol   9–10
 proprietary configuration   10
Network data analytics function
  (NWDAF)   92
Network data management
 components   78
 data ingestion module   79
 data processing/analytics module   79
 open-source frameworks   79
Network function (NF)   38

Network functions virtualization (NFV)
  36, 72, 101
 functionalities   75
 interoperability of solutions   15
 MANO   15
 VIM   76
Network Function Virtualization
  infrastructure (NFVI)   76
Network interface cards (NICs)   230
Network management algorithms   177
Network Operation Center (NOC)   4
Network slicing
 admission control   55–56
 baseband processing   44
 end-to-end (E2E) VN   43
 examples   43
 MANO   44–46
 resource allocation   56–59
Network virtualization (NV)
 AI/ML techniques   37
 architectures   35
 automated failure management   61
 dynamic slice orchestration   61
 IDS   36
 initial orchestration and configuration
   37
 intelligent monitoring   60
 network slices   36
 network slicing   43–44
 NF   38
 NFV   36
 placement   49–52
 private and secure tunnels   38
 QoS   35
 reactive human-in-the loop
   management   59
 resource partitioning   38–40
 seamless operation and maintenance
   60
 securing machine learning   62–63

sensitivity to heterogeneous hardware
62
Neural combinatorial optimization
(NCO) 51
NFV orchestrator (NFVO) 76, 83
NICs. *See* Network interface cards (NICs)
Non-graph-related problems 192
Non-RT RANIntelligent Controller
(Non-RT RIC) 74–75
Novelty and outlier detection library
279
NV. *See* Network virtualization (NV)

o

OFDMA. *See* Orthogonal
frequency-division multiple
access (OFDMA)
OpenAi gym environment structure
163
Open network automation platform
(ONAP) 45
Open shortest path first (OSPF) 8
Operational/capital expenditure
(OPEX/CAPEX) 73
Optical transport network (OTN) 78
Optimization problems
adjacency matrix 183
combinatorial optimization 183
graph 182, 183
learned algorithms *vs* mathematical
problem formulation 183
state-of-the-art technology 195
Optimizing long flows using DRL
load balancing 248, 249
lRLA 248
PG algorithm 248–249
Optimizing security chains
aggregate switches 310
capacity constraints 311
constraints 310
destination aggregates 309

merge chains 309
merge functions 309
Pyretic's operator 308
O-RAN-compliant cloud unit (O-CU)
73
Orchestration of security chains
different steps 298
elementary security rules 303
filtering traffic 305
Markov chain 300, 301
network metrics 300
network operators 301
Pokemon Go Android application
301
proposed method 297
RDAP 299
rule-based approach 302
SDN infrastructures 296, 297
security properties 306
security rules 304
states and transitions 299
Orchestration of security functions
artificial intelligence 316
learning and verification techniques
315
SDN infrastructures 316
Orthogonal frequency-division multiple
access (OFDMA) 41

p

Packet classification 228
Packet Data Convergence Protocol
(PDCP) 44
Packet routing 306
ParamILS 204
Peer-to-Peer (P2P) network protocol
321
Performance anomaly detection
AI approaches 208–210
ANNs-based 210–211
cloud systems 207

Performance anomaly detection (*contd.*)
  data-intensive technologies 207
  generic 207
  observation period 207
  traditional approaches 208
Performance evaluation
  accuracy of the generated chains
    313–314
  complexity of security chains 312
  deploying security chains 314–315
  response times 313
Peripheral system
  daemon process 239
  enforcement module 233–234,
    240–241
  monitoring module 234, 240
Personal Computer 325
p99 FCT 245, 246, 247, 249
Point data 264
Points of Presence (PoP) 39
Pointwise anomalies 265
Pokemon Go Android application 301
Policy Control Function (PCF) 44
Policy Enforcement Point (PEP) 10
Post-decision state (PDS) 215
PoW. *See* Proof-of-Work (PoW)
Power control
  Q-values 132
  state-of-the-art 131–132
  wireless networks 131
Privacy 333
Proactive VNF deployment
  challenges and use cases 112
  feature importance 113
  generalizability 114
  model training and evaluation
    112–113
  modern communication networks
    110
  prediction horizon 114
  training set size 113

VNF placement problem 111–112
Problem characteristics, anomaly
  detection 264–266
Process learning techniques
  heterogeneous datasets 298
Production-Ready Tools 277–278,
    281–282
Programmable optical switches
  OpenROADM YANG models 78
  ROADM 77
Proof-of-Authority (PoA) 332
Proof-of-Stake (POS) 332
Proof-of-Work (PoW) 321, 322, 324, 326
Prophet 278
Public BCs 323–324
Public Key (PK) 326
PyPI (for Python) 278
Python outlier detection (PyOD)
    279–280
Python programming language 253
PyTorch 232

q
Q-function 216
Q-learning agent
  exploitation 161
  exploration 161
  gym-fog environment 162
Q-learning algorithm 214
Quality of service (QoS) 35, 337
Quantized LAS (QLAS) 244
Quantized SJF (QSJF) 244
Quantum machine learning 142

r
Radio access network (RAN) 38, 73–75
Radio intelligent controller (RIC) 73
Radio link control (RLC) 44
Radio resource management (RRM)
    123
Random early detection (RED) 85
Randomized convex search (RCS) 204

RAN resource management
context-based clustering 80–81
Q-learning 81
RAN virtualization and management
near-RT RIC 75
non-RT RIC 74–75
O-RAN architecture 73, 74
RIC 73
SMO 74
Raspberry Pi 3 (RPI) 325, 326
Ray 232
Reconfigurable add drop multiplexer
(ROADM) 77
Recurrent neural network (RNN) 48,
268
Registration data access protocol (RDAP)
299
Regression algorithms 126
Regression trees 206
Reinforcement learning 274–275
model 23
spectrum of applications 24
Reinforcement learning (RL) 206
multi-armed bandit 128
Q-learning 127
stateless Q-learning 128
Relationship, instances 264
Remember patterns 268
Remote procedure call (RPC) 232
Remote radio unit (RRU) 57
Representation learning 270–271
Research alternatives 279–280
Research areas 263
Resilient distributed dataset (RDD) 202
Resource allocation
address resource management 58
E2E network 56
GAN-DDQN 59
RRU 57
slice resource allocation 58
state-of-the-art for ML-based 57

Resource partitioning
architecture 39
LR 38
PoP 39
VRF 39
Resource provisioning
centralized infrastructures 150
high-level view 149
Reward function
fog–cloud infrastructure 158
gym-fog environment 159
micro-service Ratio 160
MILP model 158
RIT. *See* Round trip time (RTT)
RL algorithms 215
RL-based auto-scaling policies 214–216
RL methods
resource allocation 154–155
Round trip time (RTT) 265
Routing Information Protocol (RIP) 8
Routing problems 178
Routing protocols
MPLS 9
OSPF 8
RIP 8
R package 278
RPI. *See* Raspberry Pi 3 (RPI)

S
Satisfiability modulo theory (SMT) 293
Scalability
AuTO's 233
BIoT 332–333
Scalable unsupervised outlier detection
(SUOD) 279
Scaling
FNN 54
proactive scaling leverages 52
state-of-the-art for ML-based scaling
53
VNFIs 52

Scaling techniques
  AI approaches   214
  traditional approaches   213
Scheduling
  DRL   134
  Markov decision process (MDP)   132
  ML-based   133
  wireless networks   132
Scikit-learn project implements   279
SCs. *See* Smart contracts (SCs)
SDN. *See* Software defined networking (SDN)
SDN-based BIoT architecture   337–338
SDN controllers   295–296
Seamless operation and maintenance   60
Seasonal Hybrid Extreme Studentized Deviate (S-H-ESD) test   278
Secret Key (SK)   326
Securing machine learning   62–63
Security, BIoT   327, 331–334
Service Data Adaptation Protocol (SDAP)   44
Service function chain (SFC)   36, 105, 147, 150
Service level agreements (SLAs)   72
Service level objective (SLO)   214
Service management and orchestration (SMO)   73, 74
Service orchestration
  5G-PPP network architecture   84
  ILP   84
  NFVO   83
  VNFs   83
Session Management Function (SMF)   44
SFC allocation
  action space   157–158
  MILP model   155
  observation space   156–157
  overall system cost   156

Q-learning agent   161–162
reward function   158–161
SHA-256   325, 326
SHA-512   325
Shadowing freedom and consistency
  guarantee shadowing freedom   307
  security function   306
  structural correctness properties   307
  symbolic model checking   308
  toy security chain in pyretic   308
Short flow Reinforcement Learning Agent (RLA) (sRLA)   234
Simple Network Management Protocol (SNMP) protocol family   5
Smart City ecosystem   328
Smart contracts (SCs)   327
Smart Grid   328
Smart Home
  IoT sensors and processing   328
SMT. *See* Satisfiability modulo theory (SMT)
Social risks   330
Softmax transfer function   210
Software defined networking (SDN)   36, 101, 289
  architecture   13
  BIoT architecture   337–338
  control plane   12
  datapath   13
  traffic processing functions   13
Sort   204
Spark   202
Spark tasks   206, 207
Spatially heterogeneous traffic   245–246
Spectrum allocation
  allocating resources   136
  control spectrum resources   139
  CRAVENET   138
  Machine learning-based   140
  sRLA   241–242

Standard Development Organizations
(SDOs) 88
State-of-the-art (SoA) 72
  fog computing (FC) 169
  load prediction
    AI approaches 212–213
    overview 211–212
    traditional approaches 212
  optimal configuration
    AI approaches 204–206
    automated parameter tuning
      methods 203
    BO 206–207
    traditional approaches 203–204
  performance anomaly detection
    AI approaches 208–210
    ANNs-based 210–211
    cloud systems 207
    data-intensive technologies 207
    generic 207
    observation period 207
    traditional approaches 208
  RL-based auto-scaling policies
    214–216
  scaling techniques
    AI approaches 214
    traditional approaches 213
  traffic optimization 231–239
Statistical anomaly detection 267
Streaming analytics and AI 278
Streaming dataflows 203
Streams 203
Subtitles
  representation learning 270
SUOD. *See* Scalable unsupervised outlier
    detection (SUOD)
Supervised learning
  If-Then-Else form 21
  knowledge extraction 22
  model 21
Supervised ML

KNN 125
logistic regression 125
regression algorithms 125–126
SVM 125
techniques 124
Supervised techniques 266
Supply chain tracing (SCT) 323
  BCs and IoT 329
Support vector machine (SVM) 125,
    205, 207, 208
Support vector regression (SVR) 205,
    212
Surprisal metric 270
SVM algorithm 209
Synthesizing security chains
  Markov chain 301
Syslog protocol 5
System overhead, AuTO module
  CS response latency
    computation overhead of DNN
      249, 250
    scaling short flows 250
    traces from four runs 250
  CS scalability 251
  PS overhead 251

t
Tachyon 202
TaskManagers 203
Telemanom 279
TelosB 326
TelosB (MSP430) 325
Temporally and spatially heterogeneous
    traffic 246–247
TensorFlow 279
The Things Network (TTN) 325
Third Generation Partnership Project
    (3GPP) 10, 71
Time division multiplexing (TDM) 41
Time-series 264, 281

TLS. *See* Transmission level security (TLS)
Topology and orchestration specification for cloud applications (TOSCA) 45
TOs. *See* Traffic optimizations (TOs)
TOSCA. *See* Topology and orchestration specification for cloud applications (TOSCA)
Traceability 327
Traffic optimizations (TOs) 223–255
  agent 224
  AuTO 224–225
  comparison targets 244
  computing overhead 252
  in Datacenter 229–230
  deep dive 247–251
  designing 223
  DRL (*see* Deep reinforcement learning (DRL))
  effectiveness 224
  homogeneous traffic 244–245
  implementation
    CS, RL agents 241–242
    peripheral system 239–241
  machine learning (ML) 224
  ML to networks 227–228
  parameter-environment mismatches 223
  process 224
  RL techniques 224
  setting 243–244
  spatially heterogeneous traffic 245–246
  State-of-the-Art (*see* State-of-the-Art)
  temporally and spatially heterogeneous traffic 246–247
  unseen environments 251–252
Traffic prediction 228
Transactions (Txs) 321, 322
Transfer learning 140

Transmission Control Protocol/Internet Protocol (TCP/IP) 3
Transmission level security (TLS) 6
Transparency
  security 333
Trust and reliability of IoT-generated data 333, 334
Trusted third parties (TTP) 322–323, 327
TTP. *See* Trusted third parties (TTP)
Twitter 278

u
UDM. *See* Unified data management (UDM)
Unified data management (UDM) 44
Universal Plug and Play (UPnP) 38
Unsupervised learning
  clustering 22
  model 22
Unsupervised ML
  clustering techniques 126–127
  hierarchical clustering 127
  soft clustering techniques 127
Unsupervised techniques 281
User association
  blackMachine-learning based 135
  cellular networks 134
  machine learning-based 137
  problem 136
  QoS specifications 136
User Datagram Protocol (UDP) 6
User plane function (UPF) 44

v
Validators 321
Verifying correctness of chains
  packet routing 306
Video, representation learning 270
Virtual infrastructure manager (VIM) 76

Virtualization   35
Virtualized Infrastructure Manager
   (VIM)   45
Virtualized network function manager
   (VNFM)   76
Virtualized network functions (VNFs)
   103
Virtual Local Area Networks (VLANs)
   37
Virtual machines (VMs)   212
Virtual network embedding (VNE)   179
   ANNs   47
   dynamic resource allocation   46
   heuristic algorithms   46
   RNNs   48
   state-of-the-art   47
Virtual network functions (VNFs)   36
   physical network equipment   40
   unikernels   40
   virtual router (vRouter)   40
Virtual private networks (VPNs)   87
Virtual Routing and Forwarding (VRF)
   39
VNE. *See* Virtual network embedding
   (VNE)
vrAIn: AI-assisted resource orchestration
   action space   82
   contextual bandit (CB) problem   81
   control loop   82
   rewards   82
   state or context space   82
   T monitoring slots   82

W
WAN. *See* Wide area network (WAN)
WAN bandwidth management   253
Wavelength allocation
   multiple network segments   86
   RSA algorithm   88
   SBVT parameters   87
   transport optical resources   87
Wavelength-division multiplexing
   (WDM)   42
WDM. *See* Wavelength-division
   multiplexing (WDM)
Wide area network (WAN)   111
WiFi   324
Wireless networks
   deep learning (DL) algorithms
      128–129
   learning accuracy   129
   RL   127–128
   supervised ML   124–126
   unsupervised ML   126–127
Wireless technologies   324
Wordcount   204
WS application   232

Y
Yahoo's EGADS   278

Z
Zero touch network and service
   management (ZSM)   70

IEEE Press Series On
Networks and Services Management

The goal of this series is to publish high quality technical reference books and textbooks on network and services management for communications and information technology professional societies, private sector and government organizations as well as research centers and universities around the world. This Series focuses on Fault, Configuration, Accounting, Performance, and Security (FCAPS) management in areas including, but not limited to, telecommunications network and services, technologies and implementations, IP networks and services, and wireless networks and services.

Series Editors:
Dr. Veli Sahin
Dr. Mehmet Ulema

1. *Telecommunications Network Management into the 21st Century*
   Edited by Thomas Plevyak and Salah Aidarous
2. *Telecommunications Network Management: Technologies and Implementations: Techniques, Standards, Technologies, and Applications*
   Edited by Salah Aidarous and Thomas Plevyak
3. *Fundamentals of Telecommunications Network Management*
   Lakshmi G. Raman
4. *Security for Telecommunications Network Management*
   Moshe Rozenblit
5. *Integrated Telecommunications Management Solutions*
   Graham Chen and Qinzheng Kong
6. *Managing IP Networks: Challenges and Opportunities*
   Edited by Thomas Plevyak and Salah Aidarous
7. *Next-Generations Telecommunications Networks, Services, and Management*
   Edited by Thomas Plevyak and Veli Sahin
8. *Introduction to IP Address Management*
   Timothy Rooney
9. *IP Address Management: Principles and Practices*
   Timothy Rooney
10. *Telecommunications System Reliability Engineering, Theory, and Practice*
    Mark L. Ayers
11. *IPv6 Deployment and Management*
    Michael Dooley and Timothy Rooney
12. *Security Management of Next Generation Telecommunications Networks and Services*
    Stuart Jacobs
13. *Cable Networks, Services, and Management*
    Mehmet Toy

14. *Cloud Services, Networking, and Management*
   Edited by Nelson L. S. da Fonseca and Raouf Boutaba
15. *DNS Security Management*
   Michael Dooley and Timothy Rooney
16. *Small Cell Networks: Deployment, Management, and Optimization*
   Holger Claussen, David Lopez-Pérez, Lester Ho, Rouzbeh Razavi, and Stepan Kucera
17. *Fundamentals of Public Safety Networks and Critical Communications Systems: Technologies, Deployment, and Management*
   Mehmet Ulema
18. *IP Address Management*
   Michael Dooley and Timothy Rooney
19. *Management of Data Center Networks*
   Nadjib Aitsaadi
20. *Communication Networks and Service Management in the Era of Artificial Intelligence and Machine Learning*
   Nur Zincir-Heywood, Marco Mellia, and Yixin Diao